Lecture Notes in Mathematics

Edited by A. Dold, B. Eckmann and F. Takens

1450

H. Fujita T. Ikebe S.T. Kuroda (Eds.)

Functional-Analytic Methods for Partial Differential Equations

Proceedings of a Conference and a Symposium held in Tokyo, Japan, July 3–9, 1989

Springer-Verlag
Berlin Heidelberg New York London
Paris Tokyo Hong Kong Barcelona

Editors
Hiroshi Fujita
Department of Mathematics, Meiji University
Higashimita, Kawasaki, 214 Japan

Teruo Ikebe
Department of Mathematics, Kyoto, University
Kyoto, 606 Japan

Shige Toshi Kuroda
Department of Mathematics, Gakushuin University
Mejiro, Tokyo, 171 Japan

Mathematics Subject Classification (1980): 35, 47, 81C, 81F

ISBN 3-540-53393-1 Springer-Verlag Berlin Heidelberg New York
ISBN 0-387-53393-1 Springer-Verlag New York Berlin Heidelberg

Printing and binding: Druckhaus Beltz, Hemsbach/Bergstr.
2146/3140-543210 – Printed on acid-free paper

Preface

In commemoration of his retirement from the University of California, Berkeley, an "International Conference on Functional Analysis and its Application in Honor of Professor Tosio Kato" was held on July 3 through 6, 1989, at Sanjo Conference Hall, University of Tokyo, the university where he began his academic career. The Organizing Committee, which consisted of Hiroshi Fujita (Meiji Univ.), S. T. Kuroda (Gakushuin Univ.), and Teruo Ikebe (Kyoto Univ., chairman), selected invited speakers mostly from among his students, students' students, and some recent collaborators. The Conference was followed by a "Symposium on Spectral and Scattering Theory" held on July 7 through 9 at Gakushuin Centennial Memorial Hall, Gakushuin University, Tokyo.

The Conference received financial supports from the Inoue Foundation for Science and the Japan Association for Mathematical Sciences, and the Symposium from Gakushuin University. We express our gratitude to these organizations.

Speakers and participants of these Conference and Symposium wish to heartily dedicate this volume to Professor Kato in celebration of his seventieth birthday.

H. Fujita
T. Ikebe
S. T. Kuroda

Programmes[1]

INTERNATIONAL CONFERENCE ON FUNCTIONAL ANALYSIS AND ITS APPLICATION IN HONOR OF PROFESSOR TOSIO KATO

MONDAY, JULY 3, 1989

James S. Howland (Univ. of Virginia)
    Quantum Stability
Peter Hess (Univ. of Zürich)
    Singular Perturbations in Periodic-Parabolic Problems
Kenji Yajima (Univ. of Tokyo)
    Smoothing Property of Schrödinger Propagators
Frank. J. Massey III (Univ. of Michigan-Dearborn)
Greg Bachelis (Wayne State Univ.)
    An Optimal Coin Tossing Problem of R. Rivest

TUESDAY, JULY 4

Takao Tayoshi (Univ. of Electro-Communications)
    Nonexistence of L^2-Eigenvalues of the Schrödinger Operator
Tosio Kato (Univ. of California, Berkeley)
    Liapunov Functions and Monotonicity for the Navier-Stokes Equation
Hiroshi Matano (Univ. of Tokyo)
    Behavior of Solutions to Elliptic Problems from the Point of View of Dynamical Systems
H. Bruce Stewart (Brookhaven National Lab.)
    Chaos, Bifurcation, and Catastrophe
    After the talk a computer-generated movie "The Lorenz System" completed by H. B. Stewart in 1984 was shown.

Conference Banquet in honor of Professor Kato

WEDNESDAY, JULY 5

Takashi Suzuki (Tokyo Metropolitan Univ.)
    Spectral Theory and Nonlinear Elliptic Equations
Rafael J. Iório, Jr. (Inst. de Mat. Pura e Aplicada)
    KdV and BO in Weighted Sobolev Spaces
Alan McIntosh (Macquarie Univ.)
    The Square Root Problem for Elliptic Operators
Gustavo Ponce (Pennsylvania State Univ.)
    The Cauchy Problem for the Generalized Korteweg-de Vries Equations
Akira Iwatsuka (Kyoto Univ.)
    On Schrödinger Operators with Magnetic Fields

[1]The titles of the papers contained in the present volume are not necessarily the same as those of talks.

THURSDAY, JULY 6

Hideo Tamura (Ibaraki Univ.)
Existence of Bound States for Double Well Potentials and the Efimov Effect
Arne Jensen (Aalborg Univ.)
Commutators and Schrödinger Operators
Charles S. Lin (Univ. of Illinois at Chicago)
On Symmetry Groups of Some Differential Equations
Takashi Ichinose (Kanazawa Univ.)
Feynman Path Integral for the Dirac Equation

SYMPOSIUM ON SPECTRAL AND SCATTERING THEORY

FRIDAY, JULY 7, 1989

Mitsuru Ikawa (Osaka Univ.)
On Poles of Scattering Matrices
Peter Hess (Univ. of Zürich)
The Periodic-Parabolic Eigenvalue Problem, with Applications
Rafael J. Iório, Jr. (Inst. de Mat. Pura e Aplicada)
Adiabatic Switching for Time Dependent Electric Fields

Short Talks

Gustavo Ponce (Pennsylvania State Univ.)
Nonlinear Small Data Scattering for Generalized KdV Equation
Tohru Ozawa (Nagoya Univ.)
Smoothing Effect for the Schrödinger Evolution Equations with Electric Fields

SATURDAY, JULY 8

Shinichi Kotani (Univ. of Tokyo)
On Some Topics of Schrödinger Operators with Random Potentials
Tosio Kato (Univ. of California, Berkeley)
Positive Commutators $i[f(P), g(Q)]$
James S. Howland (Univ. of Virginia)
Adiabatic Theorem for Dense Point Spectra
Arne Jensen (Aalborg Univ.)
High Energy Asymptotics for the Total Scattering Phase in Potential Scattering

SUNDAY, JULY 9

Yoshio Tsutsumi (Hiroshima Univ.)
L^2 Solutions for the Initial Boundary Value Problem of the Korteweg-de Vries Equation with Periodic Boundary Condition
Alan McIntosh (Macquarie Univ.)
Operator Theory for Quadratic Estimates

Contents

Papers are arranged according to the order of talks.

Spectral Concentration for Dense Point Spectrum

JAMES S. HOWLAND[1]

Division of Physics, Mathematics and Astronomy
California Institute of Technology
Pasadena, CA 91125
and
Department of Mathematics[2]
University of Virginia
Charlottesville, VA 22903

Abstract. The degree of spectral concentration at an eigenvalue λ_0 embedded in a dense point spectrum is shown to depend on the extent to which λ_0 is approximated by other eigenvalues whose eigenfunctions have appreciable overlap with the eigenvectors of λ_0. The examples considered include rank one perturbations and time-periodic perturbation of Floquet operators of discrete system.

This article is concerned with the perturbation theory of an eigenvalue λ_0 embedded in a dense point spectrum. This occurs, for example, in connection with Anderson localization or with time-periodic perturbations of discrete systems [2,3,8]. The difficulties involved may be illustrated by recalling the results of Simon and Wolff [14], who show that for certain oprators H_0 with dense pure point spectra, a rank one perturbation leads to an operator

$$H(\beta) = H_0 + \beta\langle\cdot,\varphi\rangle\varphi,$$

which is pure point for *almost every* β. This leaves open the possibility of singular continuous spectrum occurring for arbitrarily small β. The situation is reminiscent of the Stark effect, in which an (isolated) eigenvalue λ_0 disappears into an (absolutely) continuous spectrum for (all) small β.

We shall examine the problem from the point of view of spectral concentration, which was originally invented by Titchmarsh [15] to study the Stark effect. We show that the *degree of concentration depends on the extent to which* λ_0 *is approximated by other eigenvalues whose eigenfunctions have appreciable overlap with the eigenvector*

---

[1]Supported by NSF Contract DMS-8801548.
[2]Permanent address.

of λ_0. A similar phenomenon occurs in the adiabatic theorem for dense point spectrum, with regard to the degree to which the actual motion is approximated by the adiabatic motion [1].

In order to treat these problems, we must first note that the classical theory for isolated eigenvalues extends to the non-isolated case, a fact which seems to have been noted first in the literature by Greenlee [6]. We summarize the necessary results in the first section.

We then treat several examples. We first consider rank one perturbations, as discussed by Aronszajn and Donoghue [5], and Simon and Wolff [14], and then generalize to certain compact perturbations, as in [7]. Finally, we discuss the physically interesting case of a time-periodic perturbation of a discrete Hamiltonian, which has been of considerable recent interest [2,3,8].

The author wishes to thank Barry Simon and David Wales for the hospitality of Caltech, where this work was done.

§1 Spectral Concentration for Non-Isolated Eigenvalues.

We shall assume throughout this section that $H_\beta = \int \lambda dE_\beta(\lambda)$, $0 \le \beta \le \beta_0$, is a family of self-adjoint operators on a Hilbert space $\mathcal{H}$, with $H_\beta \to H_0$ in the strong resolvent sense as $\beta \to 0$; and that λ_0 is an eigenvalue of H_0 of *finite multiplicity* m. Let P_0 be the projection onto the kernel of $H_0 - \lambda_0$.

We say that the spectrum of H_β is *concentrated at* λ_0 on a family of Borel sets S_β iff

$$E_\beta[S_\beta] \to P_0 \tag{1.1}$$

strongly as $\beta \to 0$. For $p \ge 0$, we say that H_β is *concentrated to order* p at λ_0 if the Lebesgue measure

$$|S_\beta| = o(\beta^p), \qquad \text{as } \beta \to 0. \tag{1.2}$$

A *pseudoeigenvector for* H_β of *order* p, or more briefly, a *p-pair* is a family φ_β of unit vectors and a real-valued function λ_β such that

$$(H_\beta - \lambda_\beta)\varphi_\beta = o(\beta^p), \qquad \text{as } \beta \to 0. \tag{1.3}$$

An *asymptotic basis of order* p for H_β at λ_0 is a family $\left\{\varphi_\beta^{(j)}, \lambda_\beta^{(j)} : j = 1, \ldots, m\right\}$ of p-pairs, such that $\lambda_\beta^{(j)} \to \lambda_0$ and $\varphi_\beta^{(j)} \to \varphi^{(j)}$, where $\varphi^{(1)}, \ldots, \varphi^{(m)}$ is a basis of $P_0\mathcal{H}$.

There are two main results of [13]. The first is the equivalence of spectral concentration and the existence of p-pairs. The following is proved in [4], [10, p. 473], and [13] for isolated eigenvalues, and in [6] for non-isolated.

1.1 THEOREM. *If H_β has an asymptotic basis of order p at λ_0, then the spectrum of H_β is concentrated at λ_0 to order p.*

The set S_β is taken as the union of m intervals, centered at $\lambda_\beta^{(j)}$, and of width γ_β where $\gamma_\beta = o(\beta^p)$.

PROOF: The proof is exactly the same as that of Theorem 5.2 of [10, p. 473], except that since λ_0 is not isolated, it must be shown at the end that if $Q = I - P_0$, then

$$\text{s-}\lim_{\beta\to 0} E_\beta[S_\beta]Q = 0. \tag{1.4}$$

Let $J_\varepsilon = (\lambda_0 - \varepsilon, \lambda_0 + \varepsilon)$. For β small, $S_\beta \subset J_\varepsilon$ so that

$$|E_\beta[S_\beta]Qu| \leq |E_\beta[J_\varepsilon]Qu|.$$

In the limit, by [10, Theorem 1.15, p. 432], this gives

$$\overline{\lim}\, |E_\beta[S_\beta]Qu| \leq |E_0[J_\varepsilon]Qu|.$$

As $\varepsilon \to 0$, the right side converges to $|P_0 Qu| = 0$. ∎

1.2 Remark. Riddell also proves the converse result [13, p. 384], that if there is concentration to order p, then an asymptotic basis can be found. We will not need this result, since in practice concentration is usually proved by constructing p-pairs.

The second result of [13] is that p-pairs can be constructed by the perturbation method.

Assume that

$$H_\beta = H_0 + \beta V, \tag{1.5}$$

where V is H_0-bounded, which implies strong resolvent convergence. The *reduced resolvent*

$$S = (H_0 - \lambda_0)^{-1} Q \tag{1.6}$$

is a well-defined self-adjoint operator, although it is bounded only if λ_0 is isolated.

1.3 THEOREM. *Assume that for $k = 1, \cdots, p$, the operators*

$$X_1 X_2 \cdots X_k P_0 \tag{1.7}$$

are all bounded, where each X_j is either S or SV. Then H_β has an asymptotic basis of order p at λ_0.

The idea is that $X_1 \cdots X_k P$ are exactly the objects needed to be able to solve the perturbation equations out to order p. For multiplicity $m = 1$, these equations are simple, and as shown in [6], Riddell's proof [13, p. 391] works without change, if one remembers that S need not be bounded. For $m > 1$, Riddell's inductive procedure is less transparent but seems to go through as well. The author has given a different proof for the non-isolated case in [9], basing the argument on Nenciu's idea [11] of applying the adiabatic theorem.

By similar methods, one also obtains

1.4 THEOREM. *Let*

$$V(\beta) = \sum_{k=1}^{\infty} \beta^k V^{(k)} \tag{1.8}$$

be bounded and analytic for $|\beta| < \beta_0$. Assume that the operators

$$X_1 X_2 \cdots X_p P_0 \tag{1.9}$$

are all bounded, where each X_j is either S or $SV^{(k)}$ for some $k \leq p$. Then

$$H_\beta = H_0 + V(\beta)$$

is concentrated at λ_0 to order p.

§2. Applications.

We shall now give some applications of Theorem 1.3. All will be deduced from a simple corollary. As above, we let λ_0 be an eigenvalue of H_0 of finite multiplicity m, and $S = (H_0 - \lambda_0)^{-1} Q$ the reduced resolvent.

2.1 THEOREM. *Let $H_\beta = H_0 + \beta V$ where V is H_0-bounded. If $S^p V$ is bounded, then H_β is concentrated at λ_0 to order p.*

PROOF: This follows from Theorem 1.3 because the operators $X_1 X_2 \cdots X_p$ are products of operators of the form $S^k V$ with $1 \leq k \leq p$. It is necessary to show that boundedness of $S^p V$ implies boundedness of $S^k V$ for $1 \leq k \leq p$ as well. Let $E = E_0\,(\lambda_0 - 1, \lambda_0 + 1)$ and write

$$S^k V = \left(E S^{k-p}\right)\left(S^p V\right) + (1 - E) S^k V.$$

Both factors of the first term are bounded, while for the second, we have

$$(1 - E) S^k V = \lim_{z \to \lambda_0} \left[V\,(H_0 - z)^{-k}\,(1 - E)\right]^*.$$

The quantity in brackets on the right is bounded and norm analytic at $z = \lambda_0$ since V is H_0-bounded. ∎

We shall also need a real variable lemma [1,7,14].

2.2 LEMMA. *If $a_n > 0$ and $\sum_{n=1}^{\infty} a_n < \infty$ then for any sequence λ_n, and any $\alpha \geq 1$,*

$$\sum_{n=1}^{\infty} a_n^\alpha \left|\lambda - \lambda_n\right|^{-\alpha} < \infty \tag{2.1}$$

for a.e. λ.

EXAMPLE 1: RANK ONE PERTURBATIONS [7,14]. Let H_0 be pure point, of finite multiplicity, with $H_0 e_n = \lambda_n e_n$, $n \geq 0$, and e_n a complete orthonormal set. Define

$$H_\beta = H_0 + \beta\langle\cdot,\varphi\rangle\varphi \tag{2.2}$$

where $|\varphi|^2 = 1$. Since the perturbation is of rank one, we may assume that H_0 has simple multiplicity. We are, of course, thinking primarily of the case in which the eigenvalue λ_n are dense in some interval.

One has, for a fixed eigenvalue λ_0,

$$S^p V u = \sum_{n\neq 0} (\lambda_n - \lambda_0)^{-p} \langle\varphi, e_n\rangle\langle u, \varphi\rangle e_n, \tag{2.3}$$

so that $S^p V$ is bounded iff

$$\sum_{n\neq 0} (\lambda_n - \lambda_0)^{-2p} \left|\langle\varphi, e_n\rangle\right|^2 < \infty. \tag{2.4}$$

Define, for $p \geq 1$, the set

$$N_p = \left\{\lambda : \sum_{k=n}^{\infty} |\lambda_k - \lambda|^{-2p} \left|\langle\varphi, e_k\rangle\right|^2 = \infty, \text{ for every } n \geq 1\right\}, \tag{2.5}$$

and

$$N_\infty = \bigcup_{p=1}^{\infty} N_p.$$

According to Lemma 2.2, N_p has Lebesgue measure zero if

$$\sum_{n=0}^{\infty} \left|\langle\varphi, e_n\rangle\right|^{1/p} < \infty. \tag{2.6}$$

Hence, if $\langle\varphi, e_n\rangle$ decays exponentially, N_∞ has measure zero. By Theorem 2.1, we have

2.3 THEOREM. *If (2.6) holds, then N_p has measure zero. If $\lambda_0 \notin N_p$, then H_β is concentrated at λ_0 to order p.*

For $p = \infty$, this means that H_β is concentrated to order p for every finite p.

The set N_p consists of points which are well approximated by eigenvalues λ_n whose eigenvectors e_n are substantially disturbed by the perturbation $\langle\cdot,\varphi\rangle\varphi$. Thus, the degree of concentration depends on the degree to which λ_0 can be approximated by such λ_n's. Condition (2.6) assures us that N_p *depends only on the tails* $\{\lambda_k : k \geq n\}$ *of the eigenvalue sequence.*

EXAMPLE 2. COMPACT PERTURBATIONS [7]. A natural generalization of preceding example is the following. Let H_0 be as above, and let

$$H_\beta = H_0 + \beta V,$$

where V is self-adjoint and satisfies

$$\sum_{n=0}^{\infty} |Ve_n|^{1/p} < \infty. \tag{2.7}$$

This is a strong condition which even for $p = 1$ implies that V is trace class (cf. [7]). For *any* λ, define

$$S_0(\lambda) = (H_0 - \lambda)^{-1} Q(\lambda), \tag{2.8}$$

where $I - Q(\lambda)$ is the projection onto $ker(H_0 - \lambda)$ (which may be zero). Then

$$S_0(\lambda)^p V u = \sum_{\lambda_n \neq \lambda} (\lambda_n - \lambda)^{-p} \langle u, Ve_n \rangle e_n,$$

so that

$$|S_0(\lambda)^p V u|^2 \leq |u|^2 \sum_{\lambda_n \neq \lambda} (\lambda_n - \lambda)^{-2p} |Ve_n|^2$$

which is finite for a.e. λ by (2.7) and Lemma 2.2.
Define

$$N_p = \left\{ \lambda : \sum_{\lambda_n \neq \lambda} (\lambda_n - \lambda)^{-2p} |Ve_n|^2 = \infty \right\}. \tag{2.9}$$

2.4 THEOREM. *If (2.7) holds, then N_p is of mesure zero. If $\lambda_0 \notin N_p$, then H_β is concentrated to order p at λ_0.*

EXAMPLE 3. FLOQUET HAMILTONIANS [2,3,8]. Next, let H_0 be *discrete,* with eigenvalues $0 < \lambda_1 < \lambda_2 < \cdots$ of *simple multiplicity*. Let $H_0 e_n = \lambda_n e_n$, $|e_n|^2 = 1$. Let $V(t)$ be bounded, strongly C^r and 2π-periodic. We consider the time-dependent Hamiltonian

$$H_\beta(t) = H_0 + \beta V(t),$$

or, more precisely, its Floquet Hamiltonian:

$$K_\beta = i\frac{d}{dt} + H_0 + \beta V(t)$$

on $L_2(0, 2\pi) \otimes \mathcal{H}$ with periodic boundary condition $u(2\pi) = u(0)$. For $\beta = 0$, K_0 has pure point spectrum, with eigenvalues

$$\Lambda_{n,k} = n + \lambda_k$$

($n = 0, \pm 1, \pm 2, \cdots, k = 1, 2, \cdots$). We shall *assume that all $\Lambda_{n,k}$ are of finite multiplicity*, which implies that they are dense, since the spectum of K_0 is periodic.

Such operators have been of considerable recent interest as a problem in *quantum stability*, which may be said to occur when K_β has pure point spectrum. (See [3,8], and especially [2].) Under conditions weaker than those assumed below, the author [8] showed that K_β has no absolutely continuous spectrum.

Since the degree of spectral concentration can be regarded as a *measure of the stability of an eigenvalue under perturbation*, the following result seems of interest in this context.

Define, for intergers $p \geq 1$, and real $\gamma > 0$, the set

$$N(p,\gamma) = \left\{ \mu : \sum_{n,k}{}' (\lambda_k + n - \mu)^{-2p} k^{-2\gamma} = \infty \right\} \tag{2.10}$$

where the prime on the summation means that terms with $\lambda_k + n = \mu$ are omitted. By assumption, these terms are finite in number.

2.5 Lemma. *$N(p,\gamma)$ has measure zero if $1 \leq p < \gamma$. Hence for any fixed $\delta > 0$, the set*

$$N_\infty = \bigcup_{p=1}^{\infty} N(p, p+\delta)$$

has measure zero.

Proof: The set $N(p,\gamma)$ is periodic, with period 1, so it suffices to prove that $N(p,\gamma) \cap J$ is of measure zero, where $J = [0,1)$. Fix $\mu \in (0,1)$ and write the sum in (2.10) as

$$\left\{ \sum_{\lambda_k + n \in J} + \sum_{\lambda_k + n \notin J} \right\} (\lambda_k + n - \mu)^{-2p} k^{-2\gamma}.$$

The second term is analytic on $(0,1)$ and thus always finite. The first term can be treated by observing that $\lambda_k + n \in J$ for *at most one value* n_k of n. Thus the term is equal to

$$\sum_k{}' (\lambda_k + n_k - \mu)^{-2p} k^{-2\gamma} \varepsilon_k,$$

where ε_k is zero if $\lambda_k + n$ is never in J, and is one otherwise. By lemma 2.2, this sum is finite for a.e. μ if $p < \gamma$.

2.6 Theorem. *Assume that K_0 has finite multiplicity, that $V(t)$ is strongly C^∞, and that the gap*

$$\Delta\lambda_n = \lambda_{n+1} - \lambda_n$$

between eigenvalues satisfies

$$\Delta\lambda_n \geq c n^\alpha$$

for some $\alpha > 0$. Then the spectrum of K_β is concentrated at λ_0 to all orders if $\lambda_0 \notin N_\infty$.

PROOF: According to [8], K_β is unitarily equivalent to an operator of the form

$$\tilde{K}_\beta = i\frac{d}{dt} + \tilde{H} + \beta A W(t,\beta) A,$$

where $\tilde{H}$ is discrete and diagonal in the same basis e_n as H, $W(t,\beta)$ is bounded and analytic in β, and

$$A = \sum_n n^{-\gamma} \langle \cdot, e_n \rangle e_n.$$

For $V(t)$ in C^∞, γ may be taken as large as desired.

For p fixed, choose $\gamma > p + \delta$. By Lemma 2.5, $S(\lambda)^p A$ is bounded. Thus if we expand the perturbation

$$V(\beta) = AW(\beta)A = \sum_{k=1}^{\infty} \beta^k A W^{(k)} A,$$

we will have

$$S^p(\lambda) V^{(k)} = (S^p(\lambda) A)(W^k A)$$

bounded for all k. Using Theorm 1.4, and the argument in the proof of Theorem 2.1 gives the result. ∎

REMARK. It is possible to keep track of the relationship between α, the degree of smoothness of $V(t)$, and the order of concentration that can be expected.

§3 Remarks.

There are two points that need clarification. In the first place, *is concentration really relevant here?* For example, in [2,3], a KAM-type argument leads to an explicit diagonalization of H_β, so that the spectrum is concentrated on a *one point* set, the perturbed eigenvalue. Of course, [2,3] contain strong assumptions, like analyticity, but the question is still in order.

A complete answer would require a rather complete theory of these operators, which we are at present far from having. Nevertheless, the following example is instructive.

Let ν be a measure on $[0,1]$ which is singular continuous, for which the set N of λ where

(3.1) $$\int_0^1 (\lambda - t)^{-2} \nu(dt) = \infty$$

is *of measure zero, but dense.* Let the operator H_0 of multiplication by λ on $L^2(\nu)$ be perturbed by the vector 1:

$$H_\beta = H_0 + \beta \langle \cdot, 1 \rangle 1$$

(cf. [14]). Then H_β is pure point for a.e. β, but can have *no point spectrum* in N [8,14]. This means that H_β cannot have an eigenvalue $\lambda(\beta)$ which varies continuously, as in the case [2,3]. The author finds it probable that worse examples can be constructed.

In the second place, *how do we know that all the eigenvalues of* H_0 *are not in the bad set* N_p *?* In this case, our theorems would say nothing! While it might be possible for this to occur, the following result shows that it is in some sense rare.

Recall that if H is a self-adjoint, we write $P(\lambda)$ for the projection onto the kernel of $H - \lambda$ (which may be trivial), $Q(\lambda) = I - P(\lambda)$, and $S(\lambda) = (H-\lambda)^{-1}Q(\lambda)$ for the reduced resolvent. By definition, $\lambda \notin N_p(H,V)$ iff $S(\lambda)^p V$ is bounded.

3.1 THEOREM. *Let* $H = H_0 + V$, *where* V *is* H_0*-bounded, and assume that* $\lambda \notin N_p(H_0, V)$ *and that* $S_0(\lambda)$ *is compact. If* λ *is an eigenvector of* H *or* H_0, *we also assume that its multiplicity is finite.*

(a) *if* $\lambda \notin \sigma_p(H)$, *then* $\lambda \notin N_p(H,V)$.
(b) *if* $\lambda \in \sigma_p(H)$, *then* $\lambda \notin N_{p-1}(H,V)$.

Hence, in general, $N_{p-1}(H,V) \subset N_p(H_0,V)$.

If we apply this result to Example 1, where

$$H_\beta = H_0 + \beta\langle\cdot,\varphi\rangle\varphi,$$

we see that $N_{p-1}(H_\beta) \subset N_p(H_0)$. According to [8] and [14], however, for *any fixed null set* N

$$E_\beta[N] = 0$$

for a.e. β. Thus if

$$\sum_n |\langle\varphi, e_n\rangle|^{1/p} < \infty,$$

then for a.e.β, $\lambda \notin N_{p-1}(H_\beta)$ *for every eigenvalue of* H_β. This indicates that having eigenvalues in N_p is an unstable condition and does not obtain in the generic case. Thus, here, the perturbation problem

$$H'_\beta = H'_0 + \beta\langle\cdot,\varphi\rangle\varphi$$

with

$$H'_0 = H_0 + \beta_0\langle\cdot,\varphi\rangle\varphi$$

has $N_{p-1}(H'_0) \cap \sigma_p(H'_0) = \emptyset$ for a.e. β_0.

We sketch a proof, leaving some details about domains to the reader.

PROOF OF THEOREM 3.1: First, note that we can assume that $\lambda \notin \sigma_p(H_0)$ by writing

$$H = H_0 + V = (H_0 + P_0) + (V - P_0) = H'_0 + V', \tag{3.2}$$

where $P_0 = P_0(\lambda)$. Then

$$\begin{aligned} R'_0(\lambda)^p V' &= [S_0(\lambda)^p + P_0](V - P_0) \\ &= S_0(\lambda)^p V + P_0 V - P_0 \end{aligned} \tag{3.3}$$

is bounded and compact for $p=1$. Thus, replacing H_0 and V by H_0' and V', we can assume $\lambda \notin \sigma_p(H_0)$, and hence that $I + R_0(\lambda)V$ has a bounded inverse.

For part (a), observe that (suppressing λ)

$$R_0^n - R^n = \sum_{k=1}^{n} R_0^k R^{n-k} - R_0^{k-1} R^{n-k+1} \tag{3.4}$$
$$= \sum_{k=1}^{n} R_0^{k-1}[R_0 - R]R^{n-k} = \sum_{k=1}^{n} R_0^k V R^{n-k+1}$$
$$= R_0 V R^n + \sum_{k=2}^{n} R_0^k V R^{n-k+1}.$$

Solving for $\mathbf{R}^n$ gives

$$R^n V = [I + R_0 V]^{-1} R_0^n V - \sum_{k=2}^{n} R_0^k V R^{n-k+1} V. \tag{3.5}$$

It follows by induction that $R_0^p V$ bounded implies $R^p V$ bounded.

For (b), suppose that $\lambda \in \sigma_p(H)$. Then every eigenvector ψ satisfies

$$\psi = -R_0(\lambda) V \psi \tag{3.6}$$

so that

$$R_0(\lambda)^{p-1}\psi = R_0(\lambda)^p V \psi \tag{3.7}$$

or, in other words,

$$R_0(\lambda)^{p-1} P(\lambda)$$

is bounded. As above, write

$$H' = H + P = H_0 + (V + P) = H_0 + V'.$$

Then

$$R_0^{p-1} V' = R_0^{p-1} V + R_0^{p-1} P \tag{3.8}$$

is bounded, while

$$(R')^{p-1} V' = S^{p-1} V + (PV + P). \tag{3.9}$$

The last two terms are bounded, so S^{p-1} is bounded iff $(R')^{p-1}V'$ is. Applying (a) now yields the result. ∎

References

1. Averon, J. E., J. S. Howland and B.Simon, *Adiabatic theorems for dense point spectrum*, Comm. Math. Phys. (to appear).
2. Bellissard, J., *Stability and instability in quantum mechanics*, in "Trends and Developments in the Eighties," Albevario and Blanchard, eds., World Scientific, Singapore, 1985.
3. Combescure, M., *The quantum stability problem for time-periodic perturbation of the harmonic oscillator*, Ann. Inst. H. Poincaré **47** (1987), 63–84.
4. Conley, C. C. and P. A. Rejto., *Spectral concentration II*, in "Perturbation Theory and its Applications in Quantum Mechanics," C. H. Wilcox, ed., Wiley, New York, 1966, pp. 129–143.
5. Donoghue, W., *On the perturbation of spectra*, Comm. Pure Appl. Math. **18** (1965), 559–579.
6. Greenlee, W. M., *Spectral concentraion near embedded eigenvalues*, J. Math. Anal. Appl. (to appear).
7. Howland, J. S., *Perturbation theory of dense point spectra*, J. Funct. Anal. **74** (1987), 52–80.
8. Howland, J. S., *Floquet oprators with singular spectra, I and II*, Ann. Inst. H. Poincaré **50** (1989), 309–323, 325–334.
9. Howland, J. S., *A note on spectral concentration for nonisolated eigenvalues*, J. Math. Anal. Appl. (to appear).
10. Kato, T., "Perturbation Theory for Linear Operators," Springer-Verlag, New York, 1966.
11. Nenciu, G., *Adiabatic theorem and spectral concentration I*, Comm. Math. Phys. **82** (1981), 121–135.
12. Reed, M. and B. Simon, "Methods of Modern Mathematical Physics IV," Academic Press, New York, 1978.
13. Riddell, R. C., *Spectral concentration for self-adjoint operators*, Pacific J. Math. **23** (1967), 377–401.
14. Simon, B. and T.Wolff, *Singular continuous spectrum under rank one perturbations and localization for random Hamiltonians*, Comm. Pure Appl. Math. **39** (1986), 75–90.
15. Titchmarsh, E. C., *Some theorem on perturbation theory*, J. Analyse Math. **4** (1954), 187–208.

BEHAVIOUR OF A SEMILINEAR PERIODIC-PARABOLIC PROBLEM WHEN A PARAMETER IS SMALL

E.N. Dancer[1] and P. Hess[2]

[1] Department of Mathematics, University of New England, Armidale, N.S.W. 2361, Australia

[2] Mathematics Institute, University of Zurich, Rämistrasse 74, CH-8001 Zurich, Switzerland

1 Introduction

In this paper we consider the time-periodic Neumann problem

$$(*)\qquad \begin{aligned} u_t - \varepsilon^2 \Delta u &= m(x,t)h(u) && \text{in } \Omega \times (0,\infty) \\ \tfrac{\partial u}{\partial n} &= 0 && \text{on } \partial\Omega \times (0,\infty)\,. \end{aligned}$$

Here Ω is a smooth bounded domain in $I\!R^N$. We assume that h is C^1, $h(0) = h(1) = 0$, $h'(0) > 0$, $h'(1) < 0$, $h(y) > 0$ on $(0,1)$, and that m is Hölder-continuous and T-periodic in t. Of course we require that $m \not\equiv 0$.

Note that $u \equiv 0$ and $u \equiv 1$ are solutions of the equation. We refer to these as trivial T-periodic solutions. Our results concern the behaviour of other T-periodic solutions $u(x,t)$ with $0 \le u(x,t) \le 1$ on $\Omega \times [0,T]$ as $\varepsilon \to 0$. Our results improve a theorem of Alikakos and the second author [1] by removing a convexity condition on h, by allowing arbitrary space dimensions, and by permitting much more general behaviour of m. We prove uniform convergence of the solutions away from the transition layer. We note that our theorem can easily be used to provide examples where the transition layers form complicated patterns.

There has been a good deal of earlier work on this problem in the autonomous setting, for equilibrium solutions, which is discussed in [1]. Note that (*) includes the important Fisher's equation.

2 The main results

We assume that the conditions of the introduction hold. Let $\xi(x) := \int_0^T m(x,t)dt$.

Theorem *Assume that K is a compact subset of $\overline{\Omega}$ such that $\xi(x) > 0$ (resp. < 0) on K, that $\varepsilon_n \to 0$ as $n \to \infty$ (where $\varepsilon_n > 0$), and that u_{ε_n} are nontrivial T-periodic solutions of $(*)$ such that $0 \le u_{\varepsilon_n} \le 1$ on $\Omega \times [0,T]$.*

Then $u_{\varepsilon_n}(x,t) \to 1$ (resp. 0) uniformly on $K \times [0,T]$ as $u \to \infty$. Moreover, if ξ changes sign on $\overline{\Omega}$, nontrivial stable T-periodic solutions with values in $[0,1]$ exist for all small $\varepsilon > 0$.

REMARK If $\{x \in \Omega : \xi(x) = 0\}$ has measure zero, then

$$u_{\varepsilon_n}(x,t) \to \chi_{\{(x,t)\in\Omega\times[0,T]\,:\,\xi(x)>0\}}$$

in $L^p(\Omega \times [0,T])$ as $n \to \infty$ for all p with $1 \le p < \infty$. Note that we do not expect uniform convergence near where $\xi(x) = 0$ because there will be transition layers.

Corollary *If $\xi(x) > 0$ (or < 0) on $\overline{\Omega}$, (*) has no nontrivial T-periodic solution with values in $[0,1]$ for all sufficiently small positive ε.*

3 An auxiliary result

Before proving the theorem, we prove a lemma. We always assume that the hypotheses of the first part of the theorem hold. By an admissible solution we mean a nontrivial T-periodic solution $u(x,t)$ of (*) such that $0 \le u(x,t) \le 1$ on $\Omega \times [0,T]$. By the maximum principle, it follows that $0 < u(x,t) < 1$.

Lemma *Assume that K is a compact subset of $\overline{\Omega}$ such that $\xi(x) > 0$ on K. Then there is a $\delta > 0$ such that any admissible solution $u(x,t)$ of (*) satisfies $u(x,t) \ge \delta$ on $K \times [0,T]$ if ε is small.*

PROOF It suffices to assume that $h'(0) = 1$. We first consider the case where $K \subset \Omega$. Clearly it then suffices to prove that, if $x_0 \in \Omega$ and $\xi(x_0) > 0$, there is a ball B centered at x_0 such that $u(x,t) \ge \delta$ on $\overline{B} \times [0,T]$ for small ε (by using finite covers). We prove the result by constructing subsolutions and using Serrin's sweeping technique (cf. [7, Thm. 2.7.1] for the elliptic case).

If $x_0 \in \Omega$ and $\xi(x_0) > 0$, choose a ball U centered at x_0 such that $U \subset \Omega$ and $\int_0^T \min_{x\in\overline{U}} m(x,t)dt > 0$ (by continuity of m). Let $\overline{m}(t) := \min_{x\in\overline{U}} m(x,t)$. Let ϕ_1^U denote the positive first eigenfunction of $-\Delta u = \lambda u$ in U, $u = 0$ on ∂U, and let λ_1^U denote the corresponding principal eigenvalue. A simple calculation shows that

$$\tilde{\phi}_\varepsilon^U(x,t) := \phi_1^U(x)\exp[(-\varepsilon^2\lambda_1^U + \overline{\mu}_\varepsilon)t + \int_0^t \overline{m}(\tau)d\tau]$$

is a positive solution of the problem

$$\begin{cases} u_t - \varepsilon^2\Delta u - \overline{m}(t)u &= \overline{\mu}_\varepsilon u & \text{in } U\times[0,T] \\ u(x,t) &= 0 & \text{on } \partial U\times[0,T] \\ u \text{ is } T\text{-periodic in } t \end{cases}$$

provided that $\overline{\mu}_\varepsilon = \varepsilon^2\lambda_1^U - T^{-1}\int_0^T \overline{m}(\tau)d\tau$. Since $\int_0^T \overline{m}(\tau)d\tau > 0$, we see that $\overline{\mu}_\varepsilon \le -\mu < 0$ for small ε. We prove that $\delta\tilde{\phi}_\varepsilon^U$ is a periodic subsolution of (*) on $U\times[0,T]$ if δ is small and non-negative. To see this, we note by a simple calculation that $\delta\tilde{\phi}_\varepsilon^U$ is a subsolution if

(1) $$\overline{m}(t)\delta\tilde{\phi}_\varepsilon^U + \overline{\mu}_\varepsilon\delta\tilde{\phi}_\varepsilon^U \le m(x,t)h(\delta\tilde{\phi}_\varepsilon^U)$$

on $U\times[0,T]$. Here we have used the equation satisfied by $\tilde{\phi}_\varepsilon^U$ and that the condition on $\partial U\times[0,T]$ is automatically satisfied since $\tilde{\phi}_\varepsilon^U = 0$ on $\partial U\times[0,T]$. (1) is trivially satisfied if $\delta = 0$. Thus we assume $\delta > 0$. Since $\tilde{\phi}_\varepsilon^U > 0$ on $U\times[0,T]$, and since $h(y) = y + o(y)$ for small y (since $h'(0) = 1$), (1) becomes

$$\overline{m}(t) + \overline{\mu}_\varepsilon \le m(x,t) + o(1) \qquad (\text{as } \delta \to 0)$$

on $U\times[0,T]$. Note that $\tilde{\phi}_\varepsilon^U$ is bounded. Since $m(x,t) \ge \overline{m}(t)$ on $U\times[0,T]$ and $\overline{\mu}_\varepsilon \le -\mu < 0$, this is satisfied if δ is small (independent of ε for small ε).

Thus there exist $\delta_0 > 0$, $\varepsilon_0 > 0$ such that $\delta\tilde{\phi}_\varepsilon^U$ is a subsolution if $0 \le \delta \le \delta_0$ and $0 < \varepsilon \le \varepsilon_0$. Suppose that u is an admissible solution of (*). Then, if $\delta = 0$, $u \ge \delta\tilde{\phi}_\varepsilon^U$ on $U\times[0,T]$. By the Serrin sweeping principle, and since $u \ge 0$ on $\partial U\times[0,T]$ while $\delta\tilde{\phi}_\varepsilon^U = 0$ on $\partial U\times[0,T]$, we see that $u \ge \delta_0\tilde{\phi}_\varepsilon^U$ on $U\times[0,T]$ if $\varepsilon \le \varepsilon_0$. (To prove this, for fixed ε we let $\tilde{\delta} := \sup\{\delta \in [0,\delta_0] : u \ge \delta\tilde{\phi}_\varepsilon^U \text{ on } U\times[0,T]\}$, note that $u \ge \tilde{\delta}\tilde{\phi}_\varepsilon^U$ on $U\times[0,T]$, and apply the parabolic maximum principle to $u - \tilde{\delta}\tilde{\phi}_\varepsilon^U$ to deduce that this function has a positive lower bound on $\overline{U}\times[0,T]$. This contradicts the maximality of $\tilde{\delta}$ if $\tilde{\delta} < \delta_0$). Finally, if we replace U by a ball of half the radius we obtain the required estimate. (Note that $\tilde{\phi}_\varepsilon^U$ is not small away from $\partial U\times[0,T]$.) This proves the lemma for $K \subset \Omega$.

To prove the result for K intersecting $\partial\Omega$, it suffices to prove that for each $x_0 \in \partial\Omega$ with $\xi(x_0) > 0$ there is a set W relative-open in $\overline{\Omega}$, with $x_0 \in$ rel. int. K, and $\delta > 0$, such that $u(x,t) \ge \delta$ on $\overline{W} \cap \overline{\Omega}$ if ε is small. To do this, we use essentially the same argument as before except that we modify the construction of U and ϕ_1^U. We first choose W to be the intersection of a small ball centered at x_0 and $\overline{\Omega}$, except that we "round off" the corner where the boundary of the ball meets $\partial\Omega$, so that ∂W is a C^1 manifold. This is easy (but tedious) if one recalls that a small ball intersected with $\overline{\Omega}$ is nearly a hemisphere. We define ϕ_1^W as before except that the boundary condition changes to $u(x) = 0$ on $\partial W \setminus \partial\Omega$, $\frac{\partial u}{\partial n}(x) = 0$ on $\partial W \cap \partial\Omega$. (The existence follows easily from variational methods.) A result of Stampacchia [8, p. 245] ensures that ϕ_1^W is continuous on $\overline{W}$ (in fact, ϕ_1^W is smooth except on the boundary Z of $\partial W \cap \partial\Omega$ relative to $\partial\Omega$), and $\phi_1^W > 0$ except on $\overline{\partial W \setminus \partial\Omega}$. We construct $\tilde{\phi}_\varepsilon^W$ as before. The only difference is that $\tilde{\phi}_\varepsilon^W$ satisfies Dirichlet boundary conditions on $(\partial W \setminus \partial\Omega) \times [0,T]$ and Neumann boundary conditions on $(\partial W \cap \partial\Omega) \times [0,T]$. We can argue as before to deduce that $\delta\tilde{\phi}_\varepsilon^W$ are periodic subsolutions on $W \times [0,T]$ if $0 \le \delta \le \delta_0$ and $0 < \varepsilon \le \varepsilon_0$. We can then deduce much as before that $u \ge \delta_0\tilde{\phi}_\varepsilon^W$ on $\overline{W} \times [0,T]$ if $\varepsilon \le \varepsilon_0$. There are two points to be mentioned here. If we define $\tilde{\delta}$ as before, then $\tilde{\delta} > 0$ because the parabolic maximum principle applied on $\overline{\Omega} \times [0,T]$ ensures that $u(x,t) > 0$ if $(x,t) \in \overline{\Omega} \times [0,T]$. Secondly, if we look for a point where $u(x,t) - \tilde{\delta}\tilde{\phi}_\varepsilon^W(x,t)$ has a minimum value and this value equals zero, then it can not occur on $Z \times [0,T]$ (where $\tilde{\phi}_\varepsilon^W$ is zero) and hence the lack of smoothness on $Z \times [0,T]$ does not affect the use of the parabolic maximum principle. Hence the lemma follows also in this case. □

REMARKS

1. By applying the same argument to $1-u$ (which satisfies a similar equation with the same boundary conditions), we see that, if K is a compact subset of $\overline{\Omega}$ where $\xi(x) < 0$, there is a $\delta > 0$ such that $u(x,t) \le 1-\delta$ on $K \times [0,T]$ if ε is small. In this process, we construct T-periodic supersolutions close to 1 (of the form $1 - \delta\tilde{\phi}$).

2. If $\xi(x) > 0$ and $\xi(x) < 0$ somewhere in Ω, then in the proof of the lemma we constructed a T-periodic subsolution u_1 of (*) near zero on part of $\Omega \times [0,T]$ while in the remark above we constructed a T-periodic supersolution v_1 near 1 on part of $\Omega \times [0,T]$ (for ε small). As in [2] we can piece together u_1 with zero to obtain a subsolution on $\Omega \times [0,T]$, and v_1 with 1 to obtain a supersolution on $\Omega \times [0,T]$. (In both cases we do not lose continuity.) Clearly this does not affect the ordering, and hence by [3, Theorem 1] there is a stable periodic solution of (*) between u_1 and v_1 (for small ε). This proves the last statement in our theorem. (It could also be proved by using the techniques of [1].)

4 Proof of the Theorem

We have already proved the last statement. By a compactness argument (and by also using the equation for $1-u$ as in the proof of the lemma), we see that it suffices to prove that if $x_n \in \overline{\Omega}$, $x_n \to x_0$ as $n \to \infty$, $\xi(x_0) > 0$ and $\varepsilon_n \to 0$ as $n \to \infty$, then $\inf_{0 \le t \le T} u_{\varepsilon_n}(x_n,t) \to 1$ as $n \to \infty$. Suppose not, i.e. suppose that there exist $t_n \in [0,T]$ and $\alpha \in (0,1)$ such that $u_{\varepsilon_n}(x_n,t_n) \le 1-\alpha$ (at least for a subsequence). We will obtain a contradiction by using a blowing up argument.

First assume that $x_0 \in \Omega$. We define a new variable by $\boldsymbol{x} := \varepsilon_n^{-1}(x - x_n)$ and let $v_n(\boldsymbol{x}, t) := u_{\varepsilon_n}(x_n + \varepsilon_n \boldsymbol{x}, t)$. In the new variables, v_n is a solution of

$$(2) \qquad \begin{aligned} v_t - \Delta v &= m_n(\boldsymbol{x}, t)h(v) && \text{in } \tilde{\Omega}_n \times [0,T] \\ \tfrac{\partial v}{\partial n} &= 0 && \text{on } \partial\tilde{\Omega}_n \times [0,T] \\ &v \text{ is } T\text{-periodic} . \end{aligned}$$

Here $m_n(\boldsymbol{x}, t) := m(x_n + \varepsilon_n \boldsymbol{x}, t)$ and $\tilde{\Omega}_n := \{\varepsilon_n^{-1}(x - x_n) : x \in \Omega\}$. By our construction $0 \le v_n \le 1$ on $\tilde{\Omega}_n \times [0,T]$, $v_n(\mathbf{o}, t_n) \le 1 - \alpha$ and $d(\mathbf{o}, \partial\tilde{\Omega}_n) \to \infty$ as $n \to \infty$ (since x_n is not close to $\partial\Omega$). By choosing a further subsequence, we may assume $t_n \to \bar{t}$ as $n \to \infty$. Since the v_n are uniformly bounded in L^∞, [5, Theorem III 10.1] implies that v_n satisfies a Hölder condition uniformly in n on compact subsets of $\tilde{\Omega}_n \times [0,T]$. Thus, a subsequence of the v_n will converge uniformly on compact subsets of $\mathbb{R}^N \times [0,T]$ to a T-periodic function $\overline{v}(\boldsymbol{x}, t)$ such that $0 \le \overline{v} \le 1$ on $\mathbb{R}^N \times [0,T]$ and

$$(3) \qquad \overline{v}_t - \Delta\overline{v} = m(x_0, t)h(\overline{v}) \text{ in } \mathbb{R}^N \times [0,T] .$$

To see that $\overline{v}$ is a solution of (3), we note that, by the formula for m_n, $m_n(\boldsymbol{x}, t) \to m(x_0, t)$ uniformly on compact sets, and we pass to the limit in (2) by multiplying by a smooth T-periodic function ψ which is of compact support in $\boldsymbol{x}$ and integrate by part so that all the derivatives are on the ψ. Moreover, since $v_n(\mathbf{o}, t_n) \le 1 - \alpha$, $\overline{v}(\mathbf{o}, \bar{t}) \le 1 - \alpha$. Since $\overline{v} \le 1$ and $h(1) = 0$, the parabolic maximum principle implies that $\overline{v}(\boldsymbol{x}, t) < 1$ always ($\overline{v} \equiv 1$ is impossible because $\overline{v}(\mathbf{o}, \bar{t}) < 1$).

So far we have not used that $\xi(x_0) > 0$. By the lemma, there is a neighborhood U of x_0 and a $\delta > 0$ such that $u_{\varepsilon_n}(x, t) \ge \delta$ if $x \in U$, $t \in [0,T]$ and n is large. In the new variable, this implies that $v_n(\boldsymbol{x}, t) \ge \delta$ if $\boldsymbol{x}$ lies in a compact subset of $\mathbb{R}^N$ and n is large. Thus, passing to the limit, we see that $\overline{v} \ge \delta$ on $\mathbb{R}^N \times [0,T]$.

To obtain a contradiction, we use a change of variable similar to that in [4]. Choose $\alpha > \sup\{|h'(y)| \; : \; y \in [0,1]\}$ and let

$$w(\boldsymbol{x}, t) := \gamma(\overline{v}(\boldsymbol{x}, t))$$

where $\gamma(s) := \exp\,(-\alpha \int_{\frac{1}{2}}^{s} (h(\sigma))^{-1} d\sigma)$. Note that γ is positive and C^2 on $(0,1)$. Now $\gamma' = -\alpha\gamma/h < 0$ and $\gamma'' = \alpha(\alpha + h')\gamma/(h)^2 > 0$ on $(0,1)$ by our choice of α. Since γ is positive and decreasing, $\gamma(1)$ is finite. Thus w is positive, and since $\overline{v} \ge \delta$, w is bounded on $\mathbb{R}^N \times [0,T]$. Now

$$\begin{aligned} w_t - \Delta w &= \gamma'(\overline{v})(\overline{v}_t - \Delta\overline{v}) - \gamma''(\overline{v})|\nabla\overline{v}|^2 \\ &\le \gamma(\overline{v})m(x_0, t)h(\overline{v}) \\ &= -\alpha m(x_0, t)w \end{aligned}$$

by (3), the convexity of γ, and the formula for γ'. Let $z(x, t) := w(x, t)\exp(\alpha \int_0^t m(x_0, \tau) d\tau)$.

By a simple calculation,

$$z_t \le \Delta z \qquad \text{in } I\!R^N \times [0,T] \tag{4}$$

and

$$\begin{aligned} z(\boldsymbol{x},T) &= w(\boldsymbol{x},T)\exp(\alpha\xi(x_0)) \\ &= \tilde{\mu}w(\boldsymbol{x},T) = \tilde{\mu}w(\boldsymbol{x},0) = \tilde{\mu}z(\boldsymbol{x},0)\ , \end{aligned} \tag{5}$$

where $\tilde{\mu} = \exp\,(\alpha\xi(x_0)) > 1$. On the other hand since z is uniformly bounded on $I\!R^N \times [0,T]$, [6, Theorem III.10 and Remark (II) on p. 184] imply that $\sup_{\boldsymbol{x}\in I\!R^N} z(\boldsymbol{x},t)$ is not increasing in t. This contradicts (5) since z is positive in $I\!R^N \times [0,T]$ and $\tilde{\mu} > 1$. Hence we have proved uniform convergence in case $x_0 \in \Omega$.

Now suppose that $x_0 \in \partial\Omega$. The proof needs some minor changes. As before, the v_n are Hölder continuous uniformly in n (here we need the boundary estimates as in § V.7 of [5] as well as interior estimates and note that – as always in blowing up arguments – our change of variable only flattens the boundary). If $d(\mathbf{o},\partial\tilde{\Omega}_n)$ is not bounded as $n \to \infty$, we complete the argument as before. If this sequence is bounded, $\tilde{\Omega}_n$ will approach a half space $\tilde{H}$. Much as before v_n will converge uniformly on compact sets to a bounded continuous function $\overline{v}$ on $\tilde{H} \times [0,T]$ which solves (3) in int $\tilde{H} \times [0,T]$ and which is T-periodic. Moreover, the estimates in § V.7 in [5] imply that the v_n are bounded in C^1_α (in the $\boldsymbol{x}$-variables) on compact subsets of $\tilde{\Omega}_n \times [0,T]$ uniformly in n. Thus the $\boldsymbol{x}$-derivatives of the v_n converge on compact sets to those of $\overline{v}$. Since $\frac{\partial v_n}{\partial n} = 0$ on $\partial\tilde{\Omega}_n \times [0,T]$, it follows that $\frac{\partial \overline{v}}{\partial l} = 0$ on $\partial\tilde{H} \times [0,T]$, where l is the normal to $\tilde{H}$. Since $\tilde{H}$ is a half space and (3) has no explicit $\boldsymbol{x}$-dependence (and only even order and non-mixed derivatives) we can extend $\overline{v}$ to $I\!R^N \times [0,T]$ by reflecting in $\tilde{H}$ so that $\overline{v}$ is even about $\partial\tilde{H}$ and $\overline{v}$ will still be a solution of (3) (the Neumann boundary condition ensures that the extension of $\overline{v}$ is C^2). Hence we are back to a problem on $I\!R^N \times [0,T]$ and obtain a contradiction as before. This completes the proof of the theorem. □

REMARKS

1. We could replace the Laplacian by any second order uniformly elliptic operator with smooth coefficients and no constant term by essentially the same proof. We could also multiply Δ by a positive periodic function of t, but this could be eliminated by a rescaling in t.

2. It would be of interest to understand the transition layers and to cover the case where ξ vanishes identically. (Note that there always exist admissible solutions of (*) when ξ vanishes identically.) Basic to both of these questions would be a better understanding of T-periodic solutions of (3) (in particular solutions with $\inf u = 0$ and $\sup u = 1$). One difficulty with (3) is that if $\int_0^T m(x_0,t)dt = 0$, one easily sees that there always exist many spatially constant T-periodic solutions (cf. [4]). These are the only T-periodic solutions which satisfy $\inf u > 0$ or $\sup u < 1$. Note

that if there is an x_0 such that $m(x_0,t) \equiv 0$ on $[0,T]$ while ξ changes sign at x_0, there must be a transition layer near x_0 by our theorem. In this case the only T-periodic positive bounded solutions of (3) are constants by [6], and it follows that the transition layer near x_0 must have width of order larger than ε. In general, one easily sees that the width of the transition layer must be of order at least ε.

3. It would be of interest to know when the nontrivial T-periodic solution in the theorem is unique for small ε (as in [4], it is unique if h is strictly concave).

5 Proof of the Corollary

Suppose by way of contradiction that $\xi(x) < 0$ on $\overline{\Omega}$ and u_{ε_n} are admissible solutions of (*) for $\varepsilon = \varepsilon_n$ where $\varepsilon_n > 0$ and $\varepsilon_n \to 0$ as $n \to \infty$. By the theorem $u_{\varepsilon_n} \to 0$ uniformly on $\overline{\Omega} \times [0,T]$ as $n \to \infty$. Thus u_{ε_n} is a T-periodic positive solution of

$$(6)\qquad \begin{aligned} u_t - \varepsilon_n^2 \Delta u &= m_n(x,t)u && \text{in } \Omega \times [0,T] \\ \tfrac{\partial u}{\partial n} &= 0 && \text{on } \partial\Omega \times [0,T]\,, \end{aligned}$$

where $m_n(x,t) = g(u_{\varepsilon_n}(x,t))m(x,t)$. Here $g(y) = y^{-1}h(y)$ if $y \neq 0$ and $h'(0)$ if $y = 0$. Since $u_{\varepsilon_n} \to 0$ uniformly on $\overline{\Omega} \times [0,T]$, $m_n(x,t) \to h'(0)m(x,t)$ as $n \to \infty$ uniformly on $\overline{\Omega} \times [0,T]$.

Let $w_n = \frac{u_{\varepsilon_n}}{\|u_{\varepsilon_n}\|_\infty}$. Then w_n is a solution of (6) with $\|w_n\|_\infty = 1$. Choose $(x_n, t_n) \in \overline{\Omega} \times [0,T]$ such that $w_n(x_n,t_n) = 1$. By essentially the same blowing up argument as in the proof of the theorem, we obtain a nonnegative T-periodic solution $\overline{v}$ of

$$v_t - \Delta v = m(x_0,t)v \qquad \text{in } \mathbb{R}^N \times \mathbb{R}$$

such that $\|\overline{v}\|_\infty = 1$. (Here $x_n \to x_0$ as $n \to \infty$, and it is assumed for simplicity that $h'(0) = 1$ and hence $m_n \to m$ uniformly.) Note that $\|\overline{v}\|_\infty \leq 1$ since $\|w_n\|_\infty \leq 1$, and that $\overline{v}$ is nontrivial since, after we rescale w_n to v_n, we have that $v_n(\mathbf{o}, t_n) = 1$ and v_n converges uniformly to $\overline{v}$ on compact sets. We then obtain a contradiction by letting $z(x,t) = \overline{v}(x,t)\exp(-\int_0^t m(x_0,s)ds)$ and applying Protter-Weinberger's result as in the proof of the theorem. If $\xi(x) > 0$ on $\overline{\Omega}$, we obtain a similar contradiction by using the equation for $1-u$.

REMARK. Our methods here prove a general result on the spectrum of the linear problem

$$\begin{aligned} u_t - \varepsilon^2 \Delta u &= \lambda m(x,t)u && \text{in } \Omega \times \mathbb{R} \\ \tfrac{\partial u}{\partial n} &= 0 && \text{on } \partial\Omega \times \mathbb{R} \\ & u \text{ T-periodic in } t \end{aligned}$$

as $\varepsilon \to 0$. Assume that the positive principal eigenvalue $\lambda_1(m)$ exists for small positive ε. (This holds e.g. if $\int_\Omega \xi(x)dx < 0$ but $\int_0^T \max_{\overline{\Omega}} m(x,t)dt > 0$, cf. [1].) Then in case

$\xi(x_0) > 0$ for some x_0, $\lambda_1(m)$ converges to 0 as $\varepsilon \to 0$, while in case $\xi(x) < 0$ on $\overline{\Omega}$, $\lambda_1(m)$ is unbounded as $\varepsilon \to 0$.

References

[1] N.Alikakos - P. Hess : On a singularly perturbed semilinear periodic-parabolic problem. *Preprint.*

[2] H. Berestycki - P.L. Lions : Some applications of the method of sub- and supersolutions. *In : Bifurcation and nonlinear eigenvalue problems, Springer Lecture Notes in Mathematics* 782 (1980), pp. 16-41.

[3] E.N. Dancer - P. Hess : On stable solutions of quasilinear periodic-parabolic problems. *Annali Sc. Norm. Sup. Pisa* 14 (1987), pp. 123-141.

[4] P. Hess - H. Weinberger : Convergence to spatial-temporal clines in the Fisher equation with time-periodic fitnesses *J. Math. Biology* 28 (1990), pp. 83-98.

[5] O. Ladyzhenskaya, V. Solonnikov, N. Ural'ceva : Linear and quasilinear equations of parabolic type. *Amer. Math. Soc., Providence* 1968.

[6] M. Protter - H. Weinberger : Maximum principles in differential equations. *Prentice Hall, Englewood Cliffs* 1967.

[7] D. Sattinger : Topics in stability and bifurcation theory. *Springer Lectures Notes in Mathematics* 309 (1973).

[8] G. Stampacchia : Equations elliptiques du second ordre à coefficients discontinus. *Université de Montreal Presses* 1966.

Acknowledgement. This paper was written while the second author visited the University of New England at Armidale. This visit was supported by an Australian Research Council grant of E. N. Dancer.

On smoothing property of Schrödinger propagators

KENJI YAJIMA

Department of Pure and Applied Sciences
College of Arts and Sciences
University of Tokyo
3-8-1 Komaba, Meguroku, Tokyo, 153 Japan

§1. Introduction, Theorems.

In this paper, we are concerned with the smoothing property of the propagator, or the fundamental solution, for a time dependent Schrödinger equation

$$(1.1) \qquad i\partial_t u = (1/2)\sum_{j=1}^{n}(-i\partial_j - A_j(t,x))^2 u + V(t,x)u, \quad t \in \mathbf{R}^1,\ x \in \mathbf{R}^n$$

with vector and scalar potentials given by $A(t,x) = (A_1(t,x), \dots, A_n(t,x))$ and $V(t,x)$ respectively, where $\partial_j = \partial/\partial x_j$, $j = 1, \dots, n$.

Some 25 years ago, when Kato [**7**] introduced the notion of H-smooth operators, a rather astonishing fact was observed: A unitary group e^{-itH} in a Hilbert space $\mathcal{H}$ can have a small subspace $\mathcal{K}$ of $\mathcal{H}$ such that every trajectory of the group $e^{-itH}u$, $u \in \mathcal{H}$, is in $\mathcal{K}$ at $a.e.\, t \in \mathbf{R}^1$. A densely defined closed operator A from $\mathcal{H}$ to another Hilbert space $\widetilde{\mathcal{H}}$, possibly identical with $\mathcal{H}$, is said to be H-smooth if

$$\sup_{Im\,\zeta \neq 0} |((H-\zeta)^{-1}A^*\tilde{u}, A^*\tilde{u})| \le C\|\tilde{u}\|^2, \qquad \tilde{u} \in D(A^*) \subset \widetilde{\mathcal{H}}.$$

It is shown that A is H-smooth if and only if

$$\int_{-\infty}^{\infty} \|Ae^{-itH}u\|^2 dt \le C\|u\|^2, \qquad u \in \mathcal{H}$$

and, if A is H-smooth, $e^{-itH}u \in D(A)$ for $a.e.\, t \in \mathbf{R}^1$, $u \in \mathcal{H}$.

In the same paper [**7**], it was shown that the multiplication operator A in $L^2(\mathbf{R}^n)$ with $A \in L^{n-\varepsilon}(\mathbf{R}^n) \cap L^{n+\varepsilon}(\mathbf{R}^n)$, $\varepsilon > 0$ is $(-\Delta)$-smooth and $e^{it\Delta}u \in D(A)$ for $a.e.\, t \in \mathbf{R}^1$, $u \in L^2(\mathbf{R}^n)$. This suggested that $e^{it\Delta}u$ is smoother than the original function u for $a.e.\, t$ and, indeed, some ten years later it was found (cf. Strichartz [**13**]) that

$$(\int\int |e^{it\Delta}u(x)|^{2(n+2)/n} dxdt)^{n/2(n+2)} \le C\|u\|, \quad u \in L^2(\mathbf{R}^n),$$

which was subsequently extended to the form

$$(1.2) \qquad \left(\int\left\{\int |e^{it\Delta}u(x)|^p dx\right\}^{\theta/p}\right)^{1/\theta} \le C\|u\|, \quad u \in L^2(\mathbf{R}^n)$$

for $0 \le 2/\theta = n(1/2 - 1/p) < 1$ (cf. Ginibre-Velo [**6**]). The latter implies that for every $u \in L^2(\mathbf{R}^n)$, $e^{it\Delta}u \in L^2(\mathbf{R}^n) \cap L^p(\mathbf{R}^n)$, $2 \le p < 2n/(n-2)$ for *a.e.* $t \in \mathbf{R}^1$ and manifests the smoothing effect of the free Schrödinger propagator in the sense that it improves L^p-smoothness.

Again it was Kato [**8**] who discovered that a certain unitary group in L^2 can improve even the differentiability property: The solutions of the KdV equation

$$\partial_t u + \partial_x^3 u + a(u)\partial_x u = 0, \quad t > 0,\ x \in \mathbf{R}^1$$

with $u(0) = f \in L^2(\mathbf{R}^1)$ satisfy

$$\int_0^T \int_{-R}^R |\partial_x u(t,x)|^2 dx dt \le K(\|f\|, T, R)$$

if $\limsup_{|\lambda|\to\infty} |\lambda|^{-4}a(\lambda) \le 0$.

Later it was found (cf. Sjölin [**12**] and Constantin-Saut [**3**]) that such differentiability improving property is commonly possessed by the unitary groups generated by dispersive differential equations with constant coefficients and, for the free Schrödinger propagator, it is proved that

$$(1.3) \qquad \int_{\mathbf{R}^1}\int_{\mathbf{R}^n} |\phi(t,x)(1-\Delta)^{1/4}e^{it\Delta}u(x)|^2 dx dt \le C\|u\|^2, \quad u \in L^2(\mathbf{R}^n)$$

and for $1 \le q \le 2$, $r \ge 2$ and $\alpha < 1/r - n(1/q - 1/r)$,

$$(1.4) \qquad \left[\int_{\mathbf{R}^1}\int_{\mathbf{R}^n} |\phi(t,x)(1-\Delta)^{\alpha/2}e^{it\Delta}u(x)|^r dx dt\right]^{1/r} \le C\|u\|_{L^q}, \quad u \in L^q(\mathbf{R}^n),$$

where $\phi \in C_0^\infty(\mathbf{R}^{n+1})$ (see also Kato-Yajima [**9**] where ϕ in (1.3) is replaced by $\langle x\rangle^{-1/2-\varepsilon}$, $\varepsilon > 0$).

The purpose of this paper is to show that estimates of types (1.3) and (1.4) are satisfied not only by unitary groups generated by differential equations with constant coefficients but also by the propagators of Schrödinger equations of the form (1.1) with time dependent magnetic and scalar potentials which may increase at infinity $|x| \to \infty$. We assume that $A(t,x)$ and $V(t,x)$ respectively satisfy the following conditions (A.1) and (A.2). $\partial_x = (\partial_1, ..., \partial_n)$ and for multi-index $\alpha = (\alpha_1, ..., \alpha_n)$, $\partial_x^\alpha = \partial_1^{\alpha_1} \cdots \partial_n^{\alpha_n}$.

(A.1): For $j = 1, \ldots, n$, $A_j(t,x)$ is a real-valued function of $(t,x) \in \mathbf{R}^1 \times \mathbf{R}^n$ such that $\partial_x^\alpha A_j(t,x)$ is C^1 for any multi-index α. For $|\alpha| \ge 1$ we have, with some $\varepsilon > 0$,

$$(1.5) \qquad |\partial_x^\alpha B_{jk}(t,x)| \le C_\alpha(1+|x|)^{-1-\varepsilon}, \quad j,k = 1,\ldots,n,$$

$$(1.6) \qquad |\partial_x^\alpha A(t,x)| + |\partial_x^\alpha \partial_t A(t,x)| \le C_\alpha, \quad (t,x) \in \mathbf{R}^1 \times \mathbf{R}^n,$$

where $B_{jk}(t,x) = \partial_j A_k(t,x) - \partial_k A_j(t,x)$ is the strength tensor of the magnetic field.

(A.2): $V(t,x)$ is a real-valued function of $(t,x) \in \mathbf{R}^1 \times \mathbf{R}^n$ such that $\partial_x^\alpha V(t,x)$ is continuous for every α. For $|\alpha| \geq 2$ we have

$$|\partial_x^\alpha V(t,x)| \leq C_\alpha, \quad (t,x) \in \mathbf{R}^1 \times \mathbf{R}^n. \tag{1.7}$$

It is known ([**17**]) under the conditions (A.1) and (A.2) that Eqn. (1.1) generates a unique unitary propagator $\{U(t,s) : t,s \in \mathbf{R}^1\}$ in $L^2(\mathbf{R}^n)$ so that for every $s \in \mathbf{R}^1$ and

$$u_0 \in \Sigma(2) = \{f \in L^2(\mathbf{R}^n) : \|f\|^2_{\Sigma(2)} = \sum_{|\alpha+\beta|\leq 2} \|x^\alpha \partial_x^\beta f\|^2 < \infty\}, \tag{1.8}$$

$u(\cdot) = U(\cdot, s)u_0 \in C^1(\mathbf{R}^1, L^2(\mathbf{R}^n)) \cap C^0(\mathbf{R}^1, \Sigma(2))$ is a unique solution of (1.1) in $L^2(\mathbf{R}^n)$ with the initial condition $u(s) = u_0$. Here and hereafter $\| \quad \|$ and $(\ , \ )$ stand for the norm and the inner product in $L^2(\mathbf{R}^n)$. $\mathcal{S}(\mathbf{R}^n)$ is the space of rapidly decreasing functions and $\langle x \rangle = (1+x^2)^{1/2}$, $\langle D \rangle = (1-\Delta)^{1/2}$. The main results of this paper are summarized in the following four theorems.

THEOREM 1. *Suppose that (A.1) and (A.2) be satisfied. Let $T > 0$ be sufficiently small, $\mu > 1/2$ and $\rho \geq 0$. Then there exists a constant $C_{\rho\mu} > 0$ such that for $s \in \mathbf{R}^1$*

$$\int_{s-T}^{s+T} \|\langle x \rangle^{-\mu-\rho} \langle D \rangle^\rho U(t,s) f\|^2 dt \leq C_{\rho\mu} \|\langle D \rangle^{\rho - 1/2} f\|^2, \quad f \in \mathcal{S}(\mathbf{R}^n). \tag{1.9}$$

THEOREM 2. *Suppose that (A.1) and (A.2) be satisfied. Let $T > 0$ be sufficiently small, $p \geq 2$, $0 \leq 2/\theta = 2\sigma + n(1/2 - 1/p) < 1$ and $\rho \in \mathbf{R}$. Then there exists a constant $C_{p\rho\sigma} > 0$ such that for $s \in \mathbf{R}^1$*

$$\begin{aligned}(\int_{s-T}^{s+T} \{\int_{\mathbf{R}^n} |\langle x \rangle^{-2\sigma - |\rho|} \langle D \rangle^{\rho+\sigma} U(t,s) f(x)|^p dx\}^{\theta/p} dt)^{1/\theta} \\ \leq C_{p\rho\sigma} \|\langle D \rangle^\rho f\|, \quad f \in \mathcal{S}(\mathbf{R}^n).\end{aligned} \tag{1.10}$$

Theorem 1 and Theorem 2 are extensions of (1.3) and (1.4) (with $q = 2$) to differential equations with variable coefficients. We emphasize that Theorem 2 is *not* a consequence of Theorem 1 and Sobolev inequality and that it is an improvement of (1.4)even when $A(t,x) \equiv 0$ and $V(t,x) \equiv 0$. It uses different L^p-norms for space and time variables for the first place and when $\theta = p$, σ in (1.10) can be as big as $1/p - (n/2)(1/2 - 1/p)$ whereas in (1.4), $\sigma < 1/p - n(1/2 - 1/p)$.

A consequence of Theorem 1 is the following *maximal inequality* of Schrödinger type.

THEOREM 3. *Suppose that (A.1) and (A.2) be satisfied. Let $T > 0$ be sufficiently small, $\gamma > 1/2$ and $\delta > 2\gamma + 1/2$. Then there exists a constant $C > 0$ such that*

$$\int \sup_{|t-s|<T} |\langle x \rangle^{-\delta} U(t,s) f(x)|^2 dx \leq C \|f\|^2_{H^\gamma}, \quad f \in \mathcal{S}(\mathbf{R}^n). \tag{1.11}$$

The *maximal* inequality (1.11) has the following consequence, which, in the case when $A(t,x)$ and $V(t,x)$ are t-independent, will give a *summability theorem* for the (generalized) eigenfunction expansions associated with the Schrödinger operator

$$H = (1/2)\sum_{j=1}^{n}(-i\partial_j - A_j(x))^2 + V(x). \tag{1.12}$$

We shall denote by $H(t)$ the t-dependent Schrödinger operator appearing in the right hand side of (1.1).

THEOREM 4. *Suppose that (A.1) and (A.2) be satisfied and $f \in H^{\gamma}(\mathbf{R}^n) \cap L^1(\mathbf{R}^n)$ with $\gamma > 1/2$. Then for any $s \in \mathbf{R}^1$*

$$\lim_{t\to s} U(t,s)f(x) = f(x), \quad a.e.\, x \in \mathbf{R}^n. \tag{1.13}$$

REMARK: (a) The case that the potentials are singular will be treated in a forthcoming paper and we treat only smooth potentials here.
(b) The gauge transformation $Tu(t,x) = exp(-i\int_s^t V(r,x)dr)u(t,x)$ eliminates $V(t,x)$ from the equation (1.1) by changing $A(t,x)$ to $A(t,x) + \int_s^t \partial_x V(r,x)dr$. However, T will change the Sobolev norm simultaneously and for this reason we shall *not* use this transformation here (cf. [**17**]).
(c) For the free Schrödinger propagator $\exp(it\Delta)$, Carleson[**2**] proved the summability theorem, Theorem 4, for $n = 1$ with $\gamma = 1/4$, and Dahlberg and Kenig[**4**] remarked that Carleson's result is sharp. Again for $n = 1$, Kenig and Ruiz[**10**] showed that the maximal inequality (1.12) is satisfied with $\gamma = 1/4$ and that the estimate is sharp. These one-dimensional results were subsequently extended by Sjölin [**12**] to higher dimensions. It seems still open, however, if these estimates are sharp or not in general.
(d) We should mention that Constantin and Saut[**3**] treats singular scalar potentials such that $V(t,x) \in L^q(I, L^p(\mathbf{R}^n))$ for some p and q by perturbation technique.
(e) The smoothing effect of the propagator $U(t,s)$ in terms of L^p-norms has already been established in [**16**] and [**17**] for a wider class of potentials: If $V = V_0 + V_1$ and V_0 satisfies (A.2) and $V_1 \in L^{\alpha}(\mathbf{R}^1, L^p(\mathbf{R}^n)) + L^1(\mathbf{R}^1, L^{\infty}(\mathbf{R}^n))$, $p \geq 1$ with $p > n/2$, $\alpha = 2p/2p - n$, then

$$\left(\int_{s-T}^{s+T} \|U(t,s)f\|_{2p/p-1}^{4p/n} dt\right)^{n/4p} \leq C_p(1+T)^{n/4p}\|f\|, \quad f \in L^2(\mathbf{R}^n), \tag{1.14}$$

where $\| \quad \|_q$ is the L^q-norm. This is an extension of (1.2) to the case with interaction. Note that the right hand side of (1.14) increases with T in contrast to (1.2). This is natural because bound states exist in general (which are periodic (in t) solutions of (1.1) when $H(t)$ is t-independent).

The rest of this paper is devoted to proving Theorems 1 $\sim$ 4.

§2. Preliminaries.

As the proofs of Theorems 1 $\sim$ 2 will heavily rely upon the results in [**17**], in particular, the representation formula of the propagator $U(t,s)$ in the form of an oscillatory integral operator (OIO), we shall briefly recall some of its contents and derive some of their immediate consequences which will be necessary in what follows. We refer the readers to [**17**] for details and proofs which are omitted here.

We first recall how the propagator is constructed in the form of OIO. We write $H(t,x,\xi) = (1/2)\sum_{j=1}^{n}(\xi_j - A_j(t,x))^2 + V(t,x)$ for the classical mechanical Hamiltonian associated with (1.1) and denote by $(x(t,s,y,\eta),\xi(t,s,y,\eta))$ the solutions of Hamilton's equations

$$dx/dt = \partial H/\partial\xi, \quad d\xi/dt = -\partial H/\partial x \tag{2.1}$$

with the initial conditions $(x(s),\xi(s)) = (y,\eta)$. Introducing the variables of the same dimension ([**5**]), we set,

$$\tilde{x}(t,s,y,\eta) = x(t,s,y,\eta/(t-s)), \quad \tilde{\xi}(t,s,y,\eta) = (t-s)\xi(t,s,y,\eta/(t-s))$$

for $t \neq s$ and $\tilde{x}(t,t,y,\eta) = y+\eta$, $\tilde{\xi}(t,t,y,\eta) = \eta$. It is clear that the map

$$(y,\eta) \to (\tilde{x}(t,s,y,\eta),\tilde{\xi}(t,s,y,\eta)) \tag{2.2}$$

is also canonical.

LEMMA 2.1. *The solutions $(\tilde{x}(t,s,y,\eta),\tilde{\xi}(t,s,y,\eta))$ exist globally for $t,s \in \mathbb{R}^1$ and, for every α and β, $\partial_y^\alpha\partial_\eta^\beta\tilde{x}$ and $\partial_y^\alpha\partial_\eta^\beta\tilde{\xi}$ are C^1 functions of (t,s,y,η). For $|t-s| \le 1$*

$$\begin{aligned} &|\partial_y^\alpha\partial_\eta^\beta(\partial\tilde{x}_j/\partial y_k - \delta_{jk})| + |\partial_y^\alpha\partial_\eta^\beta(\partial\tilde{x}_j/\partial\eta_k - \delta_{jk})| \\ &\quad + |\partial_y^\alpha\partial_\eta^\beta(\partial\tilde{\xi}_j/\partial y_k)| + |\partial_y^\alpha\partial_\eta^\beta(\partial\tilde{\xi}_j/\partial\eta_k - \delta_{jk})| \le C_{\alpha\beta}|t-s|, \end{aligned} \tag{2.3}$$

where δ_{jk} is Kronecker's delta;

$$|\tilde{\xi}(t,x,y,\eta) - \eta| + |\tilde{x}(t,s,y,\eta) - y - \eta| \le C|t-s|(1+|y|+|\eta|). \tag{2.4}$$

It can be deduced from (2.3) and Hadamard's global implicit function theorem that there exists $T > 0$ such that the mapping $\mathbb{R}^n \ni \eta \to x = \tilde{x}(t,s,y,\eta) \in \mathbb{R}^n$ is a global diffeomorphism of $\mathbb{R}^n$ for every fixed $|t-s| < T$ and $y \in \mathbb{R}^n$. Write $\tilde{\eta}(t,s,y,x)$ for its inverse. Then

$$(x(\tau),\xi(\tau)) = (x(\tau,s,y,(t-s)^{-1}\tilde{\eta}(t,s,y,x)),\xi(\tau,s,y,(t-s)^{-1}\tilde{\eta}(t,s,y,x))) \tag{2.5}$$

is a unique solution of (2.1) such that $x(t) = x$ and $x(s) = y$. We denote by $S(t,s,x,y)$ the action integral along this path:

$$S(t,s,x,y) = \int_s^t \{(\partial_\xi H)(\tau,x(\tau),\xi(\tau))\cdot\xi(\tau) - H(\tau,x(\tau),\xi(\tau))\}d\tau. \tag{2.6}$$

LEMMA 2.2. *For $t,s \in \mathbb{R}^1$ with $0 < |t-s| < T$, $S(t,s,x,y)$ is C^1 in (t,s,x,y) and C^∞ in (x,y). $(t-s)S(t,s,x,y)$ is a generating function of the canonical map (2.2):*

$$(t-s)(\partial_x S)(t,s,x,y) = \tilde{\xi}(t,s,y,\tilde{\eta}(t,s,y,x)), \tag{2.7}$$

$$(t-s)(\partial_y S)(t,s,x,y) = -\tilde{\eta}(t,s,y,x); \tag{2.8}$$

and $S(t,s,x,y)$ satisfies Hamilton-Jacobi equations:

$$(\partial_t S)(t,s,x,y) + (1/2)((\partial_x S)(t,s,x,y) - A(t,x))^2 + V(t,x) = 0,$$
$$(\partial_s S)(t,s,x,y) - (1/2)((\partial_y S)(t,s,x,y) + A(s,y))^2 - V(s,y) = 0.$$

Moreover, we have

(2.9) $$|\partial_x^\alpha \partial_y^\beta \{S(t,s,x,y) - (x-y)^2/2(t-s)\}| \le C_{\alpha\beta}, \quad |\alpha+\beta| \ge 2.$$

Using $S(t,s,x,y)$ as a phase function, we construct the propagator $U(t,s)$ in the form of an OIO as follows:

THEOREM 2.3. *Let $T > 0$ be sufficiently small. There uniquely exists a function $e(t,s,x,y)$ with the following properties.*
(1) For any α and β, $\partial_x^\alpha \partial_y^\beta e(t,s,x,y)$ is C^1 in (t,s,x,y) for $|t-s| < T$ and we have

(2.10) $$|\partial_x^\alpha \partial_y^\beta e(t,s,x,y)| \le C_{\alpha\beta}, \quad |t-s| < T, \quad x, y \in \mathbf{R}^n.$$

(2) The family of oscillatory integral operators $\{U(t,s)\}$ defined for $0 < |t-s| < T$ by

(2.11) $$U(t,s)f(x) = (2\pi i(t-s))^{-n/2} \int e^{iS(t,s,x,y)} e(t,s,x,y) f(y)dy, \quad f \in \mathcal{S}(\mathbf{R}^n),$$

is a unique propagator for (1.1) with the following properties:
(a) For every $t \neq s$, $U(t,s)$ maps $\mathcal{S}(\mathbf{R}^n)$ into $\mathcal{S}(\mathbf{R}^n)$ continuously and extends to a unitary operator in $L^2(\mathbf{R}^n)$.
(b) If we set $U(t,t) = 1$, the identity operator, then $\{U(t,s) : |t-s| < T, t, s \in \mathbf{R}^1\}$ is strongly continuous in $L^2(\mathbf{R}^n)$ and satisfies $U(t,r)U(r,s) = U(t,s)$.
(c) For $f \in \sum(2)$, $U(t,s)f$ is a $\sum(2)$-valued continuous and $L^2(\mathbf{R}^n)$-valued C^1 function of (t,s). It satisfies $i\partial_t U(t,s)f = H(t)U(t,s)f$ and $i\partial_s U(t,s)f = -U(t,s)H(s)f$.

By a standard technique of semi-group theory, it is then easy to extend the domain of definition of the family of operators $\{U(t,s)\}$ to the whole real line $t, s \in \mathbf{R}^1$ in such a way that $(a) \sim (c)$ of Theorem 2.3 remain valid for the resulting $\{U(t,s)\}$.

When $a(t,s,x,y)$ satisfies estimates (2.10), we say $a \in Amp$ and write $I(t,s,a)$ for the OIO with the phase $S(t,s,x,y)$ and amplitude $a(t,s,x,y)$:

(2.12) $$I(t,s,a)f(x) = (2\pi i(t-s))^{-n/2} \int e^{iS(t,s,x,y)} a(t,s,x,y) f(y)dy.$$

With this notation, (2.11) may be written

(2.11′) $$U(t,s) = I(t,s,e).$$

For operators of the form (2.12) a general theory was developed by Asada and Fujiwara[**1**] and Fujiwara[**5**] and the proofs of (1) to (3) of the following lemma can be found there. We denote by Sym_0 the set of symbols $p(t,s,x,\xi,y)$ with the parameters t and s such that

$$|\partial_x^\alpha \partial_\xi^\beta \partial_y^\gamma p(t,s,x,\xi,y)| \leq C_{\alpha\beta\gamma}, \quad |t-s|<T, \quad x,y \in \mathbf{R}^n.$$

For $p \in Sym_0$ the ν-pseudo-differential operator $p(t,s,x,\nu D,x)$, $\nu = t-s$, is defined by:

$$p(t,s,x,\nu D,x)f(x) = (2\pi\nu)^{-n} \int e^{i(x-y)\xi/\nu} p(t,s,x,\xi,y) f(y)dyd\xi.$$

$\Psi DO(t,s)$ is the set of all ν-pseudo-differential operators. We denote

$$\mathcal{B}(\mathbf{R}^m) = \{f : \|f\|_{\mathcal{B}^k} = sup_{x\in\mathbf{R}^m} \sum_{|\alpha|\leq k} |\partial_x^\alpha f(x)| < \infty, \; k=0,1,2,...\},$$

which is a Fréchet space.

LEMMA 2.4. *Let $a,b \in Amp$ and $p \in Sym_0$. Then:*
(1) $I(t,s,a)$ and $p(t,s,x,\nu D,x)$ are bounded operators in $L^2(\mathbf{R}^n)$:

$$\|I(t,s,a)f\| \leq C\|a(t,s,\cdot,\cdot)\|_{\mathcal{B}^{2n+1}}\|\|f\|, \tag{2.13}$$

$$\|p(t,s,x,\nu D,x)f\| \leq C\|p(t,s,\cdot,\cdot,\cdot)\|_{\mathcal{B}^{3n+1}}\|f\|. \tag{2.14}$$

(2) There exist c,d,f and $g \in Amp$ such that

$$I(t,s,a)^* = I(s,t,c), \quad I(t,r,a)I(r,s,b) = I(t,s,d), \tag{2.15}$$

$$I(t,s,a)p(t,s,x,\nu D,x) = I(t,s,f), \quad p(t,s,x,\nu D,x)I(t,s,a) = I(t,s,g). \tag{2.16}$$

Moreover, c,d,f and g are continuously dependent upon a,b and p in the topology of $\mathcal{B}$-spaces.
(3) $I(t,s,a)I(s,t,b) \in \Psi DO(t,s)$.
(4) There exist $a_{ij,k\ell}$ and $a_{i,k} \in Amp$, $i,j=1,\ldots,n$ and $k,\ell=1,2$, such that

$$x_i I(t,s,a) = I(t,s,a)x_i + (t-s)\sum_{j=1}^n \{I(t,s,a_{ij,11})x_j + I(t,s,a_{ij,12})D_j\} + (t-s)I(t,s,a_{i,1}); \tag{2.17}$$

$$D_i I(t,s,a) = I(t,s,a)D_i + \sum_{j=1}^n \{I(t,s,a_{ij,21})x_j + (t-s)I(t,s,a_{ij,22})D_j\} + I(t,s,a_{i,2}), \tag{2.18}$$

where x_j stands for the multiplication operator by x_j and $D_j = -i\partial/\partial x_j$, $j=1,\ldots,n$.

PROOF: The lemma is a simplified version of Lemma 3.1 and Proposition 3.2 of [**17**]. We prove (2.18) here for the reader's convenience. The stationary phase method due to [**1**] implies, with $\nu = t-s$,

$$\nu D_i I(t,s,a) = I(t,s,a)\tilde{\xi}_i(t,s,y,\nu D) + \nu I(t,s,g_i), \tag{2.19}$$

where $g_i \in Amp$ and

$$\tilde{\xi}_i(t,s,y,\nu D)f(x) = (2\pi\nu)^{-n}\int e^{i(x-y)\cdot\eta/\nu}\tilde{\xi}_i(t,s,y,\eta)f(y)dyd\eta.$$

Using Taylor's formula, we write, with $\tilde{\xi}_{ij,1}(t,s,y,\eta) = \int_0^1(\partial\tilde{\xi}_i/\partial y_j)(t,s,\theta y,\theta\eta)d\theta$ and $\tilde{\xi}_{ij,2}(t,s,y,\eta) = \int_0^1(\partial\tilde{\xi}_i/\partial\eta_j)(t,s,\theta y,\theta\eta)d\theta$,

$$\tilde{\xi}_i(t,s,y,\eta) = \tilde{\xi}_i(t,s,0,0) + \sum_{j=1}^{n}\{\tilde{\xi}_{ij,1}(t,s,y,\eta)y_j + \tilde{\xi}_{ij,2}(t,s,y,\eta)\eta_j\}. \tag{2.20}$$

Inserting (2.20) into (2.19) and dividing the resulting equation by ν yield

$$\begin{aligned}D_iI(t,s,a) &= I(t,s,a)D_i + I(t,s,a)\sum_{j=1}^{n}[\nu^{-1}\tilde{\xi}_{ij,1}(t,s,y,\nu D)x_j\\ &+\nu^{-1}\{\tilde{\xi}_{ij,2}(t,s,y,\nu D)-\delta_{ij}\}\nu D_j - i(\partial\tilde{\xi}_{ij,2}/\partial y_j)(t,s,y,\nu D)]\\ &+ I(t,s,g_i+\nu^{-1}\tilde{\xi}_i(t,s,0,0)a).\end{aligned}$$

Here (2.3) and (2.4) imply $|\tilde{\xi}(t,s,0,0)| \le C|t-s|^2$ and

$$\nu^{-1}\tilde{\xi}_{ij,1}(t,s,y,\eta),\ \nu^{-1}\{\tilde{\xi}_{ij,2}(t,s,y,\eta)-\delta_{ij}\} \in Sym_0.$$

Hence, an application of (2.16) implies (2.18). The proof for (2.17) is similar. ∎

An inductive application of (2.18) shows that for any α,

$$D^\alpha I(t,s,a) = \sum_{|\beta+\gamma|\le|\alpha|} I(t,s,a_{\alpha\beta\gamma})x^\beta D^\gamma, \quad a_{\alpha\beta\gamma}\in Amp, \tag{2.21}$$

in particular, with $\Psi_{\alpha\beta\gamma} \in \Psi DO(t,s)$,

$$\begin{aligned}D^\alpha U(t,s) &= \sum_{|\beta+\gamma|\le|\alpha|} I(t,s,e_{\alpha\beta\gamma})x^\beta D^\gamma,\\ &= \sum_{|\beta+\gamma|\le|\alpha|} U(t,s)U(s,t)I(t,s,e_{\alpha\beta\gamma})x^\beta D^\gamma = U(t,s)\sum_{|\beta+\gamma|\le|\alpha|}\Psi_{\alpha\beta\gamma}(t,s)x^\beta D^\gamma.\end{aligned} \tag{2.22}$$

LEMMA 2.5. *Let $T>0$ be sufficiently small. We have, with $\psi_{\alpha\beta\gamma}(t,s)\in\Psi DO(t,s)$,*

$$D^\alpha U(t,s) = \sum_{|\beta+\gamma|\le|\alpha|} x^\beta U(t,s)\psi_{\alpha\beta\gamma}(t,s)D^\gamma, \quad |t-s|<T. \tag{2.23}$$

PROOF: We apply (2.18) to $U(s,t)$, take the adjoint and use (2.15). This gives

$$\begin{aligned}D_iU(t,s) &= U(t,s)D_i + \sum x_jI(t,s,a_{ji})\\ &+\sum(t-s)D_jI(t,s,b_{ji}) + I(t,s,c_i)\end{aligned} \tag{2.24}$$

with a_{ji}, b_{ji} and $c_i \in Amp$. Writing

$$I(t,s,b_{ji}) = U(t,s)U(s,t)I(t,s,b_{ji}) = U(t,s)\Phi_{ji}(t,s), \ \ \Phi_{ji}(t,s) \in \Psi DO(t,s),$$

we put (2.24) in a matrix form:

$$DU(t,s)(1-(t-s)\Phi(t,s)) = U(t,s)D + xI(t,s,A) + I(t,s,c), \tag{2.25}$$

where the notation should be obvious. Here $1-(t-s)\Phi(t,s)$ is invertible for small $|t-s| \leq T$ and

$$(1-(t-s)\Phi(t,s))^{-1} = 1 + (t-s)\tilde{\Phi}(t,s) = \tilde{\Psi}(t,s) \in \Psi DO(t,s),$$

(cf. Kumano-go [**11**]). Hence multiplying (2.25) by $(1-(t-s)\Phi(t,s))^{-1}$ from the right yields

$$DU(t,s) = U(t,s)D\tilde{\Psi}(t,s) + xI(t,s,\tilde{A}) + I(t,s,\tilde{c}), \tag{2.26}$$

where $\tilde{A}$ and $\tilde{c} \in Amp$. Since $D\tilde{\Psi}(t,s) - \tilde{\Psi}(t,s)D = \Psi(t,s) \in \Psi DO(t,s)$ and

$$I(t,s,\tilde{A}) = U(t,s)U(s,t)I(t,s,\tilde{A}) = U(t,s)\Xi(t,s), \ \Xi(t,s) \in \Psi DO(t,s),$$

the right hand side of (2.26) can be put in the form

$$DU(t,s) = U(t,s)\tilde{\Psi}(t,s)D + xU(t,s)\Xi(t,s) + U(t,s)\Psi_1(t,s), \tag{2.27}$$

which proves (2.23) for $|\alpha| = 1$. Here, of course, $\Psi_1(t,s) = \Psi(t,s) + U(s,t)I(t,s,\tilde{c}) \in \Psi DO(t,s)$.

Suppose now that (2.23) is satisfied for any $|\alpha| \leq k$. Then by applying (2.27) to (2.23), we have

$$\begin{aligned} D\cdot D^\alpha U(t,s) = \sum_{|\beta+\gamma|\leq|\alpha|} & \{[D,x^\beta]U(t,s)\psi_{\alpha\beta\gamma}(t,s)D^\gamma + x^\beta(U(t,s)\tilde{\Psi}(t,s)D \\ & + xU(t,s)\Xi(t,s) + U(t,s)\Psi_1(t,s))\psi_{\alpha\beta\gamma}(t,s)D^\gamma\}. \end{aligned}$$

Except for the second one, all terms in the RHS are already of the form

$$x^\beta U(t,s)\tilde{\psi}_{\alpha\beta\gamma}(t,s)D^\gamma, \quad \tilde{\psi}_{\alpha\beta\gamma}(t,s) \in \Psi DO(t,s), \quad |\beta+\gamma| \leq |\alpha|+1.$$

The second term can be written in the form

$$\begin{aligned} & x^\beta U(t,s)\tilde{\Psi}(t,s)D\cdot\psi_{\alpha\beta\gamma}(t,s)D^\gamma \\ & \qquad = x^\beta U(t,s)\tilde{\Psi}(t,s)(\psi_{\alpha\beta\gamma}(t,s)D\cdot D^\gamma + [D,\psi_{\alpha\beta\gamma}(t,s)]D^\gamma) \end{aligned}$$

and $\tilde{\Psi}(t,s)\psi_{\alpha\beta\gamma}(t,s)$ and $[D,\psi_{\alpha\beta\gamma}(t,s)] \in \Psi DO(t,s)$. Hence (2.23) holds also for $|\alpha| = k+1$ and this concludes the proof of the lemma. ∎

We shall need two more lemmas. Recall that $\tilde{\eta}(t,s,y,x)$ is the inverse of $\eta \to x = \tilde{x}(t,s,y,\eta)$ and it satisfies $|\partial_x\tilde{\eta} - 1| \leq C|t-s|$ and $|\partial_y\tilde{\eta}+1| \leq C|t-s|$, in virtue of (2.3).

LEMMA 2.6. *Let $X' = x(t,s,(y+z)/2,\xi)$ and let $X = X(t,s,z,y,\xi)$ be defined by*

$$(t-s)\xi = \int_0^1 \tilde{\eta}(t,s,\theta z + (1-\theta)y, X)d\theta \tag{2.28}$$

for sufficiently small $|t-s| < T$. Then, we have

$$|\partial_z^\alpha \partial_y^\beta \partial_\xi^\gamma X(t,s,z,y,\xi)| \le C_{\alpha\beta\gamma}, \quad |\alpha+\beta+\gamma| \ge 1 \tag{2.29}$$

and

$$C_1|t-s||y-z| \le |X - x(t,s,(y+z)/2,\xi)| \le C_2|t-s||y-z|. \tag{2.30}$$

PROOF: In virtue of (2.3), it is obvious that X is uniquely determined by (2.28) and that (2.29) is satisfied. Since $(t-s)\xi = \tilde{\eta}(t,s,(y+z)/2,X')$, a simple computation yields,

$$\begin{aligned} 0 &= \int_0^1 \tilde{\eta}(t,s,\theta z + (1-\theta)y, X)d\theta - \tilde{\eta}(t,s,(y+z)/2,X') \\ &= J_1(t,s,y,z,\xi)(X-X') - J_2(t,s,y,z,\xi)(y-z), \end{aligned} \tag{2.31}$$

where

$$J_1 = \int_0^1 (\partial_x\tilde{\eta})(t,s,(y+z)/2, \theta X + (1-\theta)X')d\theta,$$

$$J_2 = \int_0^1 (\theta - 1/2)d\theta\{\int_0^1 (\partial_y\tilde{\eta})(t,s,\mu(\theta z + (1-\theta)y) + (1-\mu)(y+z)/2, X)d\mu\}.$$

In virtue of (2.3), we have $|J_1 - 1| \le 10^{-2}$ and $|J_2| \le C|t-s|$ for sufficiently small $|t-s| < T$ and (2.30) follows immediately. ∎

LEMMA 2.7. *There exists a positive number T such that for any $\rho > 1$,*

$$\int_{s-T}^{s+T} \langle x(t,s,y,\eta)\rangle^{-\rho} dt \le C\langle\eta\rangle^{-1}, \tag{2.32}$$

where the constant C is independent of $s \in \mathbf{R}^1$ and $y, \eta \in \mathbf{R}^n$.

PROOF: It suffices to show (2.32) for large $|\eta| > 1$. It is convenient to introduce the velocity variables $v(t,s,y,\eta) = \xi(t,s,y,\eta) - A(t,x(t,s,y,\eta))$. By (2.4) we have $(1/2)|\eta| - C_1\langle y\rangle \le |v| \le 2|\eta| + C_1\langle y\rangle$ and $(1/2)|y| - C_2|t-s||\eta| \le |x| \le 2|y| + C_2|t-s||\eta|$ for sufficiently small $|t-s| \le T$. We may assume C_1, $C_2 > 1$. (2.1), (A.1) and (A.2) imply that

$$(d^2/dt^2)x^2 = 2v^2 - 2x\cdot\{B(t,x)v + F(t,x)\} \ge 2v^2 - C_3|x|(|v| + \langle x\rangle), \quad C_3 \ge 1, \tag{2.33}$$

where $B = (B_{jk})$ and $F = -(\partial_t A + \partial_x V)$. Set $T = 10^{-4}C_2^{-1}(C_1+C_3)^{-1}$ by using these constants and consider for $|\eta| \ge 10^4(C_1 + C_3)$.

If $|\eta| \leq 10^3(C_1+C_3)\langle y\rangle$, then $|x| \geq (1/2)|y| - C_2|t-s||\eta| \geq 10^{-4}(C_1+C_3)^{-1}|\eta|$ and (2.33) immediately follows.

Suppose $|\eta| \geq 10^3(C_1+C_3)\langle y\rangle$, on the contrary. Then $(1/3)|\eta| \leq |v| \leq 3|\eta|$ and $|x| \leq 400^{-1}(C_1+C_3)^{-1}|\eta|$. Hence $(d^2/dt^2)x^2 \geq 2v^2 - C_3|x|(|v|+\langle x\rangle) \geq |\eta|^2/100$. This implies $x^2(t) \geq (200)^{-1}|\eta|^2(t-\sigma)^2 + x(\sigma)^2$ for some $s-T \leq \sigma \leq s+T$. Thus we have

$$\int_{s-T}^{s+T} \langle x(t,s,y,\eta)\rangle^{-\rho} dt \leq C \int_{s-T}^{s+T} \langle (t-\sigma)|\eta|\rangle^{-\rho} dt \leq C\langle\eta\rangle^{-1}$$

and this completes the proof of the lemma. ∎

§3. Proof of Theorems.

We begin with the

PROOF OF THEOREM 1: Without losing generality, we may assume $1/2 < \mu \leq 1$. For $a \in Amp$ we write, with $\nu = t-s$,

$$\|\langle x\rangle^{-\mu} I(t,s,a)f\|^2 = |2\pi\nu|^{-n} \int e^{i(S(t,s,x,y)-S(t,s,x,z))} \langle x\rangle^{-2\mu} a(t,s,x,y)\overline{a(t,s,x,z)} f(y)\overline{f(z)}\, dy dz dx \tag{3.1}$$

and make a change of variables $x = X(t,s,y,z,\xi)$, where $X(t,s,y,z,\xi)$ is the function defined by (2.28). We denote

$$\begin{aligned} G(t,s,y,\xi,z) &= |\nu|^{-n} a(t,s,X,y)\overline{a(t,s,X,z)}|det(\partial_\xi X)| \\ &= a(t,s,X,y)\overline{a(t,s,X,z)}|det \int_0^1 (\partial_x \tilde{\eta})(t,s,\theta z+(1-\theta)y, X)d\theta|^{-1} \end{aligned}$$

and, using the fact that

$$S(t,s,X,y) - S(t,s,X,z) = (t-s)^{-1}(\int_0^1 \tilde{\eta}(t,s,\theta z+(1-\theta)y,x)d\theta)(z-y) = \xi(z-y),$$

which is an immediate consequence of (2.8) and Talyor's formula, and performing integrations by parts in ξ variables, we write the right hand side of (3.1) in the form

$$\begin{aligned} &|2\pi\nu|^{-n} \int e^{i(z-y)\cdot\xi}\langle X\rangle^{-2\mu} a(t,s,X,y)\overline{a(t,s,X,z)}|det(\partial_\xi X)| f(y)\overline{f(z)}\, dy dz d\xi \\ &= (2\pi)^{-n} \int e^{i(z-y)\cdot\xi}\langle y-z\rangle^{-2}(1-\Delta_\xi)\{\langle X\rangle^{-2\mu} G(t,s,y,\xi,z)\} f(y)\overline{f(z)}\, dy dz d\xi. \end{aligned}$$

Here $\{G(t,s,\cdot,\cdot,\cdot) : |t-s| < T\} \subset Sym_0$ is bounded in virtue of (2.3) and (2.29) and, if we write $X' = x(t,s,(y+z)/2,\xi)$,

$$|\partial_y^\alpha \partial_z^\beta \partial_\xi^\gamma \{\langle y-z\rangle^{-2}\langle X\rangle^{-2\mu}\}| \leq C\langle y-z\rangle^{-2}\langle X\rangle^{-2\mu} \leq C\langle X'\rangle^{-2\mu}$$

by (2.29) and (2.30). Hence, by (2.32),

$$F(s,y,\xi,z) = |\nu|^{-n}\int_{s-T}^{s+T}\langle y-z\rangle^{-2}(1-\Delta_\xi)\{\langle X\rangle^{-2\mu}G(t,s,y,\xi,z)\}dt,$$

is bounded in $S^{-1}_{0,0}$, that is, $\{\langle\xi\rangle F(s,\cdot,\cdot,\cdot) : s\in \mathbf{R}^1\}$ is bounded in $\mathcal{B}(\mathbf{R}^n\times\mathbf{R}^n\times\mathbf{R}^n)$, and Calderón-Vaillancourt theorem ([**14**]) implies

(3.2)
$$\int_{s-T}^{s+T}\|\langle x\rangle^{-\mu}I(t,s,a)f\|^2dt$$
$$= (2\pi)^{-n}\int e^{i(z-y)\cdot\xi}F(t,s,y,\xi,z)f(y)\overline{f(z)}dydzd\xi \le C\|\langle D\rangle^{-1/2}f\|^2.$$

This proves Theorem 1 for $\rho=0$, since $U(t,s)=I(t,s,e)$.

When $\rho=k$ is an integer, we have by Lemma 3.3 and 3.4 that

$$\langle x\rangle^{-\mu-k}D^\alpha U(t,s)f = \sum_{|\beta+\gamma\le k}\langle x\rangle^{-\mu-k}x^\gamma I(t,s,a_{\beta\gamma})D^\gamma. \tag{3.3}$$

Then (3.2) applied to (3.3) implies Theorem 1 for integral ρ. The general case then follows by the interpolation theorem (cf. Triebel [**15**], Theorem 1.11.3 and 3.4.2, Yamazaki [**18**]). ∎

For proving Theorem 2, we prepare several lemmas.

LEMMA 3.1. *Let $0<|t-s|<T$ be sufficiently small. Then for any $a\in Amp$, $2\le p<\infty$ and $\sigma\ge 0$,*

$$\|\langle D\rangle^\sigma\langle x\rangle^{-2\sigma}I(t,s,a)\langle x\rangle^{-2\sigma}\langle D\rangle^\sigma f\|_{L^p} \le C|t-s|^{-2\sigma-n(1/2-1/p)}\|f\|_{L^{p'}}, \tag{3.4}$$

where p' is the index conjugate to p: $1/p+1/p'=1$, and the constant C does not depend on t,s and $f\in\mathcal{S}(\mathbf{R}^n)$.

PROOF: We write $\nu=t-s$. By virtue of the (complex) interpolation theory (cf. Triebel [**15**], Yamazaki [**18**]) it suffices to show the lemma for integral $\sigma\ge 0$. We prove it for $\sigma=0$ and 1. The proofs for other cases are similar. When $\sigma=0$, (3.4) for $p=2$ is an immediate consequence of (2.13), and for $p=\infty$ (3.4) is obvious since a is uniformly bounded. Hence the interpolation theorem for L^p-spaces implies (3.4) for $\sigma=0$ and $p\ge 2$.

Next we prove (3.4) for $\sigma=1$. By integration by parts we have

$$I(t,s,a)\partial_k f(x) = (2\pi i\nu)^{-n/2}\int e^{iS(t,s,x,y)}\{-ia\cdot\partial_{y_k}S-\partial_{y_k}a\}f(y)dy, \tag{3.5}$$

and differentiating (3.5) by ∂_j yields

$$\begin{aligned}\partial_j I(t,s,a)\partial_k f(x) = (2\pi i\nu)^{-n/2}\int e^{iS(t,s,x,y)}\{&a\cdot\partial_{x_j}S\cdot\partial_{y_k}S\\ &-i(\partial_{x_j}a\cdot\partial_{y_k}S+\partial_{x_j}S\cdot\partial_{y_k}a+a\cdot\partial_{x_j}\partial_{y_k}S)-\partial_{x_j}\partial_{y_k}a\}f(y)dy.\end{aligned} \tag{3.6}$$

Note that

$$(3.7)\qquad \begin{aligned} &|\partial_x^\alpha \partial_y^\beta (t-s)\langle x\rangle^{-1}\langle y\rangle^{-1}\partial_y S(t,s,x,y)| \le C_{\alpha\beta},\\ &|\partial_x^\alpha \partial_y^\beta (t-s)\langle x\rangle^{-1}\langle y\rangle^{-1}\partial_x S(t,s,x,y)| \le C_{\alpha\beta},\end{aligned}$$

by (2.7), (2.3) and (2.4). Hence the argument used for proving the case $\sigma = 0$ implies for $2 \le p \le \infty$

$$(3.8)\qquad \begin{aligned} &\|\langle x\rangle^{-2} I(t,s,a)\partial_k \langle x\rangle^{-2} f\|_{L^p(\mathbf{R}^n)} + \|\langle x\rangle^{-2}\partial_j I(t,s,a)\partial_k\langle x\rangle^{-2} f\|_{L^p(\mathbf{R}^n)}\\ &\quad \le C|t-s|^{-2-n(1/2-1/p)}\|f\|_{L^{p'}(\mathbf{R}^n)},\quad j,k=1,2,...,n.\end{aligned}$$

Here it is obvious that the order of $\langle x\rangle^{-1}$ and ∂_j or ∂_k can be interchanged. Since $\langle D\rangle = \langle D\rangle^{-1} - \sum_{j=1}^n \langle D\rangle^{-1}\partial_j \cdot \partial_j$ and $\langle D\rangle^{-1}\partial_j$, $j = 1,2,...,n$, is bounded in $L^r(\mathbf{R}^n)$ for $1 < r < \infty$, (3.8) implies (3.4) for $\sigma = 1$.

Lemma 3.3. *Let $2 \le p < \infty$, $0 \le 2/\theta = 2\sigma + n(1/2 - 1/p) < 1$ and θ' and p' be the conjugate indices of θ and p, respectively. Then for any $a \in Amp$ and $0 \le \rho$,*

$$(3.9)\qquad \|\int_{t-T}^{t+T}\langle D\rangle^\rho I(t,s,a)\langle x\rangle^{-2\sigma-\rho} g(s)\,ds\| \le C(\int_{t+T}^{t-T}\|\langle D\rangle^{\rho-\sigma} g(s)\|_{L^{p'}(\mathbf{R}^n)}^{\theta'} ds)^{1/\theta'},$$

where the constant C does not depend on $t \in \mathbf{R}^1$ and $g \in C^0(I, \mathcal{S}(\mathbf{R}^n))$.

Proof: Using (2.15), we write, with $c \in Amp$,

$$\begin{aligned} &\|\int_{t-T}^{t+T} I(t,s,a)\langle x\rangle^{-2\sigma}\langle D\rangle^\sigma g(s)ds\|^2\\ &= \int_{t-T}^{t+T}(g(s), \int_{t-T}^{t+T}\langle D\rangle^\sigma\langle x\rangle^{-2\sigma} I(s,\tau,c)\langle x\rangle^{-2\sigma}\langle D\rangle^\sigma g(\tau)d\tau)ds.\end{aligned}$$

Applying (3.4) and Hardy-Littlewood-Sobolev inequality to the RHS, we estimate it by

$$C\int_{t-T}^{t+T}\{\|g(s)\|_{L^{p'}}\int_{t-T}^{t+T}|s-\tau|^{-2/\theta}\|g(\tau)\|_{L^{p'}}d\tau\} \le C(\int_{t-T}^{t+T}\|g(s)\|_{L^{p'}}^{\theta'} ds)^{2/\theta'}.$$

This proves (3.9) for $\rho = 0$.

Next let $\rho \ge 2$ be even integer. (2.21) and (2) of Lemma 2.4 show for $|\alpha| \le \rho$ that

$$\begin{aligned} D^\alpha I(t,s,a)\langle x\rangle^{-\rho-2\sigma} g(s) &= \sum_{|\beta+\gamma|\le|\alpha|} I(t,s,a_{\alpha\beta\gamma}) x^\beta D^\gamma \langle x\rangle^{-\rho-2\sigma} g(s)\\ &= \sum_{|\beta+\gamma|\le|\alpha|} I(t,s,a_{\alpha\beta\gamma})\langle x\rangle^{-2\sigma}\langle D\rangle^\sigma\cdot\langle D\rangle^{-\sigma}\langle x\rangle^{2\sigma} x^\beta D^\gamma\langle x\rangle^{-\rho-2\sigma} g(s),\end{aligned}$$

and the L^p-boundedness theorem for classical pseudo-differential operators (**[14]**) implies

$$\|\langle D\rangle^{-\sigma}\langle x\rangle^{2\sigma} x^\beta D^\gamma\langle x\rangle^{-\rho-2\sigma} g(s)\|_{L^{p'}} \le C\|\langle D\rangle^{\rho-\sigma} g(s)\|_{L^{p'}},\quad |\beta+\gamma| \le \rho.$$

Hence (3.9) for even integers can be deduced from the already proven $\rho = 0$ case. The interpolation theorem then implies (3.9) for general $\rho \ge 0$. ∎

LEMMA 3.4. *Let p, θ and σ be as in Lemma 3.3. Then for $a \in Amp$ and $\rho \leq 0$*

$$\int_{s-T}^{s+T} \|\langle D\rangle^{\rho+\sigma}\langle x\rangle^{-2\sigma+\rho} I(t,s,a)f\|_{L^p(\mathbf{R}^n)}^{\theta} dt \leq C\|\langle D\rangle^{\rho} f\|^{\theta}, \tag{3.10}$$

where the constant C is independent of $f \in \mathcal{S}(\mathbf{R}^n)$ and $s \in \mathbf{R}^1$.

PROOF: By using (2.15) and (3.9), we estimate as follows:

$$\begin{aligned} |\int_{s-T}^{s+T} (\langle x\rangle^{-2\sigma+\rho} I(t,s,a)f, g(t))\, dt| &= |(\langle D\rangle^{\rho} f, \int_{s-T}^{s+T} \langle D\rangle^{-\rho} I(t,s,a)^{*}\langle x\rangle^{-2\sigma+\rho} g(t)\, dt)| \\ &\leq \|\langle D\rangle^{\rho} f\| \cdot \|\int_{s-T}^{s+T} \langle D\rangle^{-\rho} I(s,t,c)\langle x\rangle^{-2\sigma+\rho} g(t)\, dt\| \\ &\leq C\|\langle D\rangle^{\rho} f\|(\int_{s-T}^{s+T} \|\langle D\rangle^{-\rho-\sigma} g(t)\|_{L^{p'}(\mathbf{R}^n)}^{\theta'} dt)^{1/\theta'}. \end{aligned}$$

By duality, (3.10) follows from this. ∎

LEMMA 3.5. *Let p, σ and θ be as in Lemma 3.3. Then for every $\rho \geq 0$,*

$$\int_{s-T}^{s+T} \|\langle D\rangle^{\rho+\sigma}\langle x\rangle^{-2\sigma-\rho} U(t,s)f\|_{L^p(\mathbf{R}^n)}^{\theta} dt \leq C\|\langle D\rangle^{\rho} f\|^{\theta}, \tag{3.11}$$

where the constant $C > 0$ is independent of $f \in \mathcal{S}(\mathbf{R}^n)$ and $s \in \mathbf{R}^1$.

PROOF: By virtue of the interpolation theorem, it suffices to show (3.11) for even integers $\rho \geq 0$. Thus we need estimate

$$\int_{s-T}^{s+T} \|\langle D\rangle^{\sigma}\langle x\rangle^{-2\sigma-\rho} D^{\alpha} U(t,s)f\|_{L^p(\mathbf{R}^n)}^{\theta} dt, \quad |\alpha| \leq \rho. \tag{3.12}$$

(2.11'), (2.16) and (2.23) show that

$$D^{\alpha} U(t,s)f = \sum_{|\beta+\gamma|\leq|\alpha|} x^{\beta} I(t,s,a_{\alpha\beta\gamma}) D^{\gamma}, \quad a_{\alpha\beta\gamma} \in Amp. \tag{3.13}$$

Applying (3.10) with $\rho = 0$ to (3.14), we see for $|\beta+\gamma| \leq \rho$,

$$\begin{aligned} &\int_{s-T}^{s+T} \|\langle D\rangle^{\sigma}\langle x\rangle^{-2\sigma-\rho} x^{\beta} I(t,s,a_{\alpha\beta\gamma}) D^{\gamma} f\|_{L^p(\mathbf{R}^n)}^{\theta} dt \\ &\qquad \leq C\|D^{\gamma} f\|^{\theta} \leq C\|\langle D\rangle^{\rho} f\|^{\theta}. \end{aligned} \tag{3.14}$$

Summing up (3.12) ~ (3.14), we obtain (3.11). ∎

PROOF OF THEOREM 2: When $\rho \leq 0$, Theorem 1 is implied by Lemma 3.4 and, when $\rho \geq 0$, it is implied by Lemma 3.5. ∎

Once Theorem 1 is obtained, Theorem 2 and Theorem 3 can be proved almost in the same way as in Sjölin[**12**].

PROOF OF THEOREM 3: We prove the theorem for $s = 0$. We have by Theorem 1 with $\rho = 0$ that

$$\int_{-T}^{T} \|\langle x\rangle^{-\mu} U(t,0)f\|^2 dt \le C\|\langle D\rangle^{-1/2} f\|^2. \tag{3.15}$$

Expanding the squares $(-i\partial_j - A_j(t,x))^2$ in $H(t)$, we have

$$\begin{aligned} i\partial_t U(t,0)f(x) &= -(1/2)\Delta U(t,0)f(x) + iA(t,x)\partial_x U(t,0)f(x) \\ &+\{i(div_x A(t,x)/2) + (A^2(t,x)/2) + V(t,x)\}U(t,0)f(x). \end{aligned} \tag{3.16}$$

It follows, by applying Theorem 1 with $\rho = 2$, $\rho = 1$ and $\rho = 0$ respectively to the first, second and third terms in the RHS of (3.16), that

$$\int_{-T}^{T} \|\langle x\rangle^{-2-\mu} (\partial/\partial t) U(t,0)f(x)\|^2 dt \le C\|\langle D\rangle^{3/2} f\|^2. \tag{3.17}$$

Interpolating between (3.15) and (3.17), we have for $0 \le \gamma \le 1$,

$$\int_{\mathbb{R}^n} \|\langle \partial_t\rangle^{\gamma} \langle x\rangle^{-2\gamma-\mu} U(t,0)f(x)\|^2_{L^2([-T,T])} dx \le C\|\langle D\rangle^{2\gamma-1/2} f\|^2.$$

Taking $\gamma > 1/2$ and using Sobolev's embedding theorem, we finally obtain

$$\int_{\mathbb{R}^n} \sup_{-T<t<T} |\langle x\rangle^{-\gamma-\mu} U(t,0)f(x)|^2 dx \le C\|\langle D\rangle^{2\gamma-1/2} f\|^2$$

for any $\mu > 1/2$. This proves Theorem 3. ∎

PROOF OF THEOREM 4: As the proof of Theorem 4 can be carried out in exactly the same way as in that of Theorem 5 of [**12**], we omit it here. ∎

REFERENCES

1. Asada, K. and D. Fujiwara, *On some oscillatory integral transformations in $L^2(\mathbb{R}^n)$*, Japan. J. Math. **4** (1978), 299-361.
2. Carleson, L., *Some analytical problems related to statistical mechanics*, in "Euclidean Harmonic Analysis," Lecture Notes in Math. 799, Springer-Verlag, Berlin and New York, 1979, pp. 5-45.
3. Constantin, P and J. C. Saut, *Local smoothing properties of dispersive equations*, J. Amer. Math. Soc. **1** (1988), 413-439.
4. Dahlberg, B. E. J. and C. E. Kenig, *A note on the almost everywhere behavior of solutions to the Schrödinger equation*, in "Lecture Notes in Math. 908," Springer-Verlag, Berlin and New York, 1982, pp. 205-208.

5. Fujiwara, D., *Remarks on convergence of the Feynman path integrals*, Duke Math. J. **47** (1980), 559-600.
6. Ginibre, J. and G. Velo, *The global Cauchy problem for non-linear Schrödinger equation revisited*, Ann. Inst. H. Poincaré, Anal. Nonlinéare **2** (1985), 309–327.
7. Kato, T., *Wave operators and similarity for some non-selfadjoint operators*, Math. Ann. **162** (1966), 258–279.
8. Kato, T., *On the Cauchy problem for the (generalized) Kortweg-de Vries equation*, Studies in Appl. Math. Adv. in Math. Supplementary Studies **18**, 93–128.
9. Kato, T. and K. Yajima, *Some examples of smooth operators and associated smoothing effect*, . to appear, Rev. in Math. Phys..
10. Kenig, C. E. and A. Ruiz, *A strong type* $(2,2)$ *estimate for a maximal operator associated to the Schrödinger equation*, Trans. Amer. Math. Soc. **280** (1983), 239-246.
11. Kumano-go, H., *Fundamental solution for a hyperbolic system with diagonal principal part*, Comm. in P.D.E. **4** (1979), 959-1015.
12. Sjölin, P., *Regularity of solutions to the Schrödinger equation*, Duke Math. J. **55** (1987), 699-715.
13. Strichartz, R. S., *Restrictions of Fourier transforms to quadratic surfaces and decay of solutions of wave equations*, Duke Math. J. **44**, 705–714.
14. Taylor, M. E., "Pseudo-differential operators," Princeton Univ. Press, Princeton, New Jersey, 1981.
15. Triebel, H., "Interpolation theory, function spaces, differential operators," North Holland, Amsterdam, New York, Oxford, 1978.
16. Yajima, K., *Existence of solutions to Schrödinger evolution equations*, Commun. Math. Phys. **110** (1987), 415-426.
17. Yajima, K., *Schrödinger evolution equation with magnetic fields*, to appear in J. d'Analyse Math..
18. Yamazaki, M., Unpublished notes (1989).

A COIN TOSSING PROBLEM OF R. L. RIVEST

Gregory F. Bachelis
Department of Mathematics
Wayne State University
Detroit, MI 48202

Frank J. Massey III
Department of Mathematics & Statistics
University of Michigan-Dearborn
Dearborn, MI 48128

1. Introduction.

This paper considers Rivest's coin tossing problem. It is concerned with the following game, which consists of a series of rounds; in each round a fair coin is repeatedly tossed and a count is kept of both the number of heads and tails. A round ends whenever one of the following three conditions occurs.

(1) The count of the number of heads reaches k, in which case the entire game ends.

(2) The count of the number of tails reaches m.

(3) The player elects to end the round.

The numbers k and m are given in advance. If a round ends by conditions (2) or (3) then a new round begins with both heads and tails being counted starting from 0 again. During the game a count is also kept of the total number of tosses, N, and this is not set back to 0 when a new round begins. Let V be the expected value of N when the game ends. $V = V(S)$ depends on the strategy S used to begin a new round before it is necessary. The problem is to find a strategy S^* which minimizes $V(S)$ and to find $V^* = V(S^*)$. The origin of this problem is discussed at the end of this section.

This problem may be viewed as a Markov decision problem. Let

$i = k$ - (the number of heads accumulated in the current round)

= the number of heads needed to end the game in the current round, and

$j = m$ - (the number of tails accumulated in the current round)

= the number of tails needed before one is required to begin a new round.

The states are pairs $\mathbf{x} = (i,j)$, where $0 \le i \le k$, $0 \le j \le m$, and $(i,j) \ne (0,0)$.

A strategy S consists of a set of states where one begins a new round upon reaching such a state. A strategy must include the states $(i, 0)$ for $1 \le i \le k$ and should not include the initial state (k, m) or the states in $E = \{(0, j): 1 \le j \le m\}$ where the game ends. We shall show that an optimal strategy S^* has the form $S^* = \{(i,j): j \le S_i^*\}$, where S_i^*, $1 \le i \le k$, is a non-decreasing sequence; see Proposition 2.2. The basic theory of Markov decision problems (see [K] and [R]) gives a system of difference equations (1.6) - (1.8) and inequalities (1.11), (1.12) which characterize S^*. These do not seem to lead to explicit formulas for the S_i^* or V^*, so we try to find their asymptotic behavior as $k, m \to \infty$.

Two related optimal stopping problems are considered in Section 2. One of these is a discrete optimal stopping problem, and it gives additional information about Rivest's coin tossing problem. The discrete optimal stopping problem, when appropriately scaled, formally approaches an optimal stopping problem for a particle moving according to a Brownian motion. This latter type of problem has been studied extensively (see [VMa] and the references listed there) and has been shown to lead to a free boundary problem for the diffusion equation. In particular, Rivest's problem appears to be related to the following problem:

Let $g(t,x) = 1 - t$ except on the negative x axis, where $g(0,x) = 0$. Find a continuous function $x = Y(t)$ defined for $0 \le t < 1$ and a function $u(t,x)$ defined for $0 \le t < 1$, $x \le Y(t)$ such that

$$\frac{\partial u}{\partial t} = .5\,\frac{\partial^2 u}{\partial x^2} \quad \text{and} \quad u(t,x) \le g(t,x) \qquad \text{for } x < Y(t),\ 0 < t < 1,$$

$$u(t,x) - g(t,x) = \frac{\partial u}{\partial x}(t,x) - \frac{\partial g}{\partial x}(t,x) = 0 \qquad \text{for } x = Y(t),\ 0 < t < 1, \text{ and}$$

$$u(0,x) = g(0,x) \qquad \text{for } x < Y(0).$$

At the end of Section 2 we are led to the following conjecture. Assume that the above free boundary problem has a solution. Let $\phi(z)$ be the largest number t such that $(2t)^{-1/2}\,Y(t) = z$.

(1.1) If $k, m \to \infty$ with $(k-m)(k+m)^{-1/2} \to z$ (a constant)

then $V^*/(k+m) \to 1/\phi(z)$ and, for each $\varepsilon > 0$,

$$\sup \{\ |(i-S_i^*)(2\phi(z)/(k+m))^{1/2} - Y(2i\phi(z)/(k+m))| : \varepsilon k \le i \le k\ \} \to 0.$$

I.e., if (1.1) holds then V^* is asymptotic to $(k+m)/\phi(z)$ and S_i^* is asymptotic to $i - ((k+m)/2\phi(z))^{1/2}Y(2i\phi(z)/(k+m))$.

At present we do not have a proof of this conjecture, so we consider some simple, non-optimal strategies for which one can obtain more explicit formulas for V, and we then consider strategies which are asymptotically optimal within this more restricted class. In particular, in Section 3 a formula for $V(S)$ is obtained for the strategy where one does not start a new round until the number of tails equals m. Using this, we obtain the following asymptotic formula, which we give after fixing some notation.

If $Z(t)$ denotes the standard normal density, then we let

$$(1.2) \quad R(x) = \int_x^\infty Z(t)dt, \qquad B(x) = \int_0^x Z(t)dt, \text{ and } \qquad N(x) = \int_{-\infty}^x Z(t)dt.$$

Theorem 1.1. If one does not start a new round until the number of tails equals m and if (1.1) holds, then $V(S)/(k+m) \to 1/R(z)$.

In Section 4 the same is done for strategies where one starts a new round when the number of tails exceeds the number of heads by a fixed number a, and for these strategies we show

Theorem 1.2. Suppose (1.1) holds and

(1.3) $a \to \infty$ with $a(k+m)^{-1/2} \to z + y$, where $y > 0$ is a constant.

Let $F(x) = Z(x) - xR(x)$, and let

(1.4) $W_z(y) = 2\{B(y) + yF(y)\} / \{B(z+2y) - B(z)\}$ for $y \geq -z$,

(1.5) $W_z(y) = \{B(y) + yF(y)\} / B(y)$ for $y \leq -z$.

Then $V(S)/(k+m) \to W_z(y)$. (Note: (1.5) applies only if $z < 0$.)

Thus if (1.1) holds then the asymptotically best value of a may be obtained by finding the minimum of $W_z(y)$ on $0 < y < \infty$. If $z < (\pi/8)^{1/2}$ then this minimum exists; see Proposition 4.3. In particular, the case $m = k$ corresponds to $z = 0$, and numerical techniques show that $W_0(y)$ has a minimum of approximately 1.7 when y is approximately .57. So when $k = m$ is large the best value of a is approximately $.57(2k)^{1/2}$ and it provides approximately a 15% improvement over the case where one does not start a new round until the number of tails equal m.

As noted above, Rivest's coin tossing problem is a Markov decision problem. In this case one wants to minimize the expected total cost of a finite state Markov chain with non-negative costs and no discounting. For a given strategy S, let $C = C(S)$ consist of those states where one continues the round, i.e. those states not in S or E. It is convenient to add to the set of states a "game over" state denoted by **0**. The transition matrix $\mathbf{Q} = \mathbf{Q}(S) = [Q_{\mathbf{xy}}]$ associated with a strategy is given by

$$Q_{\mathbf{xy}} = .5 \qquad \text{if } \mathbf{x} = (i, j) \in C \text{ and } \mathbf{y} = (i-1, j) \text{ or } (i, j-1),$$

$$Q_{\mathbf{xy}} = 1 \qquad \text{if } \mathbf{x} \in S \text{ and } \mathbf{y} = (k, m) \text{ or } \mathbf{x} \in E \text{ and } \mathbf{y} = \mathbf{0}, \text{ or } \mathbf{x} = \mathbf{y} = \mathbf{0},$$

$$Q_{\mathbf{xy}} = 0 \qquad \text{otherwise.}$$

The total number of tosses N is given by $N = \sum_{t=0}^{\infty} c_{\mathbf{x}(t)}$, where the cost functional $\mathbf{c} = \mathbf{c}(S) = [c_{\mathbf{x}}]$ is given by $c_{\mathbf{x}} = 1$ if $\mathbf{x} \in C$ and $c_{\mathbf{x}} = 0$ otherwise, and $\mathbf{x}(t) = (i_t, j_t)$ denotes the state after t tosses. One wants to choose S to minimize $V(S) = v_{km}(S)$, where $v_{ij} = v_{ij}(S)$ is the expected value of N given that the initial state is (i, j).

It is well known [R, p. 133] that the vector $\mathbf{v} = \mathbf{v}(S) = [v_{ij}]$ is given by $\mathbf{v} = \sum_{t=0}^{\infty} \mathbf{Q}^t\mathbf{c}$ and satisfies $\mathbf{v} = \mathbf{c} + \mathbf{Qv}$ which implies

(1.6) $\quad v_{ij} = 1 + .5(v_{i-1,j} + v_{i,j-1}) \qquad$ if $(i, j) \in C$,

(1.7) $\quad v_{ij} = v_{km} \qquad$ if $(i, j) \in S$,

(1.8) $\quad v_{0j} = 0 \qquad$ for $1 \le j \le m$.

If S has the property that it is possible for the game to end (i.e. $V(S)$ is finite) then $\mathbf{v}(S)$ is the only finite-valued solution to (1.6) - (1.8); see [K, pp. 100-102]. Furthermore, if (1.6) - (1.8) have a non-negative, finite-valued solution, then $V(S)$ is finite; see [H, p. 7].

It is known [R, p. 133] that one can find a strategy S^* which not only minimizes $V(S)$ but also $v_{ij}(S)$ for each fixed (i, j). Such an optimal strategy satisfies the dynamic programming type optimality condition

(1.9) $$v_{\mathbf{x}}(S^*) = \min_S [\, c_{\mathbf{x}}(S) + \textstyle\sum_{\mathbf{y}} Q_{\mathbf{xy}}(S) v_{\mathbf{y}}(S^*) \,]$$

for each state $\mathbf{x} = (i, j)$. In the coin tossing problem there are two possible combinations of values for $c_{\mathbf{x}}(S)$ and $[Q_{\mathbf{xy}}(S)]$ for each $\mathbf{x} = (i, j)$, where $i, j > 0$ and $(i, j) \neq (k, m)$. For such a state (1.9) becomes

(1.10) $$v_{ij}(S^*) = \min\, [v_{km}(S^*),\ 1 + .5(v_{i-1,j}(S^*) + v_{i,j-1}(S^*))]$$

Note that (1.6) and (1.7) imply that $v_{ij}(S^*)$ is equal to the first or second term in the min depending on whether (i, j) is in S^* or not. So (1.10) translates into the following two inequalities:

(1.11) $\quad v_{ij}(S^*) \le 1 + .5(v_{i-1,j}(S^*) + v_{i,j-1}(S^*)) \qquad$ if $i, j > 0$,

(1.12) $\quad v_{ij}(S^*) \le v_{km}(S^*) \qquad$ for all i, j.

If a strategy S does not satisfy (1.11) and (1.12), then one can define a new strategy S' as the set of all (i,j) in S where either (1.11) holds or $j = 0$, together with the set of all (i,j) not in S where (1.12) fails to hold. This S' will satisfy $\mathbf{c}(S') + \mathbf{Q}(S')\mathbf{v}(S) \leq \mathbf{v}(S)$ and equality does not hold. From this it is possible to show that S' is an improvement over S in the sense that $\mathbf{v}(S') \leq \mathbf{v}(S)$ and equality does not hold; see [H, p. 7]. In this way one can generate a sequence of strategies each superior to the previous. (This procedure is called policy iteration.) Since the number of strategies is finite, one eventually obtains a strategy where (1.11) and (1.12) hold. This, in fact, is an optimal strategy. This last conclusion relies on the fact that (1.9) is a sufficient condition for a strategy S^* to be optimal, as well as a necessary one. This is not always true for a Markov decision problem, but it is true under the additional assumptions that (*i*) for each strategy S the state 0 is an absorbing state, (*ii*) $c_0(S) = 0$, and (*iii*) if $R \neq \{0\}$ is a recurrence class, then $c_{\mathbf{x}}(S) > 0$ for some state $\mathbf{x}$ in R. (See [K, p. 107] where the sufficiency of (1.9) for optimality is proved under similar conditions.)

Along the same lines, one can show that if a real valued $\mathbf{u} = [u_{\mathbf{x}}]$ satisfies $\mathbf{u} \leq \mathbf{c}(S) + \mathbf{Q}(S)\mathbf{u}$ for all S and $u_0 = 0$, then it satisfies $\mathbf{u} \leq \mathbf{v}(S^*)$. These conditions translate into the following:

(1.13) $\quad u_{ij} \leq 1 + .5(u_{i-1,j} + u_{i,j-1}) \quad$ if $i,\ j > 0$,

(1.14) $\quad u_{ij} \leq u_{km} \quad$ if $i > 0$, $\quad u_{0j} = 0 \quad$ for $1 \leq j \leq m$.

As an immediate application one can see that $u_{ij} = 2i$ satisfies these conditions, so $V(S^*) \geq 2k$.

Suppose for each strategy we let $T(S)$ be the number of tosses required to reach a state which is either in E or S (i.e. to accumulate k heads or begin a new round). By conditioning on the events of either reaching E first or S first, one can show that $v_{ij}(S) = b_{ij}(S) + v_{km}(S) \cdot g_{ij}(S)$, where $b_{ij}(S)$ is the expected value of T given that the initial state is $(i,\ j)$, and $g_{ij}(S)$ is the probability of reaching S before E given that the initial state is $(i,\ j)$. In particular

(1.15) $\quad V(S) = v_{km}(S) = b_{km}(S)/h_{km}(S),$

where $h_{ij}(S) = 1 - g_{ij}(S)$ is the probability of reaching E before S starting at (i,j).

Rivest's coin tossing problem was motivated by the problem of choosing an optimal early abort strategy in certain factoring algorithms. Briefly this problem is as follows; for more details, see Pomerance [P]. In Dixon's algorithm for factoring an integer n one chooses $B < n$ and randomly picks integers r checking to see if the quantity $(r^2 \ mod \ n)$ factors completely into prime factors less than B. One does this until one has $m+1$ values of r for which this is true, where m is the number of primes less than B. From this one can often get a factorization of n. To check if $(r^2 \ mod \ n)$ has all its prime factors less than B, one successively divides by each of the primes p less than B. If it factors completely, then one adds r to the list; otherwise one discards it. In each case one picks a new r and continues until one has $m+1$ successful values. The generation of one successful r corresponds to the coin tossing game with division by a prime p corresponding to a coin toss. One difference is that the probability, $1/p$, that p divides a random number is different from $1/2$ and varies with p, and r is also varying. However, most of the p's are on the order of B and most of the $(r^2 \ mod \ n)$'s are on the order of n. So one might model the generation of one successful r by the coin tossing game with an unfair coin where the probability of heads is $1/B$ and one needs $k \approx (\log n)/(\log B)$ heads before m tails. An early abort strategy is a rule for giving up and going on to a new r at some point in the division by the primes and corresponds to a strategy in the coin tossing game. Pomerance [P] finds early abort strategies which are asymptotically optimal as $n \to \infty$. However, his results do not seem to apply directly to Rivest's problem because of the way that k, m, and the probability of heads are related.

2. Related Stopping Problems.

We shall show (see Proposition 2.1) that the optimal strategies for Rivest's problem coincide with the optimal strategies for an optimal stopping problem for a certain random walk. This stopping problem is simpler than Rivest's problem in that the procedure for constructing optimal strategies is more direct (see Proposition 2.2). For the stopping problem, $D > 0$ is given. One tosses a coin until either one accumulates i heads or j tails, or one decides to stop before either occur. However, instead of beginning a new round if one reaches j tails or decides to stop, the game ends with a cost of D in addition to T, the number of tosses incurred so far. Let $N' = T$ if one reaches i heads before one reaches j tails or decides to stop, and let $N' = T+D$ if one reaches j tails or decides to stop before reaching i heads. Now let

$w_{ij}(S) = b_{ij}(S) + Dg_{ij}(S)$ be the expected value of N' using strategy S. The stopping problem is to find a strategy $S^{\#}$ which minimizes $w_{ij}(S)$ for all i, j. This is again a Markov decision problem. The previous definitions of the transition matrix $\mathbf{Q}(S)$ and cost functional $\mathbf{c}(S)$ should be modified by letting $Q_{\mathbf{x}0} = 1$ and $c_{\mathbf{x}} = D$ if $\mathbf{x}$ is in S.

With these modifications $\mathbf{w} = \mathbf{w}(S) = [w_{ij}(S)]$ satisfies $\mathbf{w} = \mathbf{c} + \mathbf{Q}\mathbf{w}$ and $w_{\mathbf{x}} = 0$ if $\mathbf{x} = \mathbf{0}$ or $\mathbf{x}$ is in E. This translates into

(2.1) $\quad w_{ij} = 1 + .5(w_{i-1,j} + w_{i,j-1}) \quad$ if $(i, j) \in C$,

(2.2) $\quad w_{ij} = D \quad$ if $(i, j) \in S$,

(2.3) $\quad w_{0j} = 0 \quad$ for $1 \le j \le m$.

Again, there is an optimal strategy $S^{\#}$ which minimizes $w_{ij}(S)$ for each fixed (i, j), and it is characterized by

(2.4) $\quad w_{ij}(S^{\#}) \le D \quad$ for all (i,j),

(2.5) $\quad w_{ij}(S^{\#}) \le 1 + .5(w_{i-1,j}(S^{\#}) + w_{i,j-1}(S^{\#})) \quad$ if $i, j > 0$.

A $\mathbf{u}$ which satisfies (1.13) and (1.14) with u_{km} replaced by D satisfies $u_{ij} \le w_{ij}(S^{\#})$. For example $u_{ij} = \min(2i, D)$ has this property, so $w_{ij}(S^{\#}) \ge 2i$ for $0 \le i \le D/2$, and $w_{ij}(S^{\#}) = D$ for $i \ge D/2$. In particular (i, j) is in $S^{\#}$ if $i \ge D/2$.

Proposition 2.1. Consider Rivest's problem for a particular value of k and m and let S^* be an optimal strategy. Now consider the stopping problem with $D = v_{km}(S^*)$ and (i, j) restricted to $i \le k$ and $j \le m$, and let $S^{\#}$ be an optimal strategy for the stopping problem. Then S^* is also an optimal strategy for the stopping problem, $S^{\#}\backslash\{(k,m)\}$ is also an optimal strategy for Rivest's problem, and $v_{ij}(S^*) = w_{ij}(S^{\#})$.

Proof. Since $\mathbf{v}(S^*)$ satisfies (1.6) - (1.8), it follows that (2.1) - (2.3) are satisfied with $\mathbf{w}$ and S replaced by $\mathbf{v}(S^*)$ and S^* respectively. Thus $\mathbf{w}(S^*) = \mathbf{v}(S^*)$. Furthermore (1.11) and (1.12) imply (2.4) and (2.5) with $S^{\#}$ replaced by S^*. Thus S^* is an optimal strategy for the stopping problem, and $w_{ij}(S^{\#}) = v_{ij}(S^*)$ for all i, j. Since $w_{km}(S^{\#}) = v_{km}(S^*) = D$, it follows that we can add or delete (k, m) from $S^{\#}$ without affecting its optimality. Thus we can suppose $S^{\#}$ does not contain (k, m). Since $\mathbf{w}(S^{\#}) = \mathbf{v}(S^*)$ and $\mathbf{w}(S^{\#})$ satisfies (2.1) - (2.3), it follows that (1.6) - (1.8) are satisfied with $\mathbf{v}$ and S replaced by $\mathbf{w}$ and $S^{\#}$ respectively. So $\mathbf{v}(S^{\#}) = \mathbf{w}(S^{\#})$. From (2.4) and (2.5) we conclude that (1.11) and (1.12) hold for $S^{\#}$. So $S^{\#}$ is optimal for Rivest's problem.

The following proposition shows (as one would expect) that the optimal strategies for the stopping problem (and hence Rivest's problem) are of the form

(2.6) $S = \{(i,j):\ j \le S_i\}$ where $\{S_i\}$ is non-decreasing.

Proposition 2.2. An optimal strategy $S^{\#}$ for the stopping problem may be constructed inductively as follows: Let w_{0j} be defined by (2.3) for $j > 0$ and put $w_{00} = D$. For $i > 0$ let $S_i = \sup\{j:\ w_{i-1,j} \ge D-2\}$, and let w_{ij} be defined by (2.2) for $j \le S_i$ and by (2.1) for $j > S_i$. Then $S^{\#} = \{(i,j):\ 0 \le j \le S_i\}$ is an optimal strategy for the stopping problem, $w_{ij} = w_{ij}(S^{\#})$, the sequence $\{S_i\}$ is non-decreasing, and

(2.7) $\{w_{ij}\}$ is strictly decreasing in j for $j > S_i$ for each i, and if $i < D/2$ then $w_{ij} \to 2i$ as $j \to \infty$.

If it should happen that $w_{i-1,j} = D-2$ for $j = S_i$ then (i, S_i) may be omitted from $S^{\#}$ without affecting its optimality. These are the only optimal strategies.

Proof. If $D \le 2$ then all $S_i = \infty$, and the optimal strategy is to stop for all (i,j). From now on assume that $D > 2$. Note that all the other conclusions follow from (2.7). This is clearly true for (2.4). To see that this is true for (2.5), note that one only needs to show (2.5) for (i,j) in $S^{\#}$, since by (2.1) it holds with equality for (i,j) not in $S^{\#}$. If (i,j) is in $S^{\#}$ then so is $(i,j-1)$; hence $w_{ij}(S^{\#}) = w_{i,j-1}(S^{\#}) = D$, and (2.5) is equivalent to $w_{i-1,j} \ge D-2$. However, the latter follows from (2.7) and the way the S_i are defined.

We now show that (2.7) holds by induction. An explicit computation shows this for $i = 1$. Now assume that (2.7) holds for $i-1$, where $i \ge 2$. By the definition of S_i we have $w_{i-1,j} < D-2$ for $j = S_i + 1$. It follows from (2.1) that $w_{ij} < D$ for $j = S_i + 1$, i.e. w_{ij} decreases as j goes from S_i to $S_i + 1$. Then using the fact that $w_{i-1,j}$ is decreasing in j and (2.1), one can show (again by induction) that w_{ij} decreases for all $j > S_i$. The fact that $w_{ij} \to 2i$ as $j \to \infty$ follows by letting j go to infinity in (2.1).

Corollary. Let D be the smallest number such that an optimal strategy $S^{\#}$ for the stopping problem has the property that $(k,m) \in S^{\#}$. Then $V^* = D$ and $S^* \equiv S^{\#} \setminus \{(k,m)\}$ is an optimal strategy for Rivest's problem.

Proof. Note that, for a particular value of D, there is an optimal strategy containing (k,m) if and only if $w_{km}(S^{\#}) = D$. The $w_{ij}(S^{\#})$ in Proposition 2.2 are increasing functions of D and the S_i are non-increasing. The smallest value of D for which $w_{km}(S^{\#}) = D$ occurs when $w_{k-1,m}(S^{\#}) = D - 2$, so (k,m) may be omitted from $S^{\#}$ without affecting its optimality. Let $S^* \equiv S^{\#} \backslash \{(k,m)\}$. The conditions (2.1) - (2.5) for $\mathbf{w}(S^*)$ imply that $\mathbf{v}(S^*) = \mathbf{w}(S^*)$, that (1.6) - (1.8), (1.11) and (1.12) hold for $\mathbf{v}(S^*)$, and that $V^* = v_{km}(S^*) = D$. So S^* is optimal for Rivest's problem.

In view of the above proposition we can confine our attention to strategies of the form (2.6). For such a strategy

$$(2.8) \qquad g_{ij} = \sum_{\ell=1}^{i} g_{ij\ell} \qquad\qquad h_{ij} = \sum_{\ell=1}^{j} h_{ij\ell}$$

where $g_{ij\ell}$ is the probability of reaching the point (ℓ,S_ℓ) before E or any other point of S given that the initial state is (i,j), and $h_{ij\ell}$ is the probability of reaching $(0,\ell)$ before S or any other point of E. One has $b_{ij} = X_{ij} + Y_{ij}$, where X_{ij} (respectively Y_{ij}) is the expected number of heads (respectively tails) to reach $S \cup E$ given that the initial state is (i,j). Since the coin is assumed fair we have $X_{ij} = Y_{ij}$, so

$$(2.9) \qquad b_{ij} = 2X_{ij} = 2\sum_{\ell=1}^{i} (i-\ell)g_{ij\ell} + 2ih_{ij}$$

Let us introduce the variables $n = i + j$ and $s = i - j$. In terms of these variables, the Markov chain has the form of a symmetric random walk with n decreasing by one at each step and s increasing or decreasing by one with probability 1/2 in each direction. If the initial state for the stopping problem is (n_o, s_o), then the cost of stopping on the half line $s = -n$, $n > 0$, is $n_o - n$; otherwise it is $n_o - n + D$. In terms of the variables n and s, equation (2.1) becomes $w_{n+1,s} = 1 + .5(w_{n,s+1} + w_{n,s-1})$ and similarly for inequality (2.4). The boundary condition (2.3) transforms to $w_{n,-n} = 0$ for $n > 0$, and $S^{\#}$ takes the form $S^{\#} = \{(n,s)\colon s \geq f(n,D)\}$ for some function $f(n,D)$. For Rivest's problem we are looking for the smallest number D such that $k - m = f(k+m,D)$.

One obtains an equivalent problem by subtracting n_o from the stopping costs and dividing by D, i.e. the cost is $-n/D$ if one stops when $s = -n$, $n > 0$, and $1 - n/D$ otherwise. The equations and inequalities corresponding to (2.1) - (2.5) for the expected cost u_{ns} of this equivalent problem are the following:

$$u_{n+1,s} = .5(u_{n,s+1} + u_{n,s-1}) \quad \text{in } C$$
$$u_{ns} = 1 - n/D \quad \text{in } S$$
$$u_{n,-n} = -n/D \quad n > 0$$
$$u_{n+1,s} \le .5(u_{n,s+1} + u_{n,s-1}) \text{ and } u_{ns} \le 1 - n/D \quad \text{for } n > 0,\ |s| \le n-1.$$

(Note that $u_{ns} = (w_{ns} - n)/D$.)

A Brownian motion can be regarded as the limit of symmetric random walks where the time interval τ between steps and step size η approach 0 and are related by $\eta^2 = 2\tau$; see [F, Section XIV.6]. In terms of the variables $t = n\tau$ and $x = s\eta = s(2\tau)^{1/2}$, the cost is $-t(D\tau)^{-1}$ if one stops on the half line $x = t(2/\tau)^{1/2}$, $t > 0$, and $1 - t(D\tau)^{-1}$ otherwise, and $S^{\#} = \{(t,x)\colon x \ge Y(t,\tau,D)\}$, where $Y(t,\tau,D) = (2\tau)^{1/2} f(t/\tau, D)$. If one puts $\tau = 1/D$ and lets $D \to \infty$ then one obtains the following optimal stopping problem for a Brownian motion:

Suppose $Z(t)$, $t > 0$, is a standard Brownian motion with $Z(0) = 0$. Given $t_o > 0$ and x_o, consider the Brownian motion $X(t) = x_o + Z(t_o - t)$, $t \le t_o$, starting at (t_o, x_o) with time traveling backwards. Given a stopping time $T \le t_o$, let $u(t_o, x_o, T)$ be the expected value of $g(T, X(T))$, where, for the problem we are considering, $g(t,x) = 1-t$ except on the negative x axis where $g(0,x) = 0$. Let $u(t_o, x_o) = \inf_T u(t_o, x_o, T)$. One wants to find a stopping time T^* such that $u(t_o, x_o, T^*) = u(t_o, x_o)$.

As noted in the introduction, problems of this type have been extensively studied, and they lead to a free boundary problem for the diffusion equation. In the case where $g(x,0)$ is sufficiently smooth the following is known (see [VMa, VMb]).

(2.10) u is continuous.

(2.11) If we let $C = \{(t,x)\colon u(t,x) < g(t,x),\ t > 0\}$ and $S = \{(t,x)\colon u(t,x) = g(t,x),\ t \ge 0\}$, then stopping when one reaches S is optimal.

(2.12) u is C^∞ in C and $\dfrac{\partial u}{\partial t} = .5\,\dfrac{\partial^2 u}{\partial x^2}$ there.

(2.13) C is bounded by continuous curves $x = Y_i(t)$ which are continuously differentiable except for isolated values of t, and $\dfrac{\partial u}{\partial x}(t, Y_i(t)) = \dfrac{\partial g}{\partial x}(t, Y_i(t))$ except for those t.

In this case u is a solution to a free boundary problem analogous to the one mentioned in the introduction. The g in our problem does not satisfy the smoothness assumptions in [VMa, VMb], so it is not clear if the above conclusions are still true. However, it is not hard to see that in our problem, $u(t,x)$ is non-decreasing in x for each fixed t since $g(t,x)$ is. Hence C has the form $C = \{(t,x): x < Y(t), t \in I\}$, for some function $Y(t)$ defined on some set I.

Thus one might expect that

(2.14) $Y(t,1/D,D) = (2/D)^{1/2}f(tD,D)$ approaches $Y(t)$ as $D \to \infty$.

This sort of thing has been shown to be true for similar problems (see [S]), but we do not have a complete proof for the case at hand. However, suppose that (2.14) were true. Let us see how this is related to the conjecture in the introduction concerning the asymptotic behavior of V^* and S_i^*. Suppose that (1.1) holds. Then V^*, which is the smallest D such that $k-m = f(k+m,D)$, would be relatively close to the smallest D such that $(k-m)(k+m)^{-1/2} = (2(k+m)/D)^{-1/2}Y((k+m)/D)$. This in turn is relatively close to the smallest D such that $z = (2(k+m)/D)^{-1/2}Y((k+m)/D)$. This D is equal to $(k+m)/\phi(z)$, where $\phi(z)$ is the largest solution t of $(2t)^{-1/2}Y(t) = z$. So $V^*/(k+m)$ is close to $1/\phi(z)$. This argument needs to be made precise. In particular, it seems to require that $(2t)^{-1/2}Y(t)$ be strictly decreasing near $t = \phi(z)$, and we do not know this to be true for all t. Now consider the asymptotic behavior of the S_i^*. The free boundary $s = f(n,V^*)$ is relatively close to $s = (V^*/2)^{1/2}Y(n/V^*)$. The right hand side is relatively close to $((k+m)/2\phi(z))^{1/2}Y(n\phi(z)/(k+m))$. When we change back to the (i,j) coordinates, we have $(i-S_i^*) = f(n,V^*)$. This is relatively close to $((k+m)/2\phi(z))^{1/2}Y(2i\phi(z)/(k+m))$. Note that in the argument of Y one can approximate $n = i+j = i+S_i^*$ by $2i$, since $(i-S_i^*)(k+m)^{-1/2}$ is bounded, which implies in turn that $(i-S_i^*)/(k+m)$ is small.

3. The Case Where One Does Not Stop Until One Must.

Consider the case where one does not start over until one has m tails. For the sake of future reference let the $g_{ij\ell}$ and $h_{ij\ell}$ appearing in (2.8) be denoted by $G_{ij\ell}$ and $H_{ij\ell}$ for this particular strategy. They are given by $G_{i,j,i-s} = f(s;j)$ and $H_{i,j,j-s} = f(s;i)$, where $f(s;j) = C(s,s+j-1)\cdot 2^{-s-j}$ is the probability that exactly s successes precede the j-th failure in a series of Bernoulli trials in which successes and failures are equally likely

prove Theorem 1.2 we need the following proposition concerning the asymptotic behavior of E_{tn} when t is $O(n^{1/2})$. Its proof (by approximating the binomial by the normal and the sum by the integral) is given in Section 5, since we have been unable to find it in the literature. Note, however, that for fixed t it is known that $E_{tn}/[2 \cdot Z(0) \cdot n^{1/2}] \to 1$ as $n \to \infty$; see [F, Section XII.6] for the case $t = 0$ and use the fact [C, p. 23, Theorem 3] that $E_{tn}/E_{0n} \to 1$ as $n \to \infty$ to obtain this for arbitrary fixed t.

Proposition 4.2. $E_{tn}/[2n^{1/2}F(tn^{-1/2})] \to 1$ as $n \to \infty$ uniformly in t provided $tn^{-1/2}$ remains bounded, where $F(x) = Z(x) - xR(x)$.

Proof of Theorem 1.2. Recall $V(S) = b_{km} / h_{km}$. If $a > 0$ then it follows from (4.2) and (3.2) that

$$h_{ij} \left[B((j+2a-i)(i+j)^{-1/2}) - B((i-j)(i+j)^{-1/2}) \right]^{-1} \to 1 \qquad \text{as } i,j \to \infty$$

provided $(j-i)(i+j)^{-1/2}$ and $a(i+j)^{-1/2}$ are bounded and $(j+a-i)(i+j)^{-1/2}$ is bounded away from 0. Therefore, if (1.1) and (1.3) hold, then

$$(4.6) \qquad h_{km} \to B(z+2y) - B(z) \qquad \text{if } y \geq -z .$$

Similarly, if $a < 0$ one can show from (4.4) that

$$(4.7) \qquad h_{km} \to 2 \cdot B(y) \qquad \text{if } 0 < y < -z .$$

If $a > 0$ then it follows from (4.3), (3.2) and Proposition 4.2 that $b_{ij} [2(i+j)(B(y_{ij}) + y_{ij}F(y_{ij}))]^{-1} \to 1$ as $i, j \to \infty$, with $y_{ij} = (j+a-i)(i+j)^{-1/2}$, provided $(j-i)(i+j)^{-1/2}$ and $a(i+j)^{-1/2}$ are bounded and $(j+a-i)(i+j)^{-1/2}$ is bounded away from 0. Therefore, if (1.1) and (1.3) hold, then $b_{km}/(k+m) \to 2(B(y) + yF(y))$ if $y \geq -z$. A similar argument using (4.5) shows this holds if $0 < y < -z$. Combining this with (4.6) and (4.7) proves (1.4) and (1.5).

As noted in the introduction, Theorem 1.2 implies that if (1.1) holds, then the asymptotically best value of a may be obtained by finding the minimum of $W_z(y)$ on $0 < y < \infty$. The following proposition shows that the minimum exists if $z < (\pi/8)^{1/2}$, but that if $z \geq (\pi/8)^{1/2}$ then the strategies are increasingly better as $y \to 0$, although taking $y = 0$ in Theorem 1.2 is not valid. A better analysis of $V(S)$ for $y \to 0$ seems to be needed in this case.

Proposition 4.3. Let z be fixed, and consider $W_z(y)$ defined by (1.4) and (1.5) as a function of $y > 0$, and let $z^* = (\pi/8)^{1/2}$. If $z < z^*$ then there is $y^* = y^*(z) > 0$ such that $W_z(y)$ is strictly decreasing on $0 < y \leq y^*$ and strictly increasing on $y^* \leq y < \infty$. If $z \geq z^*$ then $W_z(y)$ is strictly increasing on $0 < y < \infty$. One has $W_z(y) \to 1/R(z)$ as $y \to \infty$ (the same value as in Theorem 1.1). If $z \geq 0$ then $W_z(y) \to 2 \cdot Z(0)Z(z)^{-1}$ as $y \to 0$, and if $z < 0$ then $W_z(y) \to 2$ as $y \to 0$.

Proof. The limits as y approaches 0 or ∞ are easy to verify. To verify the decreasing/increasing behavior let $J(y) = [B(y) + yF(y)]/[B(z+2y) - B(z)]$. A calculation shows that $J' = 2M(y)[B(z+2y) - B(z)]^{-2}$, where $M(y) = F(y)[B(z+2y) - B(z)] - [B(y) + yF(y)]Z(z+2y)$. It is not hard to check that $M(y)$ satisfies the differential equation $M'(y) + M(y)R(y)F(y)^{-1} = [B(y) + yF(y)]Z(z+2y)F(y)^{-1}u(y)$ where $u(y) = 2(z+2y)F(y) - R(y)$. The third derivative of u is given by $u^{(3)}(y) = (11 - 2zy - 3y^2)Z(y)$. It follows that there is $y_3 > 0$ such that $u^{(3)}$ is positive for $y < y_3$ and negative for $y > y_3$. Working backward one can show that if $z < z^*$ there is $y_o > 0$ such that $u(y)$ is negative for $y < y_o$ and positive for $y > y_o$. On the other hand if $z \geq z^*$ one can show that $u(y)$ is positive for $y > 0$. Then, in the case $z < z^*$, by using the differential equation for M and $M(0) = 0$, it is not hard to see that $M(y)$ is negative for $0 < y \leq y_o$. Also, it is not hard to show that $M(y) > 0$ for large y, which implies there is $y^* > y_o$ such that $M(y^*) = 0$. Again it follows from the differential equation that $M(y) > 0$ for $y > y^*$. In the case $z \geq z^*$ similar reasoning shows that $M(y) > 0$ for $y > 0$. Thus $J(y)$ has the decreasing/increasing behavior we want to prove for $W_z(y)$. This takes care of the case $z \geq 0$, since $W_z = 2 \cdot J$ for these z. If $z < 0$ then $W_z(y) = 2J(y)$ for $y \geq -z$ and $W_z(y) = K(y)$ for $y \leq -z$, where $K(y) = [B(y) + yF(y)]/B(y)$. An argument similar to the one given for $J(y)$ shows that $K(y)$ is decreasing for $y > 0$. Combining this with the result for $J(y)$ shows that $W_z(y)$ has the indicated decreasing/increasing behavior.

5. Proof of Proposition 4.2.

It is convenient to use the following lemma, whose proof is given for the sake of completeness.

Lemma 5.1 Suppose $A(t,x)$ and $f(t,x)$ are defined for $t = 0, 1, 2, \ldots$ and $x > 0$ and satisfy the following conditions:

(5.1) $|A(t,x)|$ and $f(t,x)$ are bounded by $Dx^{-1/2}$ where D is a constant.

(5.2) For any $a > 0$ one has $f(t,x)$ bounded below by $C_a x^{-1/2}$ in the set $\{(t,x): 0 \leq t \leq ax^{1/2}\}$ where C_a is a positive constant.

(5.3) $A(t,x)/f(t,x) \to 1$ as $x \to \infty$ uniformly in t provided $tx^{-1/2}$ remains bounded.

Then

(5.4) $$\int_0^x A(t,y)dy \Big/ \int_0^x f(t,y)dy \to 1 \qquad \text{as } x \to \infty$$

uniformly in t provided $tx^{-1/2}$ remains bounded.

Proof. Let a and r be given. It suffices to show that

$$\int_0^x |A(t,y) - f(t,y)|dy \le r\int_0^x f(t,y)dy \qquad \text{for } x \text{ large and } t \le ax^{1/2}.$$

Split up the integral on the left into a part from 0 to px and a part from px to x, where p is to be chosen. By (5.1) the integral from 0 to px is bounded by $4D(px)^{1/2}$, and by (5.2) the integral from $x/4$ to x for $f(t,y)$ is bounded below by $C_{2a}x^{1/2}$. So

$$\int_0^{px} |A(t,y) - f(t,y)|dy < (r/2)\int_0^x f(t,y)dy \qquad \text{if } p = (rC_{2a}/8D)^2.$$

If $p \ge 1$ then the proof is done. Otherwise, one can get a similar estimate for the integral from px to x as follows. Note that by (5.3) one has

(5.5) $$|A(t,y) - f(t,y)| < (r/2)\, f(t,y) \qquad \text{for } y \text{ large and } t \le a(y/p)^{1/2}.$$

So if x is large and $t \le ax^{1/2}$, then (5.5) holds for $px \le y \le x$. Integrating from px to x gives the desired inequality and finishes the proof of the lemma.

Proof of Proposition 4.2. We apply the above lemma with $f(t,x) = x^{-1/2}Z(tx^{-1/2})$ and $A(t,x)$ defined by $A(t,x) = 0$ for $x < t$ and $A(t,x) = (1/2)\, r(s,t+2s)$ for $t+2s \le x < t+2s+2$ and $s = 0, 1, 2,\ldots$. (Note $s = \lfloor (x-t)/2 \rfloor$, where $\lfloor\cdot\rfloor$ denotes the greatest integer function.) Clearly f satisfies (5.1) and (5.2). It remains to show (5.3) and (5.1) for A . By the normal approximation to the binomial [F, Section VII.2] one has

(5.6) $$r(j,n) \big/ [2n^{-1/2} Z((j-n/2)2n^{-1/2})] \to 1 \quad \text{as } n \to \infty$$

uniformly in j, provided $(j-n/2)n^{-1/2}$ is bounded. In particular $r(j,n) \le r(n/2, n) \le Dn^{-1/2}$ for some constant D. It follows that $r(s,t+2s) \big/ [2(t+2s)^{-1/2} Z(t(t+2s)^{-1/2})] \to 1$ if $(t+2s) \to \infty$ uniformly in t, provided $t(t+2s)^{-1/2}$ remains bounded. We also conclude that $r(s,t+2s) \le D(t+2s)^{-1/2}$. Letting $s = \lfloor (x-t)/2 \rfloor$ one obtains (5.3) and (5.1) for A. So the hypotheses of the lemma are satisfied, and (5.4) holds for this A and f.

If one makes the substitution $w = ty^{-1/2}$ and integrates by parts, one can show that $\int_0^n f(t,y)dy = 2n^{1/2} F(tn^{-1/2})$, where $F(x) = Z(x) - xR(x)$. Furthermore, $E_{tn} - \int_0^n A(t,y)dy$ is equal to $r((n-t)/2, n)$ or $r((n-t-1)/2, n-1)$ depending on whether $n-t$ is even or odd. Since $r(j,n) \leq Dn^{-1/2}$, this difference divided by $2n^{1/2}F(tn^{-1/2})$ goes to 0, and one obtains the assertion in Proposition 4.2.

BIBILIOGRAPHY

[C] Chung, Kai Lai. *Markov Chains with Stationary Transition Probabilities*, 2nd ed., Springer-Verlag, Berlin, 1967.

[F] Feller, William. *An Introduction to Probability Theory and Its Applications*, Vol. I, 3rd ed., Wiley, New York, 1968.

[H] Hordijk, A. *Dynamic Programming and Markov Potential Theory*, 2nd. ed., Mathematical Centre Tracts 51, Mathematisch Centrum, Amsterdam, 1977.

[K] Kushner, Harold. *Introduction to Stochastic Control*, Holt, Rinehart and Winston, New York, 1971.

[P] Pomerance, C. "Analysis and Comparison of Some Integer Factoring Algorithms," in *Computational Methods in Number Theory*, Part 1 (edited by H. W. Lenstra, Jr. and R. Tijdeman), Mathematical Centre Tracts 154, Mathematisch Centrum, Amsterdam, 1982, pp 89-139.

[R] Ross, Sheldon M. *Applied Probability Models with Optimization Applications*, Holden-Day, 1970.

[S] Shepp, L.A. "Explicit Solutions to Some Problems of Optimal Stopping," *Annals Math. Stat.* 40 (1969), pp. 992-1010.

[VMa] Van Moerbeke, P. "On Optimal Stopping and Free Boundary Problems," *Arch. Rational Mech. Anal.* 60 (1975), pp. 101-148.

[VMb] ____________. "An Optimal Stopping Problem with Linear Reward," *Acta Mathematica*, 132 (1974), pp. 111-151.

Liapunov functions and monotonicity in the Navier-Stokes equation

TOSIO KATO

Department of Mathematics, University of California, Berkeley, CA 94720, U.S.A.

1. Introduction

Consider the free Navier-Stokes equation in $\mathbf{R}^m$:

$$\partial_t u = \nu \Delta u - F(u), \qquad F(u) = P(u \cdot \partial)u, \qquad div\ u = 0. \tag{1.1}$$

Here $u = u(t)$: $\mathbf{R}^m \to \mathbf{R}^m$ is the velocity field; ν is the kinematic viscosity; ∂ =grad; and P denotes the projection operator onto solenoidal (= divergence free) vectors along gradients (so that the pressure term has been eliminated).

It is the purpose of this note to show that (1.1) has a large number of *Liapunov functions*, which decrease monotonically in time for every solution with small $L^m(\mathbf{R}^m)$-norm.

It is known (see [3]) that if $\phi \in L^m$, then a unique strong solution $u(t)$ with $u(0) = \phi$ exists for short time, and that it exists for all time if $\nu^{-1}\|\phi\|_m$ is small. ($\| \ \|_p$ denotes the L^p-norm.) For this reason, we may call $R[u] = \nu^{-1}\|u\|_m$ a *Reynolds number* for the flow u. Of course other Reynolds numbers exist with a similar property, but it seems that $R[u]$ is the simplest among them. For $m = 3$, it appears already in Leray's paper (see [5, p.231).

Actually it turns out that if $R[\phi]$ is small, then $R[u(t)]$ decreases in t *monotonically*. Therefore it may also be called a Liapunov function for (1.1). It seems that such a monotone decay has so far been known only for $\|u\|_2$. In fact we shall show that there are many other Liapunov functions. They include the L^p_s-norms $\|u\|_{s,p} = \|(1-\Delta)^{s/2}u\|_p$ for certain s and p. All $p \in (1, \infty)$ are allowed if $s = 0$. In particular, the Reynolds number $R[u] = \|u\|_m$ is at the same time a Liapunov function. It is obvious that if $\mathcal{L}[u]$ is a Liapunov function, then so is $\Phi(\mathcal{L}[u])$ for any monotone increasing function Φ.

To illustrate the situation in the simplest case, let us recall the L^2 -theory for (1.1) given some time ago [2]. If one works in the Hilbert space $\mathbf{H} = L^2(R^m; R^m)$, $-\Delta$ is a nonnegative selfadjoint operator, and P becomes an orthogonal projection of $\mathbf{H}$ onto the subspace $\mathbf{H}_\sigma$ consisting of solenoidal vectors. $-\Delta$ is reduced by $\mathbf{H}_\sigma$, so that its part A in $\mathbf{H}_\sigma$ is nonnegative selfadjoint, and (1.1) can be regarded as an abstract equation

$$\partial_t u + \nu A u = -F(u). \tag{1.2}$$

We shall now show, assuming $m = 3$ for simplicity, that $R_1(u) = \nu^{-1}\|A^{1/4}u\|$ is a Reynolds number and, at the same time, a Liapunov function. To see this we note that (see [2]),

$$\|F(u)\| \le M\|A^{3/4}u\|\ \|A^{1/2}u\|. \tag{1.3}$$

(In [2] the space domain Ω was assumed to be bounded, but (1.3) is also true when $\Omega = R^3$.) If $u = u(t)$ is a solution of (1.2), it follows, formally, that

$$\begin{aligned} \partial_t \|A^{1/4}u\|^2 &= 2(A^{1/2}u, \partial_t u) = 2(A^{1/2}u, -\nu Au - F(u)) \\ &\leq -2\nu\|A^{3/4}u\|^2 + 2M\|A^{1/2}u\|^2\|A^{3/4}u\| \\ &\leq -2\|A^{3/4}u\|^2(\nu - M\|A^{1/4}u\|), \end{aligned}$$

where we have used the inequality $\|A^{1/2}u\|^2 \leq \|A^{3/4}u\| \, \|A^{1/4}u\|$. Thus $R_1(u(t)) = \nu^{-1}\|A^{1/4}u(t)\|$ must continually decrease if it is smaller than M^{-1} at $t = 0$. This proves the assertion.

Similarly, we have

$$\begin{aligned} \partial_t \|A^{1/2}u\|^2 &\leq -2\nu\|Au\|^2 + 2M\|Au\| \, \|A^{3/4}u\| \, \|A^{1/2}u\| \\ &\leq -2\|Au\|^2(\nu - M\|A^{1/4}u\|), \end{aligned}$$

since

$$\|A^{1/2}u\| \leq \|A^{1/4}u\|^{2/3}\|Au\|^{1/3}, \qquad \|A^{3/4}u\| \leq \|A^{1/4}u\|^{1/3}\|Au\|^{2/3}.$$

This shows that $\|A^{1/2}u(t)\|$ decreases steadily if $R_1[u] = \nu^{-1}\|A^{1/4}u\| < M^{-1}$, which is the case if $R_1[u(0)] < M^{-1}$ as shown above.

In the same way one can show that $\|A^\alpha u\|$ is a Liapunov function for any $\alpha \in [0, 1/2]$, though we need an estimate different from (1.3) for $\alpha < 1/4$. Here again $R_1[u]$ serves as a Reynolds number.

Remarks. (a) In the preceding arguments we have disregarded the existence question. In fact it was shown in [2] that a local solution $u \in C([0,T); D(A^{1/4}))$ exists if $R_1[\phi] < \infty$. But this does not immediately lead to global existence for small $R_1[\phi]$, since the interval of existence given in [2] could not be determined by $\|A^{1/4}\phi\|$ alone. Existence of a global solution for small ϕ was proved in [2] directly, independently of any Liapunov function and without any more difficulty than for local existence.

(b) Since $\|u(t)\|$ is conserved for the solution of (1.2), $\|u\|_{2\alpha,2} = (\|A^\alpha u\|^2 + \|u\|^2)^{1/2}$ is also a Liapunov function with $\|A^\alpha u\|$. Moreover, it is known that $\|u\|_3 \leq \text{const}\|A^{1/4}u\|_2$ for $m = 3$. Thus $\|A^\alpha u\|$ and $R_1[u]$ are special cases of the Liapunov functions and the Reynolds number to be considered in next sections.

(c) The above results (L^2-theory) hold also in the case of a bounded domain Ω.

2. The main theorems

In what follows we shall show that if $m \geq 3$, $\|u\|_m$ is a Reynolds number and $\|u(t)\|_{s,p}$ are Liapunov functions for (1.1). Here it is assumed that $1 < p < \infty$ if $s = 0$ and $2 \leq p < \infty$ if $s > 0$; we have not been able to allow $p < 2$ when $s > 0$. (We recall that $\|f\|_p$ is the L^p-norm, and that $\|f\|_{s,p} = \|J^s f\|_p$, $J = (1 - \Delta)^{1/2}$, is the L^p_s-norm.)

More precisely, we prove

Theorem I. Let $m \geq 3$, $1 < p < \infty$. Let u be a solution of (1.1) such that

$$u \in C([0,T); L^m \cap L^p), \qquad \Delta u \in L^1_{loc}((0,T); L^p), \qquad 0 < T \leq \infty.$$

(Δu may be replaced by ∂u if $p \geq 2$.) Then

$$\partial_t \|u(t)\|_p^p \leq -c(\nu - K\|u(t)\|_m)Q_p(u(t)), \qquad 0 < t < T. \tag{2.1}$$

where K denotes a positive constant depending only on m and p, and where

$$0 \leq Q_p(\phi) = \int_{\partial\phi(x)\neq 0} |\phi(x)|^{p-2}|\partial\phi(x)|^2 dx, \qquad 1 < p < \infty. \tag{2.2}$$

Theorem II. Let $m \geq 3$, $s > 0$, $2 \leq p < \infty$. Let u be a solution of (1.1) such that

$$u \in C([0,T]; L^m \cap L_s^p), \qquad \partial u \in L^1_{loc}((0,T); L_s^p), \qquad 0 < T \leq \infty.$$

(Note that $L^m \cap$ is redundant if $s \geq m/p - 1 \geq 0$.) Then

$$\partial_t \|u(t)\|_{s,p}^p \leq -c(\nu - K\|u(t)\|_m)Q_p(J^s u(t)), \qquad 0 < t < T. \tag{2.3}$$

Remarks. (a) Setting $p = m$ in (2.1), we see that $R[u(t)] = \nu^{-1}\|u(t)\|_m$ decreases in time if $R[u(0)] < K^{-1}$. Returning to (2.1) or (2.3), we then see that $\|u(t)\|_p$ or $\|u(t)\|_{s,p}$ also decreases in time if $R[u(0)]$ is sufficiently small. (Here it is somewhat uncomfortable that K may depend on p and s). Thus $\|u\|_{s,p}$ are Liapunov functions, with $R[u]$ as the associated Reynolds number.

(b) The theorems *suggest* global existence of a small L^m-solution, but do not prove it. Also they do not give a simple answer to the question of whether $\|u(t)\|_{s,p} \to 0, t \to \infty$, for a global solution. These problems will be considered in section 4.

(c) It is not difficult to see that analogous Liapunov functions exist for many nonlinear parabolic equations. A particular interest in the Navier-Stokes equation lies in the fact that the presence of the pressure term (which is expressed by the nonlocal projection operator P) makes the problem rather nontrivial.

(d) $Q_p(\phi)$ is well defined, possibly with value $+\infty$, since the set of x with $\partial\phi(x) = 0$ has been excluded from the integration, so that expressions like $0/0$ do not occur. In fact we have

Lemma 1. If $2 \leq p < \infty$, let $\phi \in W^{1,p}$. If $1 < p < 2$, assume in addition that $\Delta\phi \in L^p$. Then

$$cQ_p(\phi) \leq -\langle |\phi|^{p-2}\phi, \Delta\phi\rangle < \infty. \tag{2.4}$$

Note that (2.4) is finite because $|\phi|^{p-2}\phi \in W^{1,p'}$ if $p \geq 2$ and $\in L^{p'}$ if $p < 2$, where $1/p + 1/p' = 1$.

Proof. First we assume that $p \geq 2$. We have, with summation convention,

$$\begin{aligned} -\langle |\phi|^{p-2}\phi, \Delta\phi\rangle &= \langle \partial_k |\phi|^{p-2}\phi_j, \partial_k\phi_j\rangle \\ &= \langle |\phi|^{p-2}(\delta_{ij} + (p-2)\phi_i\phi_j)|\phi|^{-2})\partial_k\phi_i, \partial_k\phi_j\rangle \\ &\geq \langle |\phi|^{p-2}, |\partial\phi|^2\rangle = Q_p(\phi); \end{aligned}$$

here we have used the fact that the matrix $(\phi_i\phi_j|\phi|^{-2})$ is nonnegative.

If $p < 2$, we set $\phi_\epsilon = (|\phi|^2 + \epsilon^2)^{1/2}$ for $\epsilon > 0$, and prove

$$-\langle \phi_\epsilon^{p-2}\phi, \Delta\phi\rangle \geq (p-1)\,\langle \phi_\epsilon^{p-2}, |\partial\phi|^2\rangle.$$

The proof is almost identical with the one given above; here we use the fact that the matrix $(\phi_i\phi_j\phi_\epsilon^{-2})$ is bounded by one. Letting $\epsilon \to 0$, we obtain (2.4) by monotone, and dominated, convergence theorems.

Another property of $Q_p(\phi)$ we need is given by

Lemma 2. Under the assumption of Lemma 1, we have

$$\|\phi\|_{mp/(m-2)} \leq cQ_p(\phi)^{1/p}, \qquad 1 < p < \infty. \tag{2.5}$$

Proof. If $p \geq 2$, we have $|\partial|\phi|^{p/2}| \leq (p/2)|\phi|^{(p-2)/2}|\partial\phi|$, hence $\|\partial|\phi|^{p/2}\|_2 \leq (p/2)Q_p(\phi)^{1/2}$. But

$$\|\phi\|_{mp/(m-2)}^{p/2} = \|\,|\phi|^{p/2}\|_{2m/(m-2)} \leq c\|\partial|\phi|^{p/2}\|_2$$

by the Sobolev inequality. This proves (2.5). If $p < 2$, we first estimate $\|\phi_\epsilon^{p/2} - \epsilon^{p/2}\|_{2m/(m-2)}$ in the manner described above, and then go to the limit $\epsilon \to 0$.

3. Proof of the Theorems

Proof of Theorem I. We compute, with the variable t suppressed for simplicity,

$$\begin{aligned}\partial_t\|u\|_p^p &= \partial_t\langle |u|^p, 1\rangle = p\langle |u|^{p-2}u, \partial_t u\rangle \\ &= \nu p\langle |u|^{p-2}u, \Delta u\rangle - p\langle |u|^{p-2}u, F(u)\rangle.\end{aligned} \tag{3.1}$$

According to Lemma 1, the first term on the right of (3.1) does not exceed $-c\nu Q_p(u)$. On the other hand, we have $F(u) = P(u\cdot\partial)u = P\partial(uu)$ in the symbolic notation ($\partial(uu)$ is a vector with j-component $\partial_k u_k u_j$, with summation convention). Therefore, the theorem will be proved if we show that

$$|\langle |u|^{p-2}u, P\partial(uu)\rangle| \leq c\|u\|_m Q_p(u). \tag{3.2}$$

To prove (3.2), we consider the cases $p \geq 2$ and $p < 2$ separately.

Case $p \geq 2$. We move the differentiation ∂ in (3.2) to the other side of the pairing, using the fact that P and ∂ commute. Since $|\partial|u|^{p-2}u| \leq c|u|^{p-2}|\partial u|$, and since P is bounded on any L^q-space with $1 < q < \infty$, it suffices to show that

$$\|\,|u|^{p-2}|\partial u|\,\|_q\,\|uu\|_{q'} \leq c\|u\|_m Q_p(u) \tag{3.3}$$

for *some* conjugate indices q, q' such that $1 < q < \infty$.

In the sequel we frequently use the obvious formulas such as

$$\|fg\|_{pq/(p+q)} \leq \|f\|_p\,\|g\|_q. \qquad \|f^k\|_p = \|f\|_{kp}^k. \tag{3.4}$$

To prove (3.3), we choose

$$1/q = 1/2 + (m-2)(p-2)/2mp, \qquad 1/q' = 1/m + 1/p - 2/mp; \tag{3.5}$$

note that these numbers are positive and add up to one. Then

$$\begin{aligned} \| \, |u|^{p-2} |\partial u| \, \|_q &\le \| \, |u|^{(p-2)/2} \|_{2q/(2-p)} \, \| \, |u|^{(p-2)/2} |\partial u| \, \|_2 \\ &\le c \|u\|_r^{(p-2)/2} Q_p(u)^{1/2}, \quad r = mp/(m-2), \end{aligned} \tag{3.6}$$

where we have used (3.5), (3.4), and the relation $(p-2)q/(2-q) = r$. Also we have

$$\|uu\|_{q'} \le c \|u\|_m \, \|u\|_r, \tag{3.7}$$

since $1/q' = 1/m + 1/r$. Thus the left member of (3.3) is dominated by $\|u\|_m \|u\|_r^{p/2} \cdot Q_p(u)^{1/2}$. Since

$$\|u\|_r \le c Q_p(u)^{1/p} \tag{3.8}$$

by Lemma 2, we have proved (3.3).

Case $p < 2$. We apply the same argument as above, without moving ∂ to the other side but using the fact that $\partial(uu) = u\partial u$. Thus it suffices to show that

$$\| \, |u|^{p-1} \|_q \, \|u \partial u\|_{q'} \le c \|u\|_m Q_p(u). \tag{3.9}$$

This time we choose

$$1/q = (m-2)(p-1)/mp, \qquad 1/q' = (m+2p-2)/mp.$$

Then we have

$$\| \, |u|^{p-1} \|_q = \|u\|_{(p-1)q}^{p-1} = \|u\|_r^{p-1}, \qquad r = mp/(m-2). \tag{3.10}$$

On the other hand,

$$\begin{aligned} \|u\partial u\|_{q'} &\le \| \, |u|^{(4-p)/2} |u|^{(p-2)/2} |\partial u| \, \|_{q'} \\ &\le \| \, |u|^{(4-p)/2} \|_{2q'/(2-q')} \, \| \, |u|^{(p-2)/2} |\partial u| \, \|_2 \\ &\le \|u\|_s^{(4-p)/2} Q_p(u)^{1/2}, \end{aligned}$$

where

$$1/s = (2-q')/(4-p)q' = \lambda/m + (1-\lambda)/r, \qquad \lambda = 2/(4-p).$$

Hence, by the Hölder inequality,

$$\|u\|_s \le \|u\|_m^{\lambda} \, \|u\|_r^{1-\lambda}, \qquad \|u\|_s^{(4-p)/2} \le \|u\|_m \, \|u\|_r^{(2-p)/2},$$

so that

$$\|u\partial u\|_{q'} \le \|u\|_m \, \|u\|_r^{(2-p)/2} Q_p(u)^{1/2}. \tag{3.11}$$

The required result (3.9) follows from (3.10), (3.11), and (3.8).

Proof of Theorem II. We need only minor modifications in the proof of Theorem I. We set $v = J^s u$ and estimate $\|v(t)\|_p = \|u(t)\|_{s,p}$. v satisfies the differential equation

$$\partial_t v = \nu \Delta v - J^s F(u).$$

We shall compute $\partial_t \|v(t)\|_p^p$ as above. The equality (3.1) is true with u replaced by v, except that $F(u)$ should be replaced by $J^s F(u)$. Instead of (3.3), therefore, we have only to prove

$$\| \, |v|^{p-2} |\partial v| \, \|_q \, \|uu\|_{s,q'} \leq c \|u\|_m Q_p(v).$$

For this we can use (3.6) and (3.8) with u replaced by v. The only remaining estimate needed is the modified form of (3.7), viz.

$$\|uu\|_{s,q'} \leq c\|u\|_m \, \|u\|_{s,r} = c\|u\|_m \, \|v\|_r,$$

which follows from (A.1) (Appendix), again by $1/q' = 1/m + 1/r$.

4. Existence theorems

The main object of this paper is to prove the monotonicity of various Liapunov functions. Therefore we have so far disregarded the question of the existence of solutions. To make the paper self-contained, we shall now show that (1.1) is well posed in $X = PL^m \cap PL_s^p$. We shall consider only the case $p \leq m$. The case $p > m$ is not very interesting, since $u(0) \in L^m$ with small norm already implies that $\|u(t)\|_{s,p}$ exists for $t > 0$, $p > m$ and $s \geq 0$, and decays with a definite rate (this was proved for $s = 0$ and 1 in [3]).

This section is almost independent of the previous ones, little use being made of Liapunov functions.

Theorem III. (local existence) Let $m \geq 2$, $1 < p \leq m$, $X = PL^m \cap PL_s^p$ with $\| \, \|_X = max\{\| \, \|_m, \| \, \|_{s,p}\}$. Given $\phi \in X$, there is $T > 0$ ($T = \infty$ is not excluded), depending on m, p, s, ν and ϕ, and a unique solution u to (1.1) such that

$$u \in BC([0,T); X), \quad (tA)^{1/2} u \in BC([0,T); X), \quad A = -P\Delta, \tag{4.1}$$

with $u(0) = \phi$. (BC means "bounded and continuous.")

Remarks. (a) If $s \geq m/p - 1$, then $X = PL_s^p$ by the Sobolev imbedding theorem.

(b) We note without proof that if $s > m/(p-1)$, one may take $T \geq c\nu^\lambda / \|\phi\|_X^{\lambda+1}$, where λ is a constant and c depends on m, p, s, and λ. If $s + 1 - m/p = \theta < 1$, a possible choice for λ is $2/\theta - 1$. For $m = 3$, $p = 2$, $s = 1$ this gives $\lambda = 3$, which appears in [5, p.229]. If $s \leq m/p - 1$, it seems difficult to estimate T in terms of $\|\phi\|_X$ alone.

(c) (4.1) does not include the second spatial derivatives of u, which were used in Theorem I. But it should not be difficult to show that $u(t)$ is smooth for $t > 0$.

Theorem IV. (global existence) There is $\delta > 0$, depending on m, p, and s, such that if $\nu^{-1}\|\phi\|_X < \delta$, then the solution u given by Theorem III can be continued to all $t \geq 0$ so that (4.1) holds with $T = \infty$. Moreover,

$$\|(tA)^h u(t)\|_X \to 0, \quad t \to \infty, \quad 0 \leq h \leq 1/2, \tag{4.2}$$

the convergence being monotone at least for $h = 0$.

Theorems III, IV were partially proved in [2] (for $p = 2$, $s = m/p - 1$, $m = 2, 3$) and in [3] (for $m \geq 2$, $s = 0$). (Cf. also [4], which considers the case $s > m/p + 1$ and includes the Euler equation). Here we show that the same proof works in the general case. Moreover, the following proof simplifies those of [2,3] in several respects.

If we introduce the new variables $\bar{u} = \nu^{-1}u$, $\bar{t} = \nu t$, then (1.1) is transformed into the same equation with $\nu = 1$. Therefore we may assume in the sequel that $\nu = 1$ without loss of generality. Then we write (1.1) in the form

$$\partial_t u + Au + F(u, u) = 0, \qquad A = -P\Delta, \tag{4.3}$$

which we regard as an evolution equation in the Banach space X. A generates a C_0-semigroup $U(t) = e^{-tA}$ on X, which is analytic and contractive; in fact A is identical with $-\Delta$ restricted on X, which is invariant under $e^{t\Delta}$.

Step 1. For the nonlinear operator F in X, we have the following estimate (see Lemma A.4 in Appendix).

$$\|A^{-\ell}F(u, v)\|_X \leq c\|A^k u\|_X \, \|A^k v\|_X, \tag{4.4}$$

$$1/4 < k < 1/2, \qquad 0 < \ell < 1/2, \qquad 2k + \ell = 1. \tag{4.5}$$

We fix a set of such numbers k, ℓ. In view of (4.4), the method given in [2] (especially that for $m = 2$) can be applied to (4.3) with minor modifications. We replace (4.3) by the integral equation

$$u = u_0 + \Phi(u), \quad u_0(t) = U(t)\phi, \quad \Phi(u)(t) = -\int_0^t U(t-\tau)F(u, u)(\tau)d\tau. \tag{4.6}$$

To solve this equation, we fix $0 < T \leq \infty$ and introduce, for each $0 \leq h < 1$, the Banach space $\mathcal{X}_h$ of functions $v \in C((0, T); D(A^h))$ such that $(tA)^h v \in BC([0, T); X)$ (with value zero at $t = 0$ if $h > 0$), with the norm

$$|\!|\!| v |\!|\!|_h = sup\{\|(tA)^h v(t)\|_X; 0 \leq t < T\}.$$

It is obvious that

$$u_0 \in \mathcal{X}_k, \quad \textit{with} \quad |\!|\!| u_0 |\!|\!|_k \leq c_1\|\phi\|_X. \tag{4.7}$$

Moreover we have (see Lemma A.5, Appendix)

$$|\!|\!| \Phi(u) - \Phi(v) |\!|\!|_h \leq c_2(|\!|\!| u |\!|\!|_k + |\!|\!| v |\!|\!|_k)\, |\!|\!| u - v |\!|\!|_k, \quad 0 \leq h < 1 - \ell = 2k. \tag{4.8}$$

The constants c_1, c_2 depend on h, k, ℓ but not on T.

We now set $\Sigma_K = \{v \in \mathcal{X}_k;\ |\!|\!| v |\!|\!|_k \leq K\}$; Σ_K is a complete metric space. We shall show that $u \mapsto u_0 + \Phi(u)$ maps Σ_K into itself if K is chosen appropriately. Indeed, due to (4.8) (with $v = 0, h = k$), this will be the case if $|\!|\!| u_0 |\!|\!|_k + c_2K^2 \leq K$. To satisfy this, we may set $K = /2c_2$, say, and let

$$|\!|\!| u_0 |\!|\!|_k < 1/4c_2^2. \tag{4.9}$$

This is possible by choosing T sufficiently small, if necessary, because $|||u_0|||_k$ tends to zero with T by $k > 0$ (although the smallness of T required could not be determined by $\|\phi\|_X$ alone). Moreover, (4.8) (with $h = k$) shows also that Φ is a contraction on Σ_K. Thus (4.6) has a unique fixed point u in $\Sigma_K \subset \mathcal{X}_k$. Using (4.8) again with $v = 0$, we see that $u \in \mathcal{X}_h$ for any $0 \leq h < 1 - \ell$, which includes $h = 0$ and $1/2$. This proves the existence of a solution u in the class (4.1), at least for sufficiently small T. Uniqueness can be proved by noting that (4.1) implies that $u \in \mathcal{X}_k$ for small T and using the contraction property of Φ.

Step 2. As is seen from (4.7), (4.9) is satisfied even with $T = \infty$ if $\|\phi\|_X$ is sufficiently small. The arguments given in step 1 then lead to a unique global solution.

Step 3. To prove (4.2), it suffices to modify slightly the definition of $v \in \mathcal{X}_h$ used above. In addition to the condition $(tA)^h v \in BC([0,\infty); X)$, we require that $\|(tA)^h(t)\|_X \to 0$ as $t \to \infty$. With this modification, $\mathcal{X}_h$ is still a Banach space, Σ_K is a complete metric space, and $u_0 \in \mathcal{X}_k$. The proof that Φ maps $\mathcal{X}_k$ into itself is nontrivial, however, and will be given in Appendix (Lemma A.5). The remaining arguments in step 2 are unchanged. Thus the fixed point constructed automatically satisfies (4.2), except for the monotonicity. But the latter follows from Theorem II, if $\|\phi\|_m \leq \|\phi\|_X$ is sufficiently small.

Appendix

1. *Some lemmas on L^p_s-norms.* In what follows we need the *homogeneous L^p_s-norms*, defined by

$$|f|_{s,p} = \|I^s f\|_p, \qquad I = (-\Delta)^{1/2},$$

in addition to the inhomongeneous ones $\|f\|_{s,p} = \|J^s f\|_p$. Naturally we have $|f|_{r,p} \leq c\|f\|_{s,p}$ for $1 < p < \infty$, $0 \leq r \leq s$ (as may be proved by Mihlin's theorem). Also we note

Lemma A.1. Let $1 < p < \infty$, $s \geq 0$. Then $f \in L^p_s$ if and only if f and $I^s f$ are in L^p. The norms $\|f\|_{s,p}$ and $\|f\|_p + |f|_{s,p}$ are equivalent.

Proof. The remark given above shows that $\|f\|_p + |f|_{s,p} \leq c\|f\|_{s,p}$. To prove the opposite inequality, it suffices to apply Mihlin's theorem to the symbol $(1+|\xi|^2)^{s/2}/(1+|\xi|^s)$.

Lemma A.2. For complex-valued function f, g (scalar or vector valued), we have

$$\|fg\|_{s,p} \leq c(\|f\|_{s,p_1}\|g\|_{q_1} + \|f\|_{q_2}\|g\|_{s,p_2}), \tag{A.1}$$

provided that

$$s \geq 0, \quad 1 < p_j < \infty, \quad 1 < q_j \leq \infty, \quad 1/p_j + 1/q_j = 1/p \quad (j = 1, 2). \tag{A.2}$$

Moreover, we have a homogeneous version of (A.1), in which the norms $\|\ \|_{s,p}$, etc. are replaced by the corresponding $|\ |_{s,p}$, etc. throughout.

Proof. (A.1) follows from a general result of Coifman-Meyer [1]. To deal with the homogeneous case, it suffices to apply (A.1) to f, g replaced by $f_\epsilon(x) = f(x/\epsilon)$, $g_\epsilon(x) = g(x/\epsilon)$ and let $\epsilon \searrow 0$.

Lemma A.3. Let $s_j < m/p\,(j = 1, 2)$, $s_1 + s_2 = s + m/p$, $0 < s \le min\{s_1, s_2\}$. We have

$$\|fg\|_{s,p} \le c\|f\|_{s_1,p}\,\|g\|_{s_2,p}.$$

Here again we may replace $\|\ \|$ by $|\ |$.

Proof. Set $m/p_1 = s_2 = m/p - s_1 + s$, $m/q_1 = m/p - s_2$. Then $1/p_1 + 1/q_1 = 1/p$ and $\|f\|_{s,p_1} \le c\|f\|_{s_1,p}$, $\|g\|_{q_1} \le c\|g\|_{s_2,p}$ by the Sobolev inequality; note that $1/p_1,\ 1/q_1 > 0$. Thus the first term on the right of (A.1) satisfies the required inequality. The second term can be handled similarly. The homogeneous version can be proved as in Lemma A.2.

2. *Basic estimates for the nonlinear operator.*

Lemma A.4. Let $1 < p \le m$, $s \ge 0$, $X = PL^p_s \cap PL^m$, with $\|\phi\|_X = max\{\|\phi\|_{s,p}\|\phi\|_m\}$. Let $1/4 < k < 1/2$, $0 < \ell < 1/2$, $2k + \ell = 1$. Then we have

$$\|A^{-\ell}F(u, v)\|_X \le c\|A^k u\|_X\,\|A^k v\|_X, \tag{A.3}$$

where c is a constant depending on m, p, s and k.

Proof. Set

$$1/p_1 = 1/p - \ell/m > 0, \quad 1/q_1 = \ell/m > 0, \quad 1/p_1 + 1/q_1 = 1/p. \tag{A.4}$$

Since $A = PI^2$ and $F(u, v) = P\partial(u, v)$, where P is a bounded operator, we obtain by using the *homogeneous version* of (A.1)

$$|A^{-\ell}F(u, v)|_{s,p} \le c|uv|_{s+1-2\ell,p} \le c|u|_{q_1}|v|_{s+1-2\ell,p_1} + (\ldots), \tag{A.5}$$

where (...) denotes a term equal to the preceding one with u, v interchanged. The Sobolev inequalities then give

$$|v|_{s+1-2\ell,p_1} \le c|v|_{2k+s,p} = c|A^k v|_{s,p} \le c\|A^k v\|_X, \tag{A.6}$$

$$|u|_{q_1} \le c|u|_{2k,m} = c\|A^k u\|_m \le c\|A^k u\|_X; \tag{A.7}$$

note that $(2k + s) - (s + 1 - 2\ell) = 2k + 2\ell - 1 = \ell = m/q_1 = m(1/p - 1/p_1)$, and $2k = 1 - \ell = m(1/m - 1/q_1)$. Treating the term (...) in the same way, we thus obtain from (A.5)

$$|A^{-\ell}F(u, v)|_{s,p} \le c\|A^k u\|_X\,\|A^k v\|_X. \tag{A.8}$$

This is true for $s = 0$ too, so that we have also $\|A^{-\ell}F(u, v)\|_p \le c\|f^k u\|_X\,\|A^k v\|_X$. It follows that

$$\|A^{-\ell}F(u, v)\|_{s,p} \le c\|A^k u\|_X\,\|A^k v\|_X.$$

Again, using Lemma A.3 with $p = m$, $s = 1 - 2\ell$, $s_1 = s_2 = 2k$, so that $s + m/p = s_1 + s_2$, we have

$$\|A^{-\ell}F(u,v)\|_m \le c|uv|_{1-2\ell,m} \le c|u|_{2k,m}|v|_{2k,m}$$
$$\le c\|A^k u\|_m \|A^k v\|_m \le c\|A^k u\|_X \|A^k v\|_X.$$

Summing up, we have proved (A.3).

Lemma A.5. Φ maps $\mathcal{X}_k$ into $\mathcal{X}_h$ if $0 \le k < 1/2$ and $0 \le h < 1 - \ell = 2k$, with

$$|||\Phi(u) - \Phi(v)|||_h \le c(|||u|||_k + |||v|||_k)|||u - v|||_k. \tag{A.9}$$

Proof. Let $u \in \mathcal{X}_k$. Writing $\| \ \|$ for $\| \ \|_X$ for simplicity, and letting $\ell = 1 - 2k > 0$, we have, using (A.3) and $\|A^\alpha U(t)\| \le ct^{-\alpha}$,

$$\begin{aligned}\|(tA)^h\Phi(u)(t)\| &\le \|t^h \int_0^t A^{h+\ell}U(t-\tau)A^{-\ell}F(u(\tau),u(\tau))d\tau\| \\ &\le ct^h \int_0^t (t-\tau)^{-h-\ell}\|A^k u(\tau)\|^2 d\tau \qquad \text{(A.10)} \\ &\le (ct^h \int_0^t (t-\tau)^{-h-\ell}\tau^{-2k}d\tau)|||u|||_k^2 \le c|||u|||_k^2.\end{aligned}$$

This proves the lemma, with the estimate (A.9) for $v = 0$. The general case in (A.9) can be proved in the same way.

If the definition of $v \in \mathcal{X}_h$ is modified to include the condition that $(tA^h)v(t) \to 0$, $t \to \infty$, we have to show, in addition, that (A.10) tends to zero as $t \to \infty$. To this end we split the integral into two parts, on $(0, t')$ and (t', ∞). We estimate the second part as above. Since $\|A^k u(\tau)\| = o(\tau^{-k})$ by the definition of $\mathcal{X}_k$, it follows that this part is arbitrarily small if t' is sufficiently large. On the other hand, the first part is dominated by

$$t^h \int_0^{t'} (t-\tau)^{-h-\ell}\tau^{-2k}d\tau \le ct^h(t-t')^{-h-\ell}t'^{1-2k},$$

which tends to zero as $t \to \infty$ with fixed t'. This proves the desired result.

Acknowledgment. The author is indebted to Mark A. Kon, Gustavo Ponce, and Christian G. Simader for valuable comments.

References

1. R. R. Coifman and Y. Meyer, *Nonlinear harmonic analysis, operator theory and P.D.E.*, in "Beijing Lectures in Harmonic Analysis," Princeton University Press, 1986, pp.3-45.

2. T. Kato and H. Fujita, *On the nonstationary Navier Stokes system*, Rend. Sem. Mat. Univ. Padova **32** (1962), 243-260.

3. T. Kato, *Strong L^p solutions of the Navier-Stokes equation in R^m, with applications to weak solutions*, Math. Z. **187** (1984), 471-480.

4. T. Kato and G. Ponce, *Commutator estimates and the Euler and Navier-Stokes equations*, Comm. Pure Appl. Math. **41** (1988), 891-907.

5. J. Leray, *Sur le mouvement d'un liquide visqueux emplissant l'espace*, Acta Math. **63** (1934), 193-248.

Dissipativity of the nonlinear operator (note added in proof)

We may write (1.1) in the form $\partial_t u = \mathcal{A}u$, where $\mathcal{A} : u \mapsto \mathcal{A}u = \nu\Delta u - F(u)$ is a nonlinear operator. From (3.1) it is easy to see that

$$\langle \mathcal{A}u, Gu \rangle \leq -c(\nu - K\|u\|_m)Q_p(u), \tag{1}$$

where $Gu = |u|^{p-2}u$ is the *duality map* on L^p to $L^{p'}$. Now a minor modification of the computation in section 3 shows that

$$\langle \mathcal{A}u - \mathcal{A}v, G(u-v) \rangle \leq -c(\nu - K(\|u\|_m + \|v\|_m))Q_p(u-v) \tag{2}$$

is true if $p \geq 2$. (2) is a stronger property than (1). It implies that $\langle \mathcal{A}u - \mathcal{A}v, G(u-v) \rangle \leq 0$ if $\|u\|_m, \|v\|_m$ are small. Thus $\mathcal{A}$ is "locally dissipative." (The theory of dissipative and accretive operators was studied extensively in 1960's; see, e.g., F. Browder, Nonlinear operators and nonlinear equations of evolution in Banach spaces, Proc. Symp. Pure Math. XVIII, Part 2, American Mathematical Society, 1976.) A direct consequence of (2) is that if u, v are two solutions of (1.1) with small Reynolds numbers, then $\|u(t) - v(t)\|_p$ decreases in time. It follows also that $\|\partial_t u(t)\|_p$ decreases for a solution u with small Reynolds number.

Obviously the local dissipativity (2) is true even for the operator $\mathcal{A}u + f(t)$, which includes a forcing term f. On the other hand, we have not been able to prove (2) for $p < 2$ or for the L^p_s-norm with $s > 0$.

Singular Solutions of a Nonlinear Elliptic Equation and an Infinite Dimensional Dynamical System

Hiroshi Matano

Department of Mathematics, University of Tokyo
Hongo, Tokyo 113, Japan

§1. Introduction

Some classes of partial differential equations of elliptic type can be converted into autonomous time-evolution equations by choosing one of the space variables as time variable. The resulting evolution equations are, of course, still of elliptic type, hence are ill-posed. Nonetheless, there are cases in which one can deal with the problem in the framework of dynamical systems or their analogues. Note that the dynamical systems relevant to those equations are of infinite dimensions, since none of those partial differential equations can be converted into a finite system of ordinary differential equations. In this paper we deal with singular solutions of some nonlinear elliptic equation and study their behavior near the singular point and near the infinity. The main techniques used here come from the theory of infinite dimensional dynamical systems. The advantage of using the dynamical systems point of view is that one can get a geometrical insight into the problem. Thanks to this approach, the results we present in this paper are somewhat of global nature and are perhaps difficult to obtain by other conventional methods of analysis.

We consider the following semilinear elliptic equation:

$$\Delta u - |u|^{q-1}u = 0. \tag{1.1}$$

Here q is a constant satisfying $q > 1$. Mainly we consider (1.1) in a subset of $\mathbf{R}^2$. Now let Ω be an open subset of $\mathbf{R}^2$. If a function $u(x)$ is defined on Ω except at points $a_1, a_2, \ldots, a_m \in \Omega$ and satisfies (1.1) in $\Omega \setminus \{a_1, a_2, \ldots, a_m\}$, then u can be regarded as a solution with singularities at $a_1, a_2, \ldots, a_m$. We call the points $a_1, a_2, \ldots, a_m$ the *singular points* of u. A related problem is found in the Thomas-Fermi theory in quantum mechanics, in which case the points $a_1, a_2, \ldots, a_m$ are where electrons are located. See Brezis and Lieb [BL] for the physical background of the problem and some a priori estimates near the singular points.

In an earlier paper [CMV], Chen, Matano and Véron have shown that isolated singularities of solutions of (1.1) can be classified in terms of their "asymptotic profiles". In what follows we state this result for a special case where the solution has only one singular point. Let $u(x)$ be a solution of the equation

$$\Delta u - |u|^{q-1}u = 0 \qquad \text{in} \quad \mathbf{R}^2 \setminus \{0\}. \tag{1.2}$$

The function u is defined everywhere on $\mathbf{R}^2$ except at the origin, which is the singular point. We call solutions of (1.2) "global singular solutions". By a solution of (1.2) we always mean a classical solution ; that is, $u \in C^2(\mathbf{R}^2 \setminus \{0\})$. It is easily seen that any weak solution of (1.2) that is locally bounded in $\mathbf{R}^2 \setminus \{0\}$ is a classical solution. Now let (r, θ) denote the polar coordinates in $\mathbf{R}^2 \setminus \{0\}$, where $r > 0$ and $\theta \in S^1 = \mathbf{R}/2\pi\mathbf{Z}$. Then we have the following:

Theorem 1.1 ([CMV]). *Let u be a C^2 solution of (1.2). Then both of the following limits exist:*

$$\omega_0(\theta) = \lim_{r\to 0} r^{2/(q-1)} u(r, \theta), \tag{1.3a}$$

$$\omega_\infty(\theta) = \lim_{r\to \infty} r^{2/(q-1)} u(r, \theta). \tag{1.3b}$$

The convergence takes place in the topology of $C^2(S^1)$. Moreover, both ω_0 and ω_∞ satisfy the following ordinary differential equation :

$$\frac{d^2\omega}{d\theta^2} + g(\omega) = 0 \qquad in \quad S^1, \tag{1.4}$$

where

$$g(s) = (\frac{2}{q-1})^2 s - |s|^{q-1} s. \tag{1.5}$$

We call ω_0 and ω_∞ the *asymptotic profile* of u at the origin and that at the infinity, respectively. The convergence in (1.3a) and (1.3b) can be proved by using an idea similar to what is found in [CM]. If $\omega_0 \equiv 0$, we say that u has a *weak singularity* at $x = 0$. In [CMN] we have classified all weak singularities, which we do not deal with in the present paper.

After all, what Theorem 1.1 is concerned with are the local behaviors of u near $x = 0$ and near the infinity. One may call this the "local theory".

On the other hand, some natural questions arise concerning a more global aspect of the problem. These are :

(Q1) If we choose a pair of solutions of (1.4) arbitrarily and call them ω_0 and ω_∞, can we always find a solution u of (1.2) that satisfies both (1.3a) and (1.3b) ?

(Q2) If the answer to the question (Q1) is "no", then what conditions should ω_0 and ω_∞ satisfy so that such a solution u may exist ?

(Q3) Can one determine the complete set of solutions of (1.2) ?

We have already shown in [CMV] that the answer to the question (Q1) is "no". This comes from a simple energy estimate. Namely :

Theorem 1.2 ([CMV]). *Let J be a functional on $H^1(S^1)$ defined by*

$$J(\psi) = \int_{S^1} \{\frac{1}{2}(\psi')^2 - G(\psi)\} d\theta, \tag{1.6}$$

where

$$G(s) = \int_0^s g(\sigma)d\sigma.$$

Let $u(x)$ be a solution of (1.2) and let ω_0, ω_∞ be as in (1.3a), (1.3b). Then either one of the following holds :
(a) $J(\omega_0) > J(\omega_\infty)$, or
(b) $\omega_0 = \omega_\infty$ and

$$u(r,\theta) = r^{-2/(q-1)}\omega_0(\theta). \tag{1.7}$$

The above theorem gives a necessary condition in regard to question (Q2). Note that the case (b) above corresponds to those solutions of (1.2) that are obtained by the method of separation of variables. As we will see later in this paper, (b) is a very special case, and most solutions satisfy (a) (see Proposition 3.11 and Remarks 6.3 and 6.4).

The aim of this paper is to show that conditions given in Theorem 1.2 are not only necessary for ω_0, ω_∞ to be the asymptotic profiles of some solution u of (1.2) but also are sufficient. In other words, if a pair of solutions ω_0, ω_∞ of (1.4) satisfy either (a) or (b), then there exists a solution $u(x)$ of (1.2) satisfying (1.3a) and (1.3b). To show this, we first construct a certain infinite-dimensional dynamical system — more precisely, a semiflow — on a suitable function space. A notable feature of this semiflow is that there is one-to-one correspondence between each solution of (1.2) and each point on the "global attractor" of the semiflow. We will then see that question (Q2) is equivalent to the the following question :

(Q2)′ Which pairs of equilibrium points have orbital connections, and which pairs not ?

Here a pair of equilibrium points is said to have an *orbital connection* if there exists an entire orbit —an orbit defined for all $-\infty < t < \infty$ —that connects those two points.

Analysis of the above-mentioned semiflow has already been partially done in the earlier paper [CMV], where we have used the theory of strongly order-preserving semiflows [M1] to show the existence of some connecting orbits. In the present paper we make a far more extensive study of the semiflow and derive results that are much stronger than those in [CMV]. One important observation that will play a key role in the later arguments is that the unstable manifolds and the stable manifolds of equilibrium points intersect with each other transversally (Theorem 5.3). This observation helps us find connections within various pairs of equilibrium points. The results we present in this paper answer question (Q2) and give us a certain insight into the global structure of the whole set of solutions of (1.2). We shall, however, leave question (Q3) to our future study.

Let us also mention some other recent works in which the point of view of dynamical systems — or their analogues — has been used to the study of elliptic equations. These include [AK], [AT], [Av], [K], [Mie], [S], [V1]. In particular, Kirchgässner [K] has pioneered the application of center manifold theorem to elliptic problems. While most of these works deal with the local behavior of the semiflow — such as the structure of a center manifold near an equilibrium point —, the present paper, on the other hand, studies a more global aspect of the semiflow.

Our paper is organized as follows : In the next section we state our main result (Theorem 2.1). In Section 3 we construct the semiflow associated with equation (1.2). In section 4 we discuss the smoothness of the semiflow. We prove, in Section 5, that the stable and the unstable manifolds of equilibrium points intersect transversally. More precisely, we show that the semiflow is an equivariant Morse-Smale system. Finally, in Section 6, we complete the proof of Theorem 2.1.

Due to limitation in space, we state some of the lemmas and propositions without proof, or give only an outline of the proof. Detailed proofs of those lemmas and propositions can be found in the forthcoming paper [M3], in which a full account of the theory will be given.

The author expresses his gratitude to Dr. Xu-Yan Chen and Prof. Shuichi Jimbo for their useful comments.

§2. Main Result

The main result of this paper is the following :

Theorem 2.1 (Main theorem). *Let $J : H^1(S^1) \longrightarrow \mathbf{R}$ be as in (1.6) and let $\omega, \tilde{\omega}$ be a pair of solutions of (1.4). Suppose either that $\omega = \tilde{\omega}$ or that $J(\omega) > J(\tilde{\omega})$. Then there exists a solution $u(x)$ of (1.2) satisfying (1.3a) and (1.3b) with $\omega_0 = \omega$ and $\omega_\infty = \tilde{\omega}$.*

To better understand what such a solution $u(x)$ looks like, we have to first study the structure of the set of solutions of equation (1.4). Let

$$\mathcal{E} = \{\omega \in C^2(S^1) \mid \frac{d^2\omega}{d\theta^2} + g(\omega) = 0 \quad \text{on} \quad S^1\}, \tag{2.1}$$

$$A = (\frac{2}{q-1})^{2/(q-1)}, \tag{2.2}$$

$$k_0 = \max\{k \in \mathbf{Z} \mid \quad k < \frac{2}{q-1}\}, \tag{2.3}$$

where $\mathbf{Z}$ is the set of all integers.

It is easily seen that (1.4) has three constant solutions, $A, -A$ and 0. One can also show that (1.4) has nonconstant solutions if and only if $k_0 \geq 1$. The graph of each nonconstant solution looks more or less like that of $A \sin k(\theta + \alpha)$ for some $0 \leq \alpha < 2\pi$ and $1 \leq k \leq k_0$. More precisely, we have the following :

Proposition 2.2 ([CMV]). *$\mathcal{E}$ has precisely $k_0 + 3$ connected components $\mathcal{E}^+, \mathcal{E}^-, \mathcal{E}^0$, and $\mathcal{E}_k$ $(1 \leq k \leq k_0)$, where*

(i) $\mathcal{E}^+ = \{A\}$, $\mathcal{E}^- = \{-A\}$, $\mathcal{E}^0 = \{0\}$;

(ii) $\mathcal{E}_k = \{\omega_k(\cdot + \alpha) \mid 0 \leq \alpha < 2\pi\}$ for each $1 \leq k \leq k_0$, where ω_k is a solution of (1.4) with minimal period $2\pi/k$.

Note that $\mathcal{E}^+, \mathcal{E}^-, \mathcal{E}^0$ are isolated points, while each $\mathcal{E}_k$ is homeomorphic to a circle. The definition of $\mathcal{E}_k$ does not depend on the choice of ω_k since any nonconstant solutions of (1.4) having the same minimal period are identical to one another up to the rotation — or phase shift — along S^1. Note also that $\mathcal{E}_k$'s are nonexistent if $q \geq 3$.

Proposition 2.3 ([CMV]). *Let J and $\mathcal{E}$ be as above. Then*

$$0 = J(\mathcal{E}^0) > J(\mathcal{E}_{k_0}) > \ldots > J(\mathcal{E}_1) > J(\mathcal{E}^+) = J(\mathcal{E}^-).$$

Corollary 2.4. *(i) Let u be a solution of (1.2), and ω_0, ω_∞ be as in (1.3a), (1.3b). Suppose $\omega_0 \in \mathcal{E}_k, \omega_\infty \in \mathcal{E}_{k'}$. Then either $k > k'$, or $k = k'$ and $\omega_0 = \omega_\infty$.*

(ii) Conversely, let $\omega, \tilde{\omega}$ be an arbitrary pair of solutions of (1.4) such that $\omega \in \mathcal{E}_k$ and $\tilde{\omega} \in \mathcal{E}_{k'}$ with $k > k'$. Then there exists a solution u of (1.2) satisfying (1.3a) and (1.3b) with $\omega_0 = \omega$ and $\omega_\infty = \tilde{\omega}$.

Statement (i) of the above corollary implies, roughly speaking, that a solution $u(x)$ of (1.2) tends to oscillate more frequently near $x = 0$ than near the infinity. To illustrate what statement (ii) implies, take, for example, a pair of solutions $\omega, \tilde{\omega}$ of (1.4) satisfying $\omega \in \mathcal{E}_3$ and $\tilde{\omega} \in \mathcal{E}_2$. The function ω oscillates three times on S^1, while $\tilde{\omega}$ twice. Corollary 2.4 (ii) asserts that there exists a solution $u(x)$ of (1.2) whose asymptotic profiles are precisely $\omega, \tilde{\omega}$. This means that the function $u(x)$ oscillates three times on the circle $|x| = R$ for R sufficiently small and that it oscillates twice if R is sufficiently large. Such a function u bears no global symmetry. Moreover, for any $\alpha, \beta \in S^1$, one can still find a solution of (1.2) whose asymptotic profiles are $\omega(\cdot + \alpha), \tilde{\omega}(\cdot + \beta)$. Since α and β can be chosen totally independently, we have a rather large set of nonsymmetric solutions.

Remark 2.5. Given $\omega \in \mathcal{E}_k, \tilde{\omega} \in \mathcal{E}_{k'}$ with $k > k'$, one may wonder if the solution u of (1.2) with asymptotic profiles $\omega, \tilde{\omega}$ is unique. It is, of course, not unique in the strict sense, since the following self-similar transformation with arbitrary $\lambda > 0$ produces a one-dimensional family — or a curve — of solutions of (1.2) sharing the same asymptotic profiles :

$$u(r, \theta) \longmapsto \lambda^{2/(q-1)} u(\lambda r, \theta). \tag{2.4}$$

We say that solutions u and v belong to the same *similarity class* if v is written as in the right-hand side of (2.4) for some $\lambda > 0$. The transformation (2.4) leaves u unchanged if and only if u is of the form (1.7), or, equivalently, if $\omega = \tilde{\omega}$. In this case the similarity class consists of a single point. Otherwise it is a curve. The question can thus be formulated as follows : Do solutions of (1.2) having the asymptotic profiles $\omega, \tilde{\omega}$ all belong to the same similarity class ? In the case $k = k' + 1$, we suspect that the answer is yes, though we have not yet come to a complete proof. (Thus, in this case, we suspect that the set of solutions sharing the same asymptotic profiles $\omega, \tilde{\omega}$ forms a simple curve.) In the general case, however, there is evidence that the whole set of solutions of (1.2) with the same asymptotic profiles $\omega, \tilde{\omega}$ forms a $2(k - k') - 1$ dimensional manifold. See Remark 6.3 in Section 6 for details.

§3. Construction of the Semiflow

This section is devoted to the construction of a semiflow associated with equation (1.1). The definition of the semiflow we give here is the same as that in [CMV] except for a few changes in constants that are not important. The basic idea is to consider the boundary value problem for (1.1) in an exterior region and interpret it as an initial value problem. We first construct the semiflow on the space $C^0(S^1)$, the space of real-valued continuous functions on S^1. We will later restrict it onto the space $H^1(S^1)$, that is, the Sobolev space consisting of $L^2(S^1)$ functions whose first-order derivative also belongs to $L^2(S^1)$. Since $H^1(S^1)$ is a Hilbert space, some geometric properties of the semiflow, as well as its smoothness properties, can be studied more easily in this space.

We start with the following lemma :

Lemma 3.1 ([CMV]). *For any $\psi \in C^0(S^1)$ there exists a unique $u \in C^0(\mathbf{R}^2 \setminus B_1) \cap C^2(\mathbf{R}^2 \setminus \overline{B}_1)$ such that*

$$\Delta u - |u|^{q-1}u = 0 \qquad in \quad \mathbf{R}^2 \setminus \overline{B_1}, \tag{3.1a}$$

$$u = \psi \qquad on \quad S^1 = \partial B_1, \tag{3.1b}$$

where $B_1 = \{x \in \mathbf{R}^2 | \quad |x| < 1\}$. Moreover u belongs to $C^3(\mathbf{R}^2 \setminus \overline{B_1})$ and satisfies

$$|u(x)| \le C(|x| - 1)^{-2/(q-1)} \qquad in \quad \mathbf{R}^2 \setminus \overline{B_1}, \tag{3.2}$$

where C is a constant independent of the choice of $\psi \in C^0(S^1)$.

The proof of the above lemma is based on the maximum principle and some a priori estimates; see [CMV; Proposition 3.1] for details. Estimates similar to (3.2) has been obtained earlier by [BL]; see also [LN], [O] and [V1].

As in [CMV], we make the following change of variables :

$$v(t,\theta) = r^{2/(q-1)}u(r,\theta), \quad t = \log r. \tag{3.3}$$

Then the elliptic boundary value problem (3.1) can be rewritten as

$$v_{tt} - \frac{4}{q-1}v_t + v_{\theta\theta} + g(v) = 0 \quad \text{for} \quad t > 0, \ \theta \in S^1, \tag{3.4a}$$

$$v(0,\cdot) = \psi(\cdot), \tag{3.4b}$$

where g is as in (1.5). By Lemma 3.1 and standard a priori estimates for elliptic equations, we immediately have the following :

Lemma 3.1′. *For any $\psi \in C^0(S^1)$, there exists a unique $v \in C^0([0,\infty) \times S^1) \cap C^2((0,\infty)\times S^1)$ that satisfies (3.4). Moreover v belongs to $C^3((0,\infty)\times S^1) \cap L^\infty((0,\infty)\times S^1)$ and satisfies*

$$\limsup_{t\to\infty} \|v(t,\cdot)\|_{L^\infty(S^1)} \le C, \tag{3.5}$$

where C is a constant independent of the choice of $\psi \in C^0(S^1)$. More precisely, for any $T > 0$ there exists $C_T > 0$, independent of the choice of ψ, such that

$$\sup_{t \geq T} \|v(t,\cdot)\|_{L^\infty(S^1)} \leq C_T, \tag{3.5$'$}$$

$$\sup_{t \geq T} \max_{0 \leq j \leq 3} \|\frac{\partial^j v}{\partial t^j}(t,\cdot)\|_{C^{3-j}(S^1)} \leq C_T. \tag{3.5$''$}$$

For each $t \geq 0$, let us define a map $\Phi_t : C^0(S^1) \longrightarrow C^0(S^1)$ by

$$[\Phi_t(\psi)](\cdot) = v(t, \cdot\,; \psi) \qquad (\psi \in C^0(S^1)), \tag{3.6}$$

where $v(t, \theta; \psi)$ denotes the solution of (3.4) for the specified boundary data ψ. Then

Proposition 3.2 ([CMV]). $\{\Phi_t\}_{t \geq 0}$ *is a semiflow on $C^0(S^1)$, that is,*
(i) $\Phi_0 = \mathtt{Id}$ *(identity map);*
(ii) $\Phi_{t+s} = \Phi_t \circ \Phi_s$ *for any* $t, s \geq 0$*;*
(iii) $(t, \psi) \longmapsto \Phi_t(\psi)$ *defines a continuous map from $[0, \infty) \times C^0(S^1)$ into $C^0(S^1)$.*

Remark 3.3. When we define the semiflow Φ out of the elliptic problem (3.4), we are looking at only those solutions that exist globally on $[0, \infty) \times S^1$. If we drop this global existence requirement and instead replace (3.4b) by standard initial conditions of the form $v(0, \cdot) = \psi_1, (\partial v/\partial t)(0, \cdot) = \psi_2$, then the problem would no longer be well-posed and therefore would not define a semiflow in any standard function spaces.

We now understand that (3.4) can be regarded as a well-posed initial value problem. In fact, we can rewrite (3.4) in the form of a parabolic equation. To see this, let

$$A_1 = a - (a^2 - \Delta)^{1/2}, \tag{3.7}$$

$$A_2 = a + (a^2 - \Delta)^{1/2}, \tag{3.8}$$

where

$$a = 2/(q-1), \tag{3.9}$$

$$\Delta = \frac{d^2}{d\theta^2}.$$

Since $a^2 - \Delta$ is a positive definite self-adjoint operator in $L^2(S^1)$, the operators A_1, A_2 are well defined and are both self-adjoint operators in $L^2(S^1)$, with the domain of definition being $H^1(S^1)$. The former is negative definite and the latter is positive definite. In particular, both A_1 and $-A_2$ generate an analytic semigroup on $L^2(S^1)$. Their restrictions onto the spaces $H^\gamma(S^1)$ $(\gamma \geq 0)$ also generate analytic semigroups on those spaces. Given γ, with $0 \leq \gamma < 1$, define $f : C^0(S^1) \longrightarrow H^\gamma(S^1)$ by

$$f(v) = \int_0^\infty e^{-\tau A_2} g(\Phi_\tau(v))d\tau, \tag{3.10}$$

where g is as in (1.5) and Φ as in (3.6). As is easily seen, there exists a constant $M > 0$ such that

$$\|e^{-\tau A_2}w\|_{H^\gamma(S^1)} \leq Me^{-2a\tau}\tau^{-\gamma}\|w\|_{L^2(S^1)}$$
$$\leq \sqrt{2\pi}Me^{-2a\tau}\tau^{-\gamma}\|w\|_{L^\infty(S^1)}$$

(see, for example, [He1]). Combining this inequality and (3.16), which will be given later, and also using (3.5)', one can show the following :

Lemma 3.4. *For any γ, $0 \leq \gamma < 1$, the map $f : C^0(S^1) \longrightarrow H^\gamma(S^1)$ is locally Lipschitz continuous. That is, for any $K > 0$ there exists $L > 0$ such that*

$$\|f(v) - f(\widetilde{v})\|_{H^\gamma(S^1)} \leq L\|v - \widetilde{v}\|_{L^\infty(S^1)} \tag{3.11}$$

for every $v, \widetilde{v} \in C^0(S^1)$ satisfying $\|v\|_{L^\infty}, \|\widetilde{v}\|_{L^\infty} \leq K$. Consequently, its restriction

$$f : H^\gamma(S^1) \longrightarrow H^\gamma(S^1)$$

is locally Lipschitz continuous for any γ with $1/2 < \gamma < 1$.

The proof of (3.11) is a straightforward computation, so we omit it. Note that the local Lipschitz continuity of the map $f : H^\gamma(S^1) \longrightarrow H^\gamma(S^1)$ is a consequence of the continuous embedding $H^\gamma(S^1) \subset C^0(S^1)$ for $\gamma > 1/2$.

Proposition 3.5. *The map $f : H^\gamma(S^1) \longrightarrow H^{\gamma+\beta}(S^1)$ is locally Lipschitz continuous for any $1/2 \leq \gamma \leq 1$, $0 \leq \beta < 1$. In particular, $f : H^1(S^1) \longrightarrow H^1(S^1)$ is locally Lipschitz continuous. Consequently, the following initial value problem is well-posed in $H^1(S^1)$:*

$$\frac{dv}{dt} = A_1 v + f(v) \qquad (t > 0), \tag{3.12a}$$

$$v(0) = \psi. \tag{3.12b}$$

Proposition 3.6. *Let $1/2 < \gamma < 1$ and let $\psi \in H^\gamma(S^1)$. Then $v = v(t, \theta)$ is a solution of (3.4) if and only if $v(t, \cdot)$ solves the initial value problem (3.12) in $H^\gamma(S^1)$.*

We first prove the latter :

(Proof of Proposition 3.6.) Let v be a solution of (3.4) with $\psi \in H^\gamma(S^1)$. Equation (3.4a) can be rewritten in the form

$$\frac{d^2v}{dt^2} - (A_1 + A_2)\frac{dv}{dt} + A_1 A_2 v + g(v) = 0,$$

Or equivalently,

$$\frac{d}{dt}\{e^{-tA_2}(\frac{d}{dt} - A_1)v\} + e^{-tA_2}g(v) = 0. \tag{3.13}$$

Note that $\|(\frac{d}{dt} - A_1)v\|_{H^\gamma(S^1)}$ remains bounded as $t \to \infty$ since $v_t, v_{t\theta}, v_\theta, v_{\theta\theta}$ are all bounded as $t \to \infty$ by virtue of (3.5)''. Hence $\|e^{-tA_2}(\frac{d}{dt} - A_1)v\|_{H^\gamma} \to 0$ as $t \to \infty$. In view of this, and integrating (3.13) over $t \le \tau < \infty$, we get (3.12a). One can also show that $v(t,\cdot) \to \psi$ as $t \to 0$ in the topology of $H^\gamma(S^1)$. (The details are ommitted.) Hence v solves the initial value problem (3.12) in the space $H^\gamma(S^1)$. Conversely, suppose that v is a solution to the initial value problem (3.12) for initial data $\psi \in H^\gamma(S^1)$. Let $\widetilde{v}$ be the solution of (3.4) for the same data ψ. Then, as we have shown above, $\widetilde{v}$ satisfies (3.12) and exists for all $t \ge 0$. By Lemma 3.4, the initial value problem (3.12) is well-posed in $H^\gamma(S^1)$, therefore the solution is unique for each initial data. It follows that $v = \widetilde{v}$. The proposition is proved.

(Proof of Proposition 3.5.) We first consider the case where $1/2 < \gamma < 1$. By Lemma 3.4, $f : H^\gamma(S^1) \longrightarrow H^\gamma(S^1)$ is locally Lipschitz continuous, hence the initial value problem (3.12) is well-posed in $H^\gamma(S^1)$. It follows that for any $K > 0$ and $T > 0$ there exists $C_1 > 0$ such that

$$\|\Phi_t(\psi) - \Phi_t(\widetilde{\psi})\|_{H^\gamma(S^1)} \le C_1 \|\psi - \widetilde{\psi}\|_{H^\gamma(S^1)} \tag{3.14}$$

for any $t \in [0, T]$ and $\psi, \widetilde{\psi} \in H^\gamma(S^1)$ with $\|\psi\|_{H^\gamma}, \|\widetilde{\psi}\|_{H^\gamma} \le K$. Now let $u, \widetilde{u}$ be the solutions of (3.1) with the boundary condition $u = \psi$ and $\widetilde{u} = \widetilde{\psi}$, respectively. Then

$$\|u - \widetilde{u}\|_{L^\infty(\mathbf{R}^2 \setminus B_1)} \le \|\psi - \widetilde{\psi}\|_{L^\infty(S^1)}. \tag{3.15}$$

This is a consequence of the maximum principle and the fact that both u and $\widetilde{u}$ decay as $|x| \to \infty$ by virtue of (3.2) (see [CMV; proof of Proposition 3.2]). (3.15) implies that

$$\|\Phi_t(\psi) - \Phi_t(\widetilde{\psi})\|_{L^\infty(S^1)} \le e^{at} \|\psi - \widetilde{\psi}\|_{L^\infty(S^1)} \tag{3.16}$$

for any $t \ge 0$. It follows that for each $T > 0$ there exists $C_2 > 0$ such that

$$\|\Phi_t(\psi) - \Phi_t(\widetilde{\psi})\|_{H^\alpha(S^1)} \le C_2 e^{at} \|\psi - \widetilde{\psi}\|_{L^\infty(S^1)} \tag{3.17}$$

for any $t \ge T$ and $0 \le \alpha \le 3$. Here the constant C_2 also depends on $K > 0$, but are independent of the choice of $\psi, \widetilde{\psi}$. To see that (3.17) holds, we note that $\xi(t,\cdot) = \Phi_t(\psi) - \Phi_t(\widetilde{\psi})$ satisfies the linear elliptic equation

$$\xi_{tt} - \frac{4}{q-1}\xi_t + \xi_{\theta\theta} + h(t,\theta)\xi = 0 \quad \textit{for } t > 0, \quad \theta \in S^1,$$

where $h = \{g(\Phi_t(\psi)) - g(\Phi_t(\widetilde{\psi}))\}/(\Phi_t(\psi) - \Phi_t(\widetilde{\psi}))$. Since h and its first derivatives remain bounded as $t \to \infty$ by virtue of (3.5)″, the interior L^2 estimates and (3.16) yield (3.17). Combining (3.14) and (3.17) for $\alpha = \gamma \in (1/2,\ 1)$, and considering that $H^\gamma(S^1) \subset L^\infty(S^1)$, we find that there exists $C_3 > 0$ such that

$$\|\Phi_t(\psi) - \Phi_t(\widetilde{\psi})\|_{H^\gamma(S^1)} \le C_3 e^{at} \|\psi - \widetilde{\psi}\|_{H^\gamma(S^1)}$$

for all $t \ge 0$ and $\psi, \widetilde{\psi} \in H^\gamma(S^1)$ with $\|\psi\|_{H^\gamma}, \|\widetilde{\psi}\|_{H^\gamma} \le K$. Using this inequality and (3.5)′, we get an inequality of the form

$$\|g(\Phi_t(\psi)) - g(\Phi_t(\widetilde{\psi}))\|_{H^\gamma(S^1)} \le C_4 e^{at} \|\psi - \widetilde{\psi}\|_{H^\gamma(S^1)}.$$

In view of this, and the fact that

$$\|e^{-\tau A_2}w\|_{H^{\gamma+\beta}(S^1)} \le M_\beta e^{-2a\tau}\tau^{-\beta}\|w\|_{H^\gamma(S^1)} \tag{3.18}$$

holds for any $0 \le \beta < 1$ and an appropriate constant $M_\beta > 0$, we find that

$$f : H^\gamma(S^1) \longrightarrow H^{\gamma+\beta}(S^1)$$

is locally Lipschitz continuous for any $0 \le \beta < 1$ and for any $1/2 < \gamma < 1$. In particular, $f : H^\gamma(S^1) \to H^1(S^1)$ is locally Lipschitz, hence so is the map $f : H^1(S^1) \to H^1(S^1)$. This implies that (3.12) is well-posed in $H^1(S^1)$. Therefore (3.14) holds also for $\gamma = 1$. By repeating the same argument as in (3.17) and (3.18), with the norm of $H^\gamma(S^1)$ replaced by that of $H^1(S^1)$, we see that

$$f : H^1(S^1) \longrightarrow H^{1+\beta}(S^1)$$

is locally Lipshitz continuous for any $0 \le \beta < 1$.

The well-posedness of (3.12), as a consequence of the local Lipschitz continuity of f, can be shown by a standard argument (see, for example, Henry [He1]). The proof of Proposition 3.5 is complete.

Remark 3.7. We see from the above propositions not only that initial value problem (3.12) is well-posed, but also that its solutions can always be continued over the whole interval $0 \le t < \infty$.

Remark 3.8. Since (3.12) is well-posed in $H^1(S^1)$, it is clear that Proposition 3.6 holds also for $\gamma = 1$. Thus Proposition 3.2 remains true if the space $C^0(S^1)$ is replaced by $H^1(S^1)$ (or any $H^\gamma(S^1)$ with $1/2 < \gamma \le 1$).

Proposition 3.9. *The semiflow Φ defined in (3.6) is "point dissipative". In other words, the following hold :*

(i) Φ is "compact", that is, given any bounded set B in $C^0(S^1)$ and any $t > 0$, $\Phi_t(B)$ is relatively compact in $C^0(S^1)$;

(ii) there exists $C > 0$ such that $\limsup_{t\to\infty} \|\Phi_t(\psi)\|_{L^\infty(S^1)} \le C$ for any $\psi \in C^0(S^1)$.

The statements (i) and (ii) remain true if we replace the space $C^0(S^1)$ by $H^\gamma(S^1)$ and the norm $\|\cdot\|_{L^\infty}$ by $\|\cdot\|_{H^\gamma}$ for any $1/2 < \gamma \le 1$.

Corollary 3.10. *The semiflow Φ has a global attractor that is compact. More precisely,*

$$\mathcal{A} = \{\psi \in C^0(S^1) \mid \text{ there exists an entire orbit passing through } \psi\} \tag{3.19}$$

is the maximal compact invariant set in $C^0(S^1)$ (and also in $H^\gamma(S^1), 1/2 < \gamma \le 1$). Moreover, for any neighborhood $U \supset \mathcal{A}$ and any bounded set B (in the space $C^0(S^1)$ or $H^\gamma(S^1), 1/2 < \gamma \le 1$), $\Phi_t(B) \subset U$ holds for all sufficiently large $t > 0$.

In the definition of $\mathcal{A}$ in (3.19), by an *entire orbit* we mean an orbit of the semiflow Φ that is defined for all $-\infty < t < \infty$. We say a set S is *invarinat* if $\Phi_t(S) = S$ for all $t \ge 0$. Proposition 3.9 is an immediate consequence of Lemma 3.1′. The existence of a compact global attractor stated in Corollary 3.10 is a well-known established property of point dissipative systems; see [HMO], [Ha]. (Note that any entire orbit of Φ is bounded by virtue of Theorem 1.1, hence the definition of the set $\mathcal{A}$ in (3.19) agrees with that in [HMO], [Ha].)

Now let us study the relation between the global attractor $\mathcal{A}$ and the set of solutions of (1.2). Roughly, to each solution of (3.1) corresponds a solution of (3.4), hence it can be identified with a positive semi-orbit of Φ, that is, $O^+(\psi) = \{\Phi_t(\psi) | \, t \geq 0\}$. Similarly, to each solution of (1.2) corresponds an entire orbit, $\{\Phi_t(\psi) | -\infty < t < \infty\}$. As a matter of fact, this statement is somewhat misleading, since each solution of (3.4) is a function of t, while an orbit is its image. In order to avoid such ambiguity, we look at the problem in the following way: Recall that each solution of (3.1) is uniquely determined by the data ψ in (3.1b). This makes it possible to identify the set of solutions of (3.1) with the space $C^0(S^1)$. Thus the set of solutions of (1.2) can be identified with a subset of $C^0(S^1)$:

$$\widetilde{\mathcal{A}} = \{\psi \in C^0(S^1) | \quad \text{solution of (3.1) can be extended to a solution of (1.2)}\}. \tag{3.20}$$

It is not difficult to see, as we will show below, that $\widetilde{\mathcal{A}} = \mathcal{A}$. Thus the set of solutions of (1.2) can be identified with the compact global attractor of Φ.

To see that $\widetilde{\mathcal{A}} = \mathcal{A}$, we first note that if $u(r, \theta)$ is a solution of (3.1) for boundary data ψ, then for each $\lambda \geq 1$ the function $\lambda^{2/(q-1)} u(\lambda r, \theta)$ is a solution of (3.1) for boundary data $\Phi_t(\psi)$, where $t = \log \lambda$. Therefore, if we identify each solution of (3.1) with its boundary data in (3.1b), then the self-similar transformation (2.4) is interpreted as the "time shift":

$$u(r, \theta) \longmapsto \lambda^{2/(q-1)} u(\lambda r, \theta) \quad \Longleftrightarrow \quad \psi \longmapsto \Phi_t(\psi) \tag{3.21}$$

Let u be a solution of (1.2). Then $\lambda^{2/(q-1)} u(\lambda r, \theta)$ is a solution of (3.1) not only for every $\lambda \geq 1$ but for every $\lambda > 0$. The converse is also true. The assertion $\widetilde{\mathcal{A}} = \mathcal{A}$ now follows easily from this and (3.21).

It is also clear that, if $u, \widetilde{u}$ are solutions of (1.2) with $u|_{|x|=1} = \psi$, $\widetilde{u}|_{|x|=1} = \widetilde{\psi}$, then they belong to the same similarity class if and only if ψ and $\widetilde{\psi}$ lie on the same orbit. Consequently, there is one-to-one correspondence between each entire orbit of Φ and each similarity class of solutions of (1.2).

In the context of the semiflow Φ, Theorem 1.1 can be interpreted that any entire orbit converges to equilibrium points, say ω_0, ω_∞, as $t \to -\infty$ and as $t \to \infty$, respectively. Moreover, if $\omega_0 = \omega_\infty$, then Theorem 1.2 states that the orbit stays at ω_0 for all $-\infty < t < \infty$, since solutions of the form (1.7) correspond to time-independent solutions of (3.4). This means that there exists no *homoclinic* orbit. Thus we have

Proposition 3.11. *Any entire orbit of Φ is either*
(a) an equilibrium point, or
(b) a heteroclinic orbit, that is, an orbit connecting a pair of distinct equilibrium points.
The case (a) corresponds to solutions of (1.2) of the form (1.7), while the case (b) corresponds to the rest of solutions of (1.2).

§4. Differentiability of the Semiflow

In this section we show that the map $f : H^1(S^1) \longrightarrow H^1(S^1)$ defined in (3.10) is differentiable, or more precisely, of class $C^{1+\alpha}$ for some $0 < \alpha < 1$. This implies that the map Φ_t is also differentiable, and makes it possible to construct the stable and the unstable manifolds of each equilibrium point. Before stating the main results of this section (Propositions 4.3 ~ 4.5), we need some preliminary lemmas. In what follows the constant a and the operators A_1, A_2 will be as in (3.9), (3.7), (3.8).

Lemma 4.1. *Let $u(x)$ be a solution of (3.1). Then for any $\varphi \in C^0(S^1)$ there exists a solution $w(x)$ to the problem (4.1) that is bounded as $|x| \to \infty$. Moreover, such a bounded solution is unique.*

$$\Delta w - q|u|^{q-1}w = 0 \qquad \text{in} \quad \mathbf{R}^2 \setminus \overline{B_1}, \tag{4.1a}$$

$$w = \varphi \qquad \text{on} \quad S^1 = \partial B_1. \tag{4.1b}$$

Corollary 4.2. *Let a be as in (3.9) and g be as in (1.5). Let v be an arbitrary solution of (3.4). Then for any $\varphi \in C^0(S^1)$ there exists a solution $\xi \in C^0([0,\infty) \times S^1) \cap C^2((0,\infty) \times S^1)$ to the problem*

$$\xi_{tt} - \frac{4}{q-1}\xi_t + \xi_{\theta\theta} + g'(v)\xi = 0 \qquad \text{for} \quad t > 0,\ \theta \in S^1, \tag{4.2a}$$

$$\xi(0,\cdot) = \varphi(\cdot). \tag{4.2b}$$

satisfying

$$\sup_{t>0,\theta\in S^1} |e^{-at}\xi(t,\theta)| < \infty. \tag{4.3}$$

Moreover, such a solution of (4.2) is unique.

(Proof of Lemma 4.1.) First we prove the existence. Let $K = \|\varphi\|_{L^\infty}$. Then the constant function $w_1(x) \equiv K$ is a supersolution to the problem (4.1) while $w_2(x) \equiv -K$ is a subsolution. It follows that there exists at least one exact solution of (4.1) that satisfies $w_1 \geq w \geq w_2$.

Next we show the uniqueness. Since (4.1a) is a linear equation, it suffices to consider the case where $\varphi = 0$. Let $t = \log r$. Then a simple computation shows that

$$w_{tt} + w_{\theta\theta} - q|v|^{q-1}w = 0 \qquad \text{for} \quad t > 0,\ \theta \in S^1, \tag{4.4}$$

where $v(t,\theta)$ is as in (3.3). Take $0 < \delta < T$. Multiplying (4.4) with w and integrating it over $\delta \leq t < T$, $\theta \in S^1$, we get

$$\int_\delta^T \int_{S^1} (w_t^2 + w_\theta^2 + q|v|^{q-1}w^2)\,d\theta dt$$

$$= \left[\int_{S^1} w(t,\theta)w_t(t,\theta)d\theta\right]_\delta^T. \tag{4.5}$$

Letting $\delta \to 0$ and $T \to \infty$, we obtain

$$\int_0^\infty \int_{S^1} (w_t^2 + w_\theta^2 + q|v|^{q-1} w^2) d\theta dt = 0. \tag{4.6}$$

In deriving (4.6) from (4.5), we have used the fact that $w(0, \cdot) = 0$, that w_t is continuous at $t = 0$, and that w remains bounded as $t \to \infty$. (The continuity of w_t near the boundary of the domain follows from the standard L^p estimates, the Sobolev embedding theorem and the fact that $\varphi = 0$.) It follows from (4.6) that $w_t \equiv 0$ and $w_\theta \equiv 0$. Hence $w \equiv 0$. The lemma is proved.

(Proof of Corollary 4.2.) Simply set $\xi = e^{at} w$, where w is as in (4.1).

The following are the main results in this section :

Proposition 4.3 (Differentiability). *Let $0 < \alpha < \min\{1, q-1\}$. Then the map $f : H^1(S^1) \longrightarrow H^1(S^1)$ is of class $C^{1+\alpha}$. More precisely,*
(i) f is Fréchet differentiable, and its derivative Df at $\psi \in H^1(S^1)$ is expressed as

$$[Df(\psi)](\varphi) = \int_0^\infty e^{-\tau A_2} g'(\Phi_\tau(\psi)) \xi(\tau) d\tau, \qquad \varphi \in H^1(S^1), \tag{4.7}$$

where $\xi(t, \cdot)$ is the solution of (4.2).
(ii) $Df(\psi) : H^1(S^1) \longrightarrow H^1(S^1)$ is a bounded linear operator and

$$\|Df(\psi) - Df(\widetilde{\psi})\|_{\mathcal{L}} \le C \|\psi - \widetilde{\psi}\|_{H^1}^\alpha$$

for some constant $C > 0$, where $\|\cdot\|_{\mathcal{L}}$ denotes the norm in $\mathcal{L}(H^1(S^1); H^1(S^1))$, the space of bounded linear operators on $H^1(S^1)$.

Proposition 4.4 (The linearized equation). *Let v be a solution of (3.4). Let $\varphi \in H^1(S^1)$. Then $\xi \in C([0, \infty); H^1(S^1))$ is a solution to the problem*

$$\frac{d\xi}{dt} = A_1 \xi + Df(v)\xi, \qquad t > 0, \tag{4.8a}$$

$$\xi(0) = \varphi \tag{4.8b}$$

if and only if $\xi(t, \theta)$ is a solution of (4.2) satisfying (4.3). In particular, any solution of (4.8) satisfies the condition (4.3).

Since A_1 is a negative definite self-adjoint operator in $H^1(S^1)$, the operator $A_1 + Df(v)$ is also self-adjoint and bounded from above. It is also easily seen that the spectra of $A_1 + Df(v)$ consist only of real eigenvalues. Let $\omega \in \mathcal{E}$, where $\mathcal{E}$ is as in (2.1), and consider the eigenvalue problem

$$A_1 \varphi + Df(\omega)\varphi = \lambda \varphi. \tag{4.9}$$

Let $\lambda_1(\omega) \ge \lambda_2(\omega) \ge \lambda_3(\omega) \ge \ldots$ be the eigenvalues of (4.9) and $\varphi_1, \varphi_2, \varphi_3, \ldots$ be the corresponding eigenfunctions. Then we have the following :

Proposition 4.5 (Eigenvalue problem). *Let $\omega \in \mathcal{E}$, and let $\lambda_j(\omega), \varphi_j$ $(j = 1, 2, \ldots)$ be as above. Then*

(i) $\lambda_{2j-1}(\omega) > \lambda_{2j}(\omega) \geq \lambda_{2j+1}(\omega)$ $(j = 1, 2, \ldots)$. *Furthermore, the equality in the second inequality always holds if $\omega \in \mathcal{E}^+ \cup \mathcal{E}^- \cup \mathcal{E}^0$.*

(ii)

$$(\lambda_{2j+1}(\omega) - j) \to 0, \quad (\lambda_{2j}(\omega) - j) \to 0 \qquad as \quad j \to \infty. \tag{4.10}$$

(iii) The function $\varphi_j(\theta)$ has only simple zeroes on S^1. Moreover, the number of zeroes of φ_j, denoted by $z[\varphi_j]$, satisfies

$$z[\varphi_{2j+1}] = j \qquad (j = 0, 1, 2, \ldots), \tag{4.11a}$$

$$z[\varphi_{2j}] = j \qquad (j = 1, 2, \ldots). \tag{4.11b}$$

(iv) For any constants c_1, c_2 with $c_1^2 + c_2^2 \neq 0$,

$$z[c_1\varphi_{2j} + c_2\varphi_{2j+1}] = j \qquad (j = 1, 2, \ldots). \tag{4.12}$$

(v) If $\omega \in \mathcal{E}^+ \cup \mathcal{E}^-$, then $0 > \lambda_1(\omega) > \lambda_2(\omega) \geq \cdots$;
if $\omega \in \mathcal{E}_k$, then $\lambda_1(\omega) > \cdots \geq \lambda_{2k-1}(\omega) > 0 = \lambda_{2k}(\omega) > \lambda_{2k+1}(\omega) > \cdots$;
if $\omega \in \mathcal{E}^0$, and if $2/(q-1) \notin \mathbf{Z}$ then

$$\lambda_1(\omega) > \cdots \geq \lambda_{2k_0+1}(\omega) > 0 > \lambda_{2k_0+2}(\omega) \geq \cdots.$$

We omit the proof of Propositions 4.3 and 4.4. See [M3] for details.

(Outline of the proof of Proposition 4.5.) It is not difficult to see that φ is an eigenfunction of (4.9) if and only if it is an eigenfunction of the following eigenvalue problem:

$$\frac{d^2\varphi}{d\theta^2} + g'(\omega)\varphi = \mu\varphi \quad \text{on} \quad S^1. \tag{4.13}$$

Let $\mu_1(\omega) \geq \mu_2(\omega) \geq \mu_3(\omega) \geq \ldots$ be the engenvalues of (4.13). Then

$$\lambda_j(\omega) = a - \sqrt{a^2 - \mu_j(\omega)} \qquad (j = 1, 2, \ldots). \tag{4.14}$$

In particular, $\lambda_j(\omega) > 0$ (resp. $= 0, < 0$) if and only if $\mu_j(\omega) > 0$ (resp. $= 0, < 0$). In view of these, one can easily prove (i) and (iii). In fact, those are consequences of the Sturm-Liouville theory for the case where the domain is S^1. Statement (iv) follows from (i) and (iii). (There are a number of different ways to derive (iv) from (i) and (iii); Henry [He2; Lemma1] uses the parabolic maximum principle to show a result similar to (4.12) above.) Statement (ii), for the special case where $\omega = 0$, can be shown by an easy computation and by using (4.14). In the general case, we note that $\mu_j(\omega) - \mu_j(0)$ remains bounded as $j \to \infty$. In view of this, and using (4.10) for $\omega = 0$ and (4.14), we obtain (4.10) for a general $\omega \in \mathcal{E}$. Statement (v) also follows easily from (4.13) and (4.14).

§5. Morse-Smale Property

Let k be an integer satisfying $1 \le k \le k_0$, where k_0 is as in (2.3). Proposition 4.5 states that for each $\omega \in \mathcal{E}_k$ the operator $A_1 + Df(\omega)$ has precisely one eigenvalue at the origin, and that all other eigenvalues are away from the origin (hence from the imaginary axis). This implies that the one-dimensional manifold of equilibria, $\mathcal{E}_k$, is "normally hyperbolic". Because of this, each point $\omega \in \mathcal{E}_k$ has the unstable and the stable manifolds, the dimension (or the codimension) of which is determined by the number of positive eigenvalues of $A_1 + Df(\omega)$. Throughout this section, the underlying space on which the semiflow is defined will be $H^1(S^1)$.

Definition 5.1. Given an equilibrium point $\omega \in \mathcal{E}$, we define the *unstable set* and the *stable set* of ω by

$$W^u(\omega) = \{\psi \in H^1(S^1) \mid \quad \alpha(\psi) = \omega\}, \tag{5.1a}$$

$$W^s(\omega) = \{\psi \in H^1(S^1) \mid \quad \lim_{t\to\infty} \Phi_t(\psi) = \omega\}. \tag{5.1b}$$

Here $\alpha(\psi)$ denotes the α-limit set of ψ; the right-hand side of (5.1a) is equivalent to saying that there exists a solution to (3.4a) for $t < 0$ satisfying (3.4b) and converging to ω as $t \to -\infty$. We call $W^u(\omega)$, $W^s(\omega)$ the *unstable manifold* and the *stable manifold*, if they are manifolds. The restrictions of $W^u(\omega)$ and $W^s(\omega)$ onto a neighborhood (which is not necessarily specified but somehow determined in the context) of ω are denoted by $W^u_{loc}(\omega)$ and $W^s_{loc}(\omega)$, respectively.

Definition 5.2. A set $W \subset H^1(S^1)$ is called a C^1 *manifold* if it is connected and if for each point $p \in W$ there exists a closed subspace T_pW such that $W - p = \{\psi - p \mid \psi \in W\}$ is expresed locally as a graph of a C^1 mapping $h : T_pW \longrightarrow (T_pW)^{\perp}$ satisfying $h(0) = 0$, $Dh(0) = 0$. T_pW is called the *tangent space* of W at p. The *dimension* of W, $\dim W$, and its *codimension*, $\mathrm{codim} W$, are defined as those of T_pW.

Proposition 5.3. *Let $1 \le k \le k_0$, and let $\omega \in \mathcal{E}_k$. Then $W^s(\omega)$ is a C^1 manifold, at least in a neighborhood of ω, while $W^u(\omega)$ is globally a C^1 manifold. Furthermore, it holds that*

$$\dim W^u(\omega) = 2k - 1, \tag{5.2a}$$

$$\mathrm{codim} W^s_{loc}(\omega) = 2k. \tag{5.2b}$$

The tangent space $T_\omega W^u(\omega)$ is spanned by $\varphi_1, \varphi_2, \dots, \varphi_{2k-1}$, while the space $T_\omega W^s_{loc}(\omega)$ is spanned by $\varphi_{2k+1}, \varphi_{2k+2}, \dots$, where φ_j $(j = 1, 2, \dots)$ are eigenfunctions of (4.9) corresponding to $\lambda_j(\omega)$.

The smoothness of $W^u_{loc}(\omega)$, $W^s_{loc}(\omega)$ can be proved by using the general results in [CLL]. See also [CL]. That $W^u(S^1)$ is globally a C^1 manifold follows from the backward uniqueness for the parabolic problem (3.12) (cf. [He1]). [HPS] deals with a related problem for diffeomorphisms in finite dimensions, in which case the arguments are simpler. For details of the proof of Proposition 5.3, see [M3].

The following is the main result of this section :

Theorem 5.4 (Transversality). *Let $\omega \in \mathcal{E}_k, \widetilde{\omega} \in \mathcal{E}_{k'}$ with $k > k'$. Then $W^u(\omega)$ and $W^s(\widetilde{\omega})$ intersect transversally, at least in a neighborhood of $\widetilde{\omega}$. More precisely, either*
(i) $W^u(\omega) \cap W^s(\widetilde{\omega}) = \emptyset$, or
(ii) for each point $p \in W^u(\omega) \cap W^s(\widetilde{\omega})$ sufficiently close to $\widetilde{\omega}$, it holds that

$$T_p W^u(\omega) + T_p W^s_{loc}(\widetilde{\omega}) = H^1(S^1).$$

Remark 5.5. As a matter of fact, case (i) above never occurs, as we will see in Section 6. The conclusion of Theorem 5.4 remains true even if $\omega \in \mathcal{E}^0$ or $\widetilde{\omega} \in \mathcal{E}^+ \cup \mathcal{E}^-$.

Remark 5.6. For each $\beta \in S^1$, denote by σ_β the rotation operator on $H^1(S^1)$, namely, $[\sigma_\beta \psi](\theta) = \psi(\theta + \beta)$. It is clear that the semiflow Φ is equivariant with respect to σ_β, that is,

$$\Phi_t \circ \sigma_\beta = \sigma_\beta \circ \Phi_t \qquad \text{for any} \quad t \geq 0, \ \beta \in S^1.$$

In view of this, one easily finds that

$$\sigma_\beta W^u(\omega) = W^u(\sigma_\beta \omega), \tag{5.3a}$$

$$\sigma_\beta W^s(\omega) = W^s(\sigma_\beta \omega). \tag{5.3b}$$

In particular, if we move ω smoothly on $\mathcal{E}_k$, then both $W^u(\omega)$ and $W^s(\omega)$ vary smoothly.

Remark 5.7. The equivariance of Φ with respect to σ_β and the transversality properties, together with the normal hyperbolicity of $\mathcal{E}$, imply that the semiflow Φ is an *equivariant Morse-Smale system* (see [C], [Ha] and [HMO]).

The proof of Theorem 5.4 can be carried out in the same spirit as in [He2] and [An], which deal with one-dimensional semilinear parabolic equations. There are, however, some additional difficulties we have to cope with in the present problem. First, the equation (3.12) that generates the semiflow Φ is not a partial differential equation, but a pseudodifferential equation, so the parabolic maximum principle used in [He2] and [An] fail to apply. Secondly, the equilibrium points are not necessarily isolated, but many of them form manifolds of equilibria. Nonetheless, much of the idea in this section comes from [He2] and [An].

Lemma 5.8. *Let $\omega, \widetilde{\omega}$ be as in Theorem 5.4 and suppose $W^u(\omega) \cap W^s(\widetilde{\omega}) \neq \emptyset$. Let ψ be any point on $W^u(\omega) \cap W^s_{loc}(\widetilde{\omega})$ and let v be a solution of*

$$v_{tt} - \frac{4}{q-1} v_t + v_{\theta\theta} + g(v) = 0, \qquad \text{for} \quad t \in \mathbf{R}, \ \theta \in S^1, \tag{5.4a}$$

$$v(0, \cdot) = \psi, \tag{5.4b}$$

where g is as in (1.5). Then
(i) $\varphi \in T_\psi W^u(\omega)$ if and only if there exists a solution ξ to the problem

$$\xi_{tt} - \frac{4}{q-1} \xi_t + \xi_{\theta\theta} + g'(v)\xi = 0 \qquad \text{for} \quad t \in \mathbf{R}, \ \theta \in S^1, \tag{5.5a}$$

$$\xi(0,\cdot) = \varphi \tag{5.5b}$$

satisfying (4.3) and

$$\lim_{t\to-\infty} \|\xi(t,\cdot)\|_{L^\infty(S^1)} = 0. \tag{5.6}$$

(ii) $\varphi \in T_\psi W^s_{loc}(\widetilde{\omega})$ *if and only if there exists a solution* ξ *to the problem (4.2) satisfying*

$$\lim_{t\to\infty} \|\xi(t,\cdot)\|_{L^\infty(S^1)} = 0. \qquad (5.6)'$$

Lemma 5.9. *Let* $\omega, \widetilde{\omega}, \psi$ *and* v *be as in Lemma 5.8. Let* φ_j $(j = 1,2,\ldots)$ *be the eigenfunctions of (4.9). Also, let* $\widetilde{\varphi}_j$ $(j = 1,2,\ldots)$ *be the eigenfunctions of (4.9) with* ω *replaced by* $\widetilde{\omega}$. *For each* $j = 1,2,\ldots$, *define* X_j *to be the space spanned by* $\varphi_{2j-2}, \varphi_{2j-1}$ *and* $\widetilde{X}_j$ *to be the space spanned by* $\widetilde{\varphi}_{2j-2}, \widetilde{\varphi}_{2j-1}$, *where we understand that* $\varphi_0 = \widetilde{\varphi}_0 = 0$. *Finally, denote by* $P_j, \widetilde{P}_j$ *the orthogonal projections associated with the subspaces* $X_j, \widetilde{X}_j$, *respectively, and let* $Q_j = I - P_j$ *and* $\widetilde{Q}_j = I - \widetilde{P}_j$. *Then the following hold:*

(i) If $\varphi \in T_\psi W^u(\omega)$, $\varphi \neq 0$, *and if* ξ *denotes the solution of (5.5) satisfying (5.6), then there exists an integer* j, $1 \le j \le k$, *such that*

$$\lim_{t\to-\infty} \frac{\|Q_j\xi(t,\cdot)\|_{H^1}}{\|P_j\xi(t,\cdot)\|_{H^1}} = 0. \tag{5.7a}$$

(ii) If $\varphi \in T_\psi W^s_{loc}(\widetilde{\omega})$, $\varphi \neq 0$, *and if* ξ *denotes the solution of (4.2) satisfying (5.6)′, then there exists an integer* $j \ge k' + 1$ *such that*

$$\lim_{t\to\infty} \frac{\|\widetilde{Q}_j\xi(t,\cdot)\|_{H^1}}{\|\widetilde{P}_j\xi(t,\cdot)\|_{H^1}} = 0. \tag{5.7b}$$

Remark 5.10. Lemma 5.9 implies that the solution curve $\xi(t,\cdot)$, which converges to 0 as $t \to -\infty$ (resp. $t \to \infty$), is tangential to the space X_j (resp. $\widetilde{X}_j$) at 0. Lemma 5.9 can be proved by using arguments similar to those in [He2] and [An]. In particular, the statement (ii) follows from [An; Lemma 7] and (4.10). Lemma 5.8 is a consequence of Proposition 4.4. The details are ommitted.

Lemma 5.11. *Let* $\varphi \in T_\psi W^u(\omega) \cap T_\psi W^s_{loc}(\widetilde{\omega}) \setminus \{0\}$ *and let* ξ *be a solution of (4.2) satisfying (4.3). Then* ξ *can be extended to a solution of (5.5) and satisfies (5.6) and (5.6)′. Moreover,* $z[\xi(t,\cdot)]$ *converges as* $t \to -\infty$ *and also as* $t \to \infty$, *and*

$$\lim_{t\to-\infty} z[\xi(t,\cdot)] \ge \lim_{t\to\infty} z[\xi(t,\cdot)], \tag{5.8}$$

where z *is as in Proposition 4.5 (ii).*

(Proof). That ξ satisfies (5.6) and (5.6)′ follows from Lemma 5.8 and Corollary 4.2. The existence of $\lim z[\xi(t,\cdot)]$ is a consequence of Lemma 5.9 and (4.12). In fact, (5.7a) implies

$$\lim_{t\to-\infty} z[\xi(t,\cdot)] = j - 1, \tag{5.9a}$$

while (5.7b) implies

$$\lim_{t\to\infty} z[\xi(t,\cdot)] = j-1. \tag{5.9b}$$

The inequality (5.8) can be shown by using the fact that both of the limits in (5.8) exist, and by applying the maximum principle to the equation

$$\Delta w - q|u|^{q-1}w = 0 \qquad \text{in} \quad \mathbf{R}^2 \setminus \{0\}, \tag{5.10}$$

which is equivalent to (5.5). See [CMV; proof of Lemma 1.6] for a similar usage of the maximum principle.

(Outline of the Proof of Theorem 5.4.) By Proposition 5.2, we have

$$\dim T_\psi W^u(\omega) = 2k-1, \tag{5.11a}$$

$$\operatorname{codim} T_\psi W^s_{loc}(\widetilde{\omega}) = 2k'. \tag{5.11b}$$

Let

$$Y = \text{the space spanned by} \quad \widetilde{\varphi}_{2k}, \widetilde{\varphi}_{2k+1}, \ldots .$$

Clearly Y is a subspace of $T_{\widetilde{\omega}} W^s_{loc}(\widetilde{\omega})$. Define

$$Z_1 = \{\varphi \in T_\psi W^u(\omega) \mid \ (5.7a) \text{ holds for some } 1 \le j \le k\},$$

$$Z_2 = \{\varphi \in T_\psi W^s_{loc}(\widetilde{\omega}) \mid \ (5.7b) \text{ holds for some } j \ge k+1\}.$$

By Lemma 5.8 (i), it holds that $Z_1 = T_\psi W^u(\omega)$, which is a $(2k-1)$-dimensional subspace of $H^1(S^1)$. On the other hand, using Lemma 5.8 (ii), one sees that Z_2 is a linear subspace and that $\varphi \in Z_2$ if and only if the solution of (4.2) is tangential to Y as $t \to \infty$. Furthermore, $\operatorname{codim} Z_2 = 2k-1$. Note also that $Z_1 \cap Z_2 = \{0\}$, which is a consequence of (5.8) and (5.9). Combinig these, one gets $Z_1 + Z_2 = H^1(S^1)$, hence

$$T_\psi W^u(\omega) + T_\psi W^s_{loc}(\widetilde{\omega}) = H^1(S^1). \tag{5.12}$$

The theorem is proved.

§6. Proof of Theorem 2.1.

To prove Theorem 2.1, the main result of this paper, we only have to consider the case where both ω and $\widetilde{\omega}$ belong to $\mathcal{E} \setminus (\mathcal{E}^+ \cup \mathcal{E}^- \cup \mathcal{E}^0)$ and satisfy $\omega \ne \widetilde{\omega}$. In fact, if $\omega = \widetilde{\omega}$, then the function in (1.7) with $\omega_0 = \omega$ is the desired solution. On the other hand, if $\omega \ne \widetilde{\omega}$ and $\omega, \widetilde{\omega} \in \mathcal{E}^+ \cup \mathcal{E}^- \cup \mathcal{E}^0$, then by the energy inequality $J(\omega) > J(\widetilde{\omega})$ we have $\omega = 0$ or $\widetilde{\omega} \in \mathcal{E}^+ \cup \mathcal{E}^-$. In both of the cases, it is shown in [CMV] that such a solution u exists.

Now let $\omega, \widetilde{\omega} \in \mathcal{E} \setminus (\mathcal{E}^+ \cup \mathcal{E}^- \cup \mathcal{E}^0)$ and $\omega \ne \widetilde{\omega}$. Then by the inequality $J(\omega) > J(\widetilde{\omega})$, we have

$$\omega \in \mathcal{E}_k \, , \; \widetilde{\omega} \in \mathcal{E}_{k'} \qquad \text{for some} \quad k > k'.$$

As shown in Section 3, each solution of (1.2) — more precisely, each similarity class of solutions of (1.2) — can be identified with an entire orbit of the semiflow Φ, and the asymptotic profiles of each solution of (1.2) correspond to the pair of equilibrium points that are connected by an entire orbit. Therefore, all we have to show is that there exists an entire orbit of the semiflow Φ connecting ω and $\widetilde{\omega}$.

In [CMV], we have applied the theory of strongly order-preserving semiflows [M1], [M2] to show the existence of such a connecting orbit for the special case where the integer k is a multiple of k'. If k is not a multiple of k', the same argument does not work. Moreover, the method in [CMV] proves only the existence of a connecting orbit between some point in $\mathcal{E}_k$ and some point in $\mathcal{E}_{k'}$, therefore, one could not choose $\omega \in \mathcal{E}_k$ and $\widetilde{\omega} \in \mathcal{E}_{k'}$ arbitrarily. What we will prove in this section will fill all those gaps.

Lemma 6.1. *Let k be an integer such that $2 \leq k \leq k_0$. Then there exists an entire orbit connecting some point in $\mathcal{E}_k$ and some point in $\mathcal{E}_{k-1}$.*

(Outline of Proof.) Part of the arguments given here are similar to those in [He2], which deals with one-dimensional semilinear heat equations. Define a subspace of $H^1(S^1)$ by

$$\widetilde{H}^1(S^1) = \{\psi \in H^1(S^1) \mid \psi(0) = 0,\ \psi(\cdot + \pi) = -\psi(\cdot)\}. \tag{6.1}$$

Clearly $\widetilde{H}^1(S^1)$ is a closed subspace of $H^1(S^1)$, and it is not difficult to see that $\widetilde{H}^1(S^1)$ is invariant under the semiflow Φ. Denote by $\widetilde{\Phi}$ the restriction of Φ onto the space $\widetilde{H}^1(S^1)$. Also, set

$$\widetilde{\mathcal{E}} = \mathcal{E} \cap \widetilde{H}^1(S^1), \tag{6.2a}$$

$$\widetilde{\mathcal{E}}_k = \mathcal{E}_k \cap \widetilde{H}^1(S^1). \tag{6.2b}$$

The set of equilibrium points of the semiflow $\widetilde{\Phi}$ is $\widetilde{\mathcal{E}}$, and it holds that

$$\widetilde{\mathcal{E}} = \mathcal{E}^0 \cup \mathcal{E}^+ \cup \mathcal{E}^- \cup \widetilde{\mathcal{E}}_{k_0} \cup \cdots \cup \widetilde{\mathcal{E}}_1.$$

The set $\widetilde{\mathcal{E}}_k$ consists of precisely two points, ω_k and $-\omega_k$. For each $\omega \in \widetilde{\mathcal{E}}$, consider the follwing eigenvalue problem:

$$A_1\varphi + Df(\omega)\varphi = \widetilde{\lambda}\varphi, \quad \varphi \in \widetilde{H}^1(S^1). \tag{6.3}$$

The eigenvalues of (6.3) are all simple. In fact, if we denote by $\widetilde{\mu}_1(\omega) > \widetilde{\mu}_2(\omega) > \widetilde{\mu}_3(\omega) > \ldots$ and $\widetilde{\varphi}_1, \widetilde{\varphi}_2, \widetilde{\varphi}_3, \ldots$ the eigenvalues and the corresponding eigenfunctions of the eigenvalue problem

$$\frac{d^2\varphi}{d\theta^2} + g'(\omega)\varphi = \widetilde{\mu}\varphi, \qquad 0 < \theta < \pi, \tag{6.4a}$$

$$\varphi(0) = \varphi(\pi) = 0, \tag{6.4b}$$

then the eigenvalues of (6.3) are given by

$$\widetilde{\lambda}_j(\omega) = a - \sqrt{a^2 - \widetilde{\mu}_j(\omega)} \qquad (j = 1, 2, \ldots) \tag{6.5}$$

and the corresponding eigenfunctions coincide with $\widetilde{\varphi}_1, \widetilde{\varphi}_2, \ldots$. As in Proposition 4.5, we see that if $\omega \in \mathcal{E}^+ \cup \mathcal{E}^-$, then

$$0 > \widetilde{\lambda}_1(\omega) > \widetilde{\lambda}_2(\omega) > \cdots; \tag{6.6a}$$

if $\omega \in \widetilde{\mathcal{E}}_k$, then

$$\widetilde{\lambda}_1(\omega) > \cdots > \widetilde{\lambda}_{k-1}(\omega) > 0 > \widetilde{\lambda}_k(\omega) > \cdots; \tag{6.6b}$$

if $\omega \in \mathcal{E}^0$, and if $2/(q-1) \notin \mathbf{Z}$, then

$$\widetilde{\lambda}_1(\omega) > \cdots > \widetilde{\lambda}_{k_0}(\omega) > 0 > \widetilde{\lambda}_{k_0+1}(\omega) > \cdots, \tag{6.6c}$$

while if $\omega \in \mathcal{E}^0$, and if $2/(q-1) \in \mathbf{Z}$, then

$$\widetilde{\lambda}_1(\omega) > \cdots > \widetilde{\lambda}_{k_0}(\omega) > 0 = \widetilde{\lambda}_{k_0+1}(\omega) > \cdots. \tag{6.6c$'$}$$

As in Theorem 5.4 and Remark 5.5, one can show that the stable and the unstable manifolds of equilibria intersect transversally.

Now take $a = 2/(q-1)$ as a parameter and vary it smoothly from $a = 0$ upward. As the value of a passes through each positive integer, k, a new pair of equilibria, namely $\widetilde{\mathcal{E}}_k = \{\omega_k, \ -\omega_k\}$, bifurcates from 0. Let $\widetilde{W}^s(\omega)$ and $\widetilde{W}^u(\omega)$ be the stable and the unstable manifold of $\omega \in \widetilde{\mathcal{E}}$. To clarify the dependence of those manifolds on the parameter a, in what follows we will use such notation as $\widetilde{W}^s(\omega; a)$, $\widetilde{W}^u(\omega; a)$ and $\omega_k(a)$. Now fix $k \geq 2$ and let ϵ be a sufficiently small positive number. By (6.6b) and (6.6c)$'$, we find that

$$\dim \widetilde{W}^u(\omega_k(k+\epsilon); \ k+\epsilon) = \dim \widetilde{W}^u(0; \ k) = k-1.$$

One can further show that the manifold $\widetilde{W}^u(\omega_k(k+\epsilon); \ k+\epsilon)$ converges as $\epsilon \to 0$ to the manifold $\widetilde{W}^u(0; \ k)$ in the C^1 sense in a neighborhood of every point on $\widetilde{W}^u(0; \ k)$. By a similar argument, one finds that $\widetilde{W}^s(\omega_{k-1}(k+\epsilon); \ k+\epsilon)$ converges as $\epsilon \to 0$ to $\widetilde{W}^u(\omega_{k-1}(k); \ k)$ in the C^1 sense in a small neighborhood of $\omega_{k-1}(k)$. Since 0 and ω_{k-1} are connected by an entire orbit of $\widetilde{\Phi}$ ([CMV; Theorem 4.1]), we have

$$\widetilde{W}^s(\omega_{k-1}(k); \ k) \cap \widetilde{W}^u(0; \ k) \neq \emptyset. \tag{6.7}$$

Choose a point ψ in the above intersection that is sufficiently close to $\omega_{k-1}(k)$. Then the manifold $\widetilde{W}^u(\omega_k(k+\epsilon); \ k+\epsilon)$ is close to $\widetilde{W}^u(0; \ k)$ in the C^1 sense, while $\widetilde{W}^s(\omega_{k-1}(k+\epsilon); \ k+\epsilon)$ is close to $\widetilde{W}^u(\omega_{k-1}(k); \ k)$, at least on a small neighborhood of ψ. In view of this, and considering that the intersection in (6.7) is transversal, we find that

$$\widetilde{W}^s(\omega_{k-1}; \ k+\epsilon) \cap \widetilde{W}^u(\omega_k; \ k+\epsilon) \neq \emptyset \tag{6.8}$$

if ϵ is sufficiently small. Now let ϵ tend to ∞ smoothly. Then the intersection in (6.8) does not vanish since it remains transversal for every $\epsilon > 0$. Thus we see that $\widetilde{\mathcal{E}}_k$ and $\widetilde{\mathcal{E}}_{k-1}$ are connected by an entire orbit so long as $2/(q-1) > k$. Since an entire orbit of $\widetilde{\Phi}$ is also an entire orbit of Φ, the conclusion of the lemma follows.

Before going into the proof of Theorem 2.1, we need to define the following functional on $H^1(S^1)$:

$$\hat{J}(\psi) = J(\psi) - \frac{1}{2}\int_{S^1} v_t(0,\theta)^2 d\theta, \tag{6.9}$$

where J is as in (1.6). Note that $v_t(0,\theta)$ is well-defined, since $\psi \in H^1(S^1)$ implies $v \in H^{3/2}_{loc}([0,\infty) \times S^1)$. It is easily seen that $\hat{J}(\omega) = J(\omega)$ for any $\omega \in \mathcal{E}$.

Lemma 6.2. *The functional $\hat{J}$ defined in (6.9) is a Liapunov functional for the semiflow Φ. More precisely, $\hat{J} : H^1(S^1) \longrightarrow \mathbf{R}$ is continuous, and $\hat{J}(\Phi_t(\psi))$ is nonincreasing in $t \geq 0$ for any $\psi \in H^1(S^1)$. Moreover, it is strictly decreasing unless ψ is an equilibrium point.*

(Proof.) The proof of the continuity of $\hat{J}$ is omitted. Now observe that

$$\frac{d}{dt}\hat{J}(\Phi_t(\psi)) = -\frac{4}{q-1}\int_{S^1} v_t(0,\theta)^2 d\theta,$$

where v_t is as in (6.9). The conclusion of Lemma 6.2 follows immediately from this.

(Outline of Proof of Theorem 2.1.) As mentioned at the beginning of this section, it suffices to prove Corollary 2.4 (ii). Let k, k' be integers satisfying $k_0 \geq k > k' \geq 1$ and let $\omega \in \mathcal{E}_k$, $\widetilde{\omega} \in \mathcal{E}_{k'}$.

First we consider the case where $k' = k-1$. By Lemma 6.1, there exists an entire orbit connecting a point, say ω_1, of $\mathcal{E}_k$ and a point, say $\widetilde{\omega}_1$, of $\mathcal{E}_{k-1}$. This entire orbit clearly lies in the intersection of $W^u(\omega_1)$ and $W^s(\widetilde{\omega}_1)$ (see Definition 5.1). Since $W^u(\omega_1)$ and $W^s(\widetilde{\omega}_1)$ intersect transversally by Theorem 5.3, their intersection does not vanish if we perturb these manifold smoothly. In view of this and (5.3), we see that the intersection of $W^u(\omega_1)$ and $W^s(\widetilde{\omega}_1)$ does not vanish if we shift ω_1 slightly within $\mathcal{E}_k$, or $\widetilde{\omega}_1$ within $\mathcal{E}_{k-1}$ (see Remark 5.5). Thus

$$S = \{(\omega,\widetilde{\omega}) \in \mathcal{E}_k \times \mathcal{E}_{k-1} |\quad \omega_1 \quad \text{and} \quad \widetilde{\omega}_1 \quad \text{have a connection}\}$$

is a relatively open subset of $\mathcal{E}_k \times \mathcal{E}_{k-1}$. On the other hand, one can also show that S is a closed set. To see this, let $(\omega_j, \widetilde{\omega}_j) \in S$ $(j = 1, 2, \ldots)$ and suppose that $\omega_j \to \omega_*$, $\widetilde{\omega}_j \to \widetilde{\omega}_*$ as $j \to \infty$. Denote by $v_j(t,\cdot)$ a solution of (5.4a) that connects ω_j and $\widetilde{\omega}_j$. By replacing $v_j(t,\cdot)$ by its time-shift $v_j(t+T,\cdot)$ if necessary, we can assume without loss of generality that

$$\hat{J}(v_j(0,\cdot)) = \{\hat{J}(\mathcal{E}_k) + \hat{J}(\mathcal{E}_{k-1})\}/2.$$

By Proposition 3.9 or the estimate (3.5)″, we can choose a subsequence of $\{v_j\}$, denoted again by $\{v_j\}$, converging as $j \to \infty$ locally uniformly in $t \in \mathbf{R}$. Denote by v_* the limit function. Clearly v_* is a solution of (5.4a). Moreover, it satisfies

$$J(\mathcal{E}_k) = \hat{J}(\mathcal{E}_k) \geq \hat{J}(v_*(t,\cdot)) \geq \hat{J}(\mathcal{E}_{k-1}) = J(\mathcal{E}_{k-1}), \quad t \in \mathbf{R}, \tag{6.10}$$

$$\hat{J}(v_*(0,\cdot)) = \{\hat{J}(\mathcal{E}_k) + \hat{J}(\mathcal{E}_{k-1})\}/2. \tag{6.11}$$

By Proposition 3.11, v_* connects a pair of points ω_{**} , $\widetilde{\omega}_{**} \in \mathcal{E}$. In view of (6.10), (6.11) and $J(\omega_{**}) \geq J(\widetilde{\omega}_{**})$, and considering that there exists no point $\bar{\omega} \in \mathcal{E}$ satisfying $J(\mathcal{E}_k) > J(\bar{\omega}) > J(\mathcal{E}_{k-1})$, we find that $J(\omega_{**}) = J(\mathcal{E}_k)$, $J(\widetilde{\omega}_{**}) = J(\mathcal{E}_{k-1})$, which implies $\omega_{**} \in \mathcal{E}_k$, $\widetilde{\omega}_{**} \in \mathcal{E}_{k-1}$. To see that $\omega_{**} = \omega_*$ and $\widetilde{\omega}_{**} = \widetilde{\omega}_*$, we recall that both $\mathcal{E}_k$ and $\mathcal{E}_{k-1}$ are normally hyperbolic. This implies that for any $\epsilon > 0$ there exists $\delta > 0$ such that for each point $\psi \in W^u(\mathcal{E}_k) \cap U_\delta(\mathcal{E}_k)$ (resp. $\psi \in W^s(\mathcal{E}_{k-1}) \cap U_\delta(\mathcal{E}_{k-1})$) there exists $\bar{\omega} \in \mathcal{E}_k$ (resp. $\bar{\omega} \in \mathcal{E}_{k-1}$) satisfying $\psi \in W^u(\bar{\omega})$ (resp. $\psi \in W^s(\bar{\omega})$) and $\|\psi - \bar{\omega}\|_{H^1(S^1)} < \epsilon$. Here $U_\delta(K)$ denotes a δ- neighborhood of a set K. From this it easily follows that $\omega_{**} = \omega_*$, $\widetilde{\omega}_{**} = \widetilde{\omega}_*$. Thus it is shown that S is a closed set. Combining these, and recalling that $\mathcal{E}_k \times \mathcal{E}_{k-1}$ is connected, we see that $S = \mathcal{E}_k \times \mathcal{E}_{k-1}$. In particular, $(\omega, \widetilde{\omega}) \in S$.

Next we consider the general case where $k > k'$. Choose a sequence of points $\omega_j \in \mathcal{E}_j$ $(k \geq j \geq k')$ such that $\omega_k = \omega$ and $\omega_{k'} = \widetilde{\omega}$. By what we have shown above, there exists a connecting orbit between each pair ω_j, ω_{j-1} :

$$\omega = \omega_k \to \omega_{k-1} \to \cdots \to \omega_{k'} = \widetilde{\omega}. \tag{6.12}$$

By using what is called the *transition property* of connections (cf. [PdM]), which holds true in (equivariant) Morse Smale systems and can be proved by using the *inclination lemma* (see [W]), we see that (6.12) implies the existence of an entire orbit connecting ω and $\widetilde{\omega}$. This completes the proof of Theorem 2.1.

Remark 6.3. Given $\omega, \widetilde{\omega} \in \mathcal{E}$, how many solutions of (1.2) exist whose asymptotic profiles are precisely $\omega, \widetilde{\omega}$? This question can be answered, roughly, by computing the dimension of

$$W^u(\omega) \cap W^s(\widetilde{\omega}),$$

since each point on this set corresponds to a solution of (1.2) with asymptotic profiles $\omega, \widetilde{\omega}$. If $\omega \in \mathcal{E}_k$ and $\widetilde{\omega} \in \mathcal{E}_{k'}$ with $k > k'$, then

$$\dim(T_\psi W^u(\omega) \cap T_\psi W^s_{loc}(\widetilde{\omega})) = 2(k - k') - 1, \tag{6.13}$$

where $T_\psi W^u(\omega)$ and $T_\psi W^s_{loc}(\widetilde{\omega})$ denote the tangent spaces of those manifolds at each point $\psi \in W^u(\omega) \cap W^s_{loc}(\widetilde{\omega})$. This is a consequence of (5.10) and (5.11). (6.13) suggests that $W^u(\omega) \cap W^s(\widetilde{\omega})$ is a $2(k - k') - 1$ dimensional manifold.

Remark 6.4. As mentioned in the remark following (3.20), the whole set of solutions of (1.2) can be identified with the compact global attractor $\mathcal{A}$ defined in (3.19). From Corollary 3.10 it is clear that

$$\mathcal{A} = \bigcup_{\omega \in \mathcal{E}} W^u(\omega). \tag{6.14}$$

(6.14) can also be written as

$$\mathcal{A} = W^u(\mathcal{E}^+) \cup W^u(\mathcal{E}^-) \cup W^u(\mathcal{E}^0) \cup W^u(\mathcal{E}_{k_0}) \cup \cdots \cup W^u(\mathcal{E}_1).$$

It follows that the Hausdorff dimension of $\mathcal{A}$ satisfies

$$\dim_H \mathcal{A} = \max\{\dim W^u(\mathcal{E}^\pm), \dim W^u(\mathcal{E}^0), \dim W^u(\mathcal{E}_{k_0}), \ldots, \dim W^u(\mathcal{E}_1)\}$$

$$= \dim W^u(\mathcal{E}^0) = 2k_0 + 1.$$

We do not know if $\mathcal{A}$ is a manifold, though some formal arguments suggest that $\mathcal{A}$ is homeomorphic to a $(2k_0+1)$–dimensional closed ball. This, however, still remains an open question.

References

[AK] C.J. Amick and K. Kirchgässner : Solitary water waves in the presence of surface tension, Arch. Rat. Mech. Anal., to appear.

[AT] C.J. Amick and R.E.L. Turner : Small internal waves in two-fluid systems, Univ. Wisc. Tech. Summary Report No 89-4.

[An] S.B. Angenent : The Morse-Smale property for a semi-linear parabolic equation, J. Differential Equations, **62** (1986), 427-442.

[AF] S.B. Angenent and B. Fiedler : The dynamics of rotating waves in scalar reaction diffusion equations, Trans. Amer. Math. Soc., **307** (1988), 545-568.

[Av] P. Aviles : On isolated singularities in some nonlinear partial differential equations, Indiana Univ. Math. J., **32** (1983), 773-790.

[BL] H. Brezis and E.H. Lieb : Long range atomic potentials in Thomas-Fermi theory, Comm. Math. Phys., **65** (1979), 231-246.

[C] K.-C. Chang : Infinite dimensional Morse theory and its applications, Seminaire de Math. Superieures, **97**, Presses de l'Université de Montreal, Quebec, 1985.

[CM] X.-Y. Chen and H. Matano : Convergence, asymptotic periodicity, and finite-point blow-up in one-dimensional semilinear heat equations, J. Differential Equations, **78** (1989), 160-190.

[CMV] X-Y. Chen, H. Matano and L. Véron : Anisotropic singularities of solutions of nonlinear elliptic equations in $\mathbf{R}^2$, J. Differential Functional Analysis, **83** (1989), 50-97.

[CLL] S.-N. Chow, X.-B. Lin and K. Lu : Smooth foliations for flows in a Banach space, preprint.

[CL] S.-N. Chow and K. Lu : C^k centre unstable manifold, Proc. Royal Soc. Edinburgh, **108A** (1988), 303-320.

[GL] D. Gilbarg and N.S. Trüdinger : Elliptic partial differential equations of second order (2nd ed.), Springer-Verlag, Berlin/New York, 1977.

[Ha] J.K. Hale : Asymptotic behavior of dissipative systems, Math. Surveys and Monographs, **25**, Amer. Math. Soc., Providence, R. I., 1988.

[HMO] J.K. Hale, L.T. Magalhães and W.M. Oliva : An introduction to infinite dimensional dynamical systems — Geometric theory, Appl. Math. Sci., **47**, Springer Verlag, New York, 1984.

[He1] D. Henry : Geometric theory of semilinear parabolic equations, Lecture Notes in Math., **840**, Springer Verlag, New York, 1981.

[He2] D. Henry : Some infinite-dimensiona Morse-Smale systems defined by parabolic differential equations, J. Differential Equations, **59** (1985), 165-205.

[HPS] M.W. Hirsh, C.C. Pugh and M. Shub : Invariant manifolds, Lecture Notes in Math., **583**, Springer Verlag, New York, 1977.

[K] K. Kirchgässner : Nonlinear wave motion and homoclinic bifurcation, Theoretical and Applied Mechanics (edits, F. Noirdson and N. Olhoff), Elsevier Science Publishers B. V. (North-Holland), 1985.

[LN] C. Loewner and L. Nirenberg : Partial differential equations invariant under conformal or projective transformations, Contributions to Analysis (edits, L.V. Ahlfors *et al*), Academic Press, Orlando, 1974, 245-272.

[M1] H. Matano : Existence of nontrivial unstable sets for equilibriums of strongly order-preserving systems, J. Fac. Sci. Univ. Tokyo, **30** (1983), 645-673.

[M2] H. Matano : Correction to: Existence of nontrivial unstable sets for equilibriums of strongly order preserving systems, J. Fac. Sci. Univ. Tokyo, **34** (1987) 853-855.

[M3] H. Matano : Nonlinear elliptic equations and infinite-dimensional dynamical systems, in preparation.

[Mie] A. Mielke : A reduction principle for nonautonomous systems in infinite dimensional spaces, J. Differential Equations, **65** (1986), 68-88.

[O] R. Osserman : On the inequality $\Delta u \geq f(u)$, Pacific J. Math., **7** (1957), 1641-1647.

[PdM] J. Palis and W. de Melo : Geometric theory of dynamical systems, Springer Verlag, New York, 1980.

[S] L. Simon : Asymptotics for a class of nonlinear evolution equations with applications to geometric problems, Ann. Math., **118** (1983), 525-571.

[V1] L. Véron : Singular solutions of some nonlinear elliptic equations, Nonlinear Anal, **5** (1981), 225-242.

[V2] L. Véron : Global behaviour and symmetry properties of singular solutions of nonlinear elliptic equations, Ann. Fac. Sci. Toulouse Math., **6** (1984), 1-31.

[W] H.-O. Walther : Inclination lemmas with dominated convergence, Z. Angew. Math. Phys., **38** (1987), 327-337.

Introduction to Geometric Potential Theory

Takashi SUZUKI

Department of Mathematics, Tokyo Metropolitan University,
Fukasawa 2-1-1, Setagayaku, Tokyo, 158, JAPAN

1 Introduction

In [8] we have done the asymptotic analysis for two-dimensional elliptic eigenvalue problem (P):

$$-\Delta u = \lambda f(u), \ u > 0 \qquad (in\ \Omega \subset R^2) \tag{1}$$

$$u = 0 \qquad (on\ \partial\Omega), \tag{2}$$

where Ω is a bounded domain with smooth boundary $\partial\Omega$, λ is a positive constant, and $u \in C^2(\Omega) \cap C^0(\overline{\Omega})$ is a classical solution.

We supposed that the semilinear term $f(u)$ is exponentially dominated, that is,

$$f(t) = e^t + g(t) \tag{3}$$

with

$$| g(t) | \in o(e^t) \qquad as\ \ t \to \infty, \tag{4}$$

where

$$| g'(t) - g(t) | \leq G(t) \tag{5}$$

and

$$G(t) + | G'(t) | \in O(e^{\gamma t}) \qquad as\ \ t \to \infty \tag{6}$$

for some $\gamma < 1/4$. Furthermore we supposed that

$$(T) \qquad \Omega \subset R^2 \ is\ convex \quad or \quad f(t) \geq 0\ for\ t \geq 0.$$

Then we have the following theorems:

Theoem 0.1. *The value* $\Sigma = \int_\Omega \lambda f(u) dH^2$ *accumulates to* $8\pi k$ *for some* $k = 0, 1, 2, ..., +\infty$ *as* $\lambda \downarrow 0$. *The solutions* $\{u\}$ *behave as follows:*

(a) In the case $k = 0$, $\|u\|_{L^\infty} \to 0$, *i.e., uniform convergence to zero*

(b) In the case $0 < k < +\infty$, $u_{|S} \to +\infty$ *and* $\|u\|_{L^\infty_{loc}(\overline{\Omega}\setminus S)} \in O(1)$ *for some set* $S \subset \Omega$ *of k-points, i.e., k point blow-up*

(c) In the case $k = +\infty$, $u(x) \to +\infty$ *for any* $x \in \Omega$, *i.e., entire blow-up*

For the case (*b*) we can classify the limiting function $u_0 = u_0(x)$, which may be called the singular limit, and also the location of the blow-up points as follows.

Theorem 0.2. *In the case (b), the blow-up points*

$$S = \{x_1, ..., x_k\} \subset \Omega$$

satisfy

$$\frac{1}{2}\nabla R(x_j) + \sum_{j \neq \ell} \nabla_x G(x_\ell, x_j) = 0 \quad (1 \leq j \leq k) \tag{7}$$

and the singular limit $u_0 = u_0(x)$ *must be of the form*

$$u_0(x) = 8\pi \sum_{j=1}^{k} G(x, x_j), \tag{8}$$

where $G = G(x, y)$ *denotes the Green function for* $-\Delta$ *in* Ω *under the Dirichlet condition and* $R = R(x)$ *denotes the Robin function:*

$$R(x) = [G(x, y) + \frac{1}{2\pi} \log | x - y |]_{y=x}.$$

In deducing these theorems, we have utilized the complex structure of the equation

$$-\Delta u = \lambda e^u \qquad (in\ \Omega \subset R^2) \tag{9}$$

Namely, $u \in C^2(\Omega)$ solves (9) if and only if there exists an analytic function $F = F(z)$ in Ω such that

$$\rho(F) \equiv \frac{| F' |}{1 + | F |^2} \tag{10}$$

is positively single-valued and satisfies

$$(\frac{\lambda}{8})^{1/2} e^{u/2} = \rho(F). \tag{11}$$

This fact was essentially discovered by J. Liouville[7]. We refer to [9] for the proof.

The relation (11) suggests a fine analogy between the nonlinear equation (9) and the linear equation

$$-\Delta u = 0 \qquad (in\ \Omega \subset R^2). \tag{12}$$

In fact the latter has the integral

$$u = Re\ F, \tag{13}$$

where $F = F(z)$ is an analytic function in Ω.

Here, we realize that the Harnack principle for harmonic functions, i.e., solutions for (12), is very similar to our Theorem 0.1. In fact the Harnack principle assures us of the alternatives for monotone harmonic functions between the uniform convergence on every compact set and the entire blow-up. Our purpose is to show that Theorem 0.1 may be actually regarded as a natural extension of the Harnack principle for solutions of the linear equation (12) to those of the nonlinear equation (9). We can derive a mean value theorem for sub-solutions of (9). Then a variant of Harnack's inequality follows immediately to establish a kind of the Harnack principle. We call the study geometric potential theory because solutions for (9) are related to conformal mappings in Ω valued into the two-dimensional round sphere and hence have a geometric meaning. Extensions to higher-dimensional cases and/or taking up other principles of the standard potential theory might become interesting themes in future.

This article is composed of four sections. Next section, §2, is devoted to some technical lemmas to develop the potential theory for (9) in §3. The results in §3 are applied to study the blow-up set for parabolic and elliptic equations and the Harnack principle for those problems will be established in §4.

2 Fundamental lemma

Let $\Omega \subset R^m$ $(m > 1)$ be a bounded domain and $B = B_R(0) \subset R^m$ be a ball with radius $R > 0$ and center origin. The function $\phi : \overline{R}_+ = [0, +\infty) \to R = (-\infty, +\infty)$ is absolutely continuous and strictly increasing. The function $p \in C^1(\Omega) \cap C^0(\overline{\Omega})$ satisfies

$$-\Delta\phi(p) \leq p, \quad p \geq 0 \quad (in\ \Omega), \qquad p \leq a \quad (on\ \partial\Omega) \tag{14}$$

in the distributional sense, where a is a nonnegative constant. The comparison function $q = q(r)$ is strictly decreasing and continuous in $r = | x |$ on $\overline{B}$ satisfying

$$-\Delta\phi(q) = q \quad (in\ B), \qquad q = a \quad (on\ \partial B). \tag{15}$$

Then we have the following

Lemma. *If*

$$\int_{\{p>a\}} p dH^m \leq \int_B q dH^m, \tag{16}$$

then

$$\int_{\{p>t\}} p dH^m \leq \int_{\{q>t\}} q dH^m \qquad for\ any\ t \geq a. \tag{17}$$

In particular,

$$t_0 \equiv max_{\overline{\Omega}}\, p \ \leq \ \overline{t}_0 \equiv q(0). \tag{18}$$

Proof: For each $t \in I \equiv [a, t_0]$, we put $\Omega_t = \{p > t\}$ and $\Gamma_t = \partial\Omega_t = \{p = t\}$. Then $\mu(t) = H^m(\Omega_t)$ denotes the distribution function of p, which is right continuous and strictly decreasing so that has a continuous nonincreasing (hence absolutely continuous) left inverse $t = t(\mu)$ satisfying

$$t(\mu) = t \quad and \quad \mu(t(\mu)) \geq \mu.$$

Co-area formula indicates

$$-\mu'(t) = \int_{\Gamma_t} \frac{dH^{m-1}}{|\nabla p|} \qquad (a.e.\ t \in I), \tag{19}$$

while Sard's lemma and Green's formula imply

$$D(t) \equiv \int_{\Omega_t} p dH^m \geq -\int_{\Gamma_t} \frac{\partial}{\partial n}\phi(p) dH^{m-1} \tag{20}$$

$$= \phi'(t) \int_{\Gamma_t} |\nabla p| \, dH^{m-1} \quad (a.e.\ t \in I),$$

where n denotes the outer unit normal vector. Therefore we have

$$\int_{\Gamma_t} |\nabla p| \, dH^{m-1} \leq \frac{D(t)}{\phi'(t)} \qquad (a.e.\ t \in I)$$

and

$$\int_{\Gamma_t} \frac{dH^{m-1}}{|\nabla p|} = -\frac{D'(t)}{t} \qquad (a.e.\ t \in I)$$

because $\phi' > 0$ and

$$D(t) = \int_t^\infty r d(-\mu(r)). \tag{21}$$

Schwarz's inequality now implies

$$H^{m-1}(\Gamma_t)^2 \leq -\frac{D(t)D'(t)}{t\phi'(t)} = -\frac{1}{2t\phi'(t)}\{D(t)^2\}' \qquad (a.e.\ t \in I), \tag{22}$$

which can be combined with the isoperimetric inequality

$$H^m(\Omega_t)^{\gamma_m} \leq c_m H^{m-1}(\Gamma_t)^2 \qquad (a.e.\ t \in I), \tag{23}$$

where $\gamma_m = \frac{2(m-1)}{m} \in [1,2)$ and $c_m = m^{-2}\omega_m^{-2/m}$ with $\omega_m = \pi^{m/2}/\Gamma(1+m/2)$ denoting the volume of $m-$dimensional unit ball.

Noting

$$H^m(\Omega_t) = \mu(t) = \int_t^\infty \frac{1}{r} d(-D(r)) = \frac{D(t)}{t} - \int_t^\infty \frac{D(r)}{r^2} dr, \tag{24}$$

which follows from (21) and an integration by parts, we obtain

$$-\frac{c_m}{2}\{D(t)^2\}' \geq t\phi'(t)\{\frac{D(t)}{t} - \int_t^\infty \frac{D(r)}{r^2} dr\}^{\gamma_m} \qquad (a.e.\ t > a) \tag{25}$$

and also

$$D(a) = \int_{\{p>a\}} p dH^m, \quad D(t) > 0 \ (a \leq t < t_0), \quad D(t_0) = 0. \tag{26}$$

We follow the same procedure for the continuous strictly decreasing function $q = q(r)$ $(r = |x|)$. We have the equalities for

$$\overline{D}(t) \equiv \int_{\{q>t\}} q dH^m \tag{27}$$

in (20) because of those for (15), in (22) because $|\nabla q| = -q'(r)$ is constant on $\{q = t\}$, and in (23) because $\{q > t\}$ is a ball. Therefore,

$$-\frac{c_m}{2}\{\overline{D}(t)^2\}' = t\phi'(t)\{\frac{\overline{D}(t)}{t} - \int_t^\infty \frac{\overline{D}(r)}{r^2} dr\}^{\gamma_m} \qquad (a.e.\ t > a) \tag{28}$$

and

$$\overline{D}(a) = \int_B q dH^m, \quad \overline{D}(t) > 0\ (a \leq t < \bar{t}_0), \quad \overline{D}(\bar{t}_0) = 0. \tag{29}$$

Now we can perform the comparison between $D(t)$ and $\overline{D}(t)$ through (25) with (26) and (28) with (29). Introducing the error function

$$e(t) = \overline{D}(t) - D(t), \tag{30}$$

we get

$$\frac{c_m}{2}\{e(t)(D(t) + \overline{D}(t))\}' \tag{31}$$

$$\geq t\phi'(t)[\{\frac{D(t)}{t} - \int_t^\infty \frac{D(r)}{r^2} dr\}^{\gamma_m} - \{\frac{\overline{D}(t)}{t} - \int_t^\infty \frac{\overline{D}(r)}{r^2} dr\}^{\gamma_m}]$$

$$= -t\phi'(t)k(t)\{\frac{e(t)}{t} - \int_t^\infty \frac{e(r)}{r^2} dr\} \qquad (a.e.\ t > a)$$

with a right continuous nonnegative function $k = k(t)$. In fact the nonnegativity of k holds by $\gamma_m \geq 1$.

In other words, we obtain

$$e'(t) \geq K(t)e(t) + \int_t^\infty L(r,t)e(r)dr \qquad (a.e.\ t \in I \equiv [a, t_0]) \tag{32}$$

with

$$e(a) \geq 0, \tag{33}$$

where

$$K(t) = -\{D'(t) + \overline{D}'(t) + \frac{2}{c_m}\phi'(t)k(t)\}/\{D(t) + \overline{D}(t)\} \tag{34}$$

and

$$L(r,t) = \frac{2}{c_m}\phi'(t)k(t)t/\{D(t) + \overline{D}(t)\}r^2 \geq 0. \tag{35}$$

Therefore, introducing the right continuous function

$$E(t) = e(t)\exp\{-\int_0^t K(t')dt'\} \tag{36}$$

as usual, we have

$$E'(t) \geq \int_t^\infty M(r,t)E(r)dr \qquad (a.e.t > a) \tag{37}$$

with

$$E(a) \geq 0, \tag{38}$$

where $M(r,t) = L(r,t)\exp\{\int_t^r K(t')dt'\}$. Hence

$$\int_a^t E'(t)dt + E(a) \geq \int_a^t N(r,t)E(r)dr \qquad (t \geq a)$$

for $N(r,t) = \int_a^{r\wedge t} L(r,t')dt' \geq 0$. Here, noting that $D = D(t)$ and $\overline{D} = \overline{D}(t)$ are nonincreasing and continuous,respectively, we have

$$E(t-0) = e(t-0)\exp(-\int_0^t K(t')dt') \leq e(t)\exp(-\int_0^t K(t')dt') = E(t)$$

for each $t \geq a$ so that

$$E(t) \geq \int_a^t E'(t)dt + E(a) \tag{39}$$

$$\geq \int_a^t N(r,t)E(r)dr \qquad (t \geq a).$$

Hence

$$E(t) \geq 0 \qquad (t \geq a). \tag{40}$$

This implies (17).

3 Spherically subharmonic functions

The structure of solutions $\mathsf{C} = \{H =^t (V,\Lambda)\}$ for

$$-\Delta V = \Lambda e^V \qquad (in\ B_1(0) \subset R^2) \tag{41}$$

with

$$V = 0 \qquad (on\ \partial B_1(0)) \tag{42}$$

is well-known. Every solution V for (41) with (42) is radial: $V = V(r)$ $(r =\mid x \mid)$ from the Gidas-Ni-Nirenberg theory[5], while radial solutions can be calculated explicitly. Furthermore, we have $V_r < 0$ $(0 < r < 1)$. See [9] for instance.

In $\Lambda - \Sigma$ plane C forms a one-dimensional manifold connecting the trivial solution $H \equiv^t (V,\Lambda) =^t (0,0)$ and the singular limit $H \equiv^t (V,\Lambda) =\ ^t(4\log\frac{1}{|x|},0)$ bending just once at $\overline{\Lambda} = 2$. Furthermore, the solutions are parametrized by

$$\Sigma = \int_{B_1(0)} \Lambda e^V dH^2 \in [0, 8\pi). \tag{43}$$

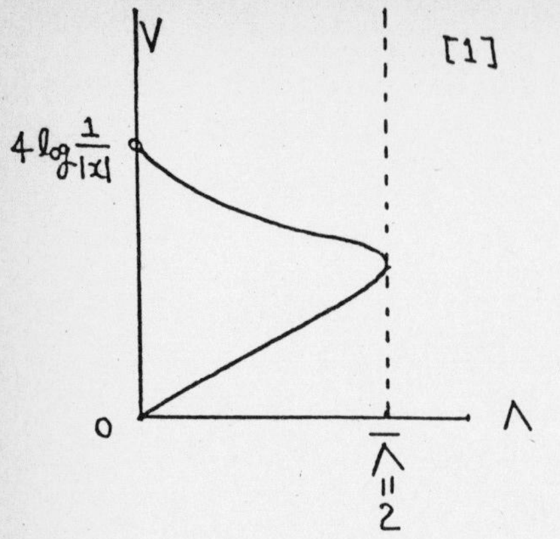

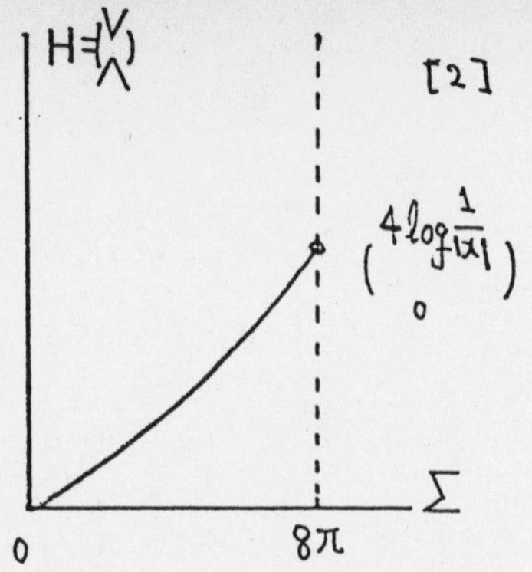

See [6] for geometrical meaning of these figures. Here we note only that under the scaling transformation $x' = Rx$ $(R > 0)$, Figure 1 is transformed so that $\Lambda' = R^2\Lambda$, while Figure 2 is invariant. We also recall the fact that

$$V(0) = -2\log(1 - \frac{\Sigma}{8\pi})$$

for the solution $H =^t (V, \Lambda)$ with

$$\int_{B_1(0)} \Lambda e^V dH^2 = \Sigma \in [0, 8\pi).$$

We call a C^2-function u is spherically subharmonic in $\Omega \subset R^2$ when

$$-\Delta u \leq \lambda e^u \qquad (in\ \Omega) \tag{44}$$

holds for some positive constant $\lambda > 0$.

Proposition 1 (Maximal principle). *If a spherically subharmonic function* $u \in C^0(\overline{\Omega})$ *satisfies*

$$\Sigma = \int_\Omega \lambda e^u dH^2 < 8\pi, \tag{45}$$

then

$$\max_{\overline{\Omega}} u \leq \max_{\partial\Omega} u - 2\log(1 - \frac{\Sigma}{8\pi}). \tag{46}$$

Proof: Setting $p = \lambda e^u$ and $a = \max_{\partial\Omega} p$, we have

$$-\Delta \log p \leq p, \quad p \geq 0 \qquad (in\ \Omega) \tag{47}$$

and

$$p \leq a \qquad (on\ \partial\Omega). \tag{48}$$

On the other hand we have a ball $B = B_R(0)$ and the strictly decreasing radial function $v = v(r)$ of $r = |x|$ on B satisfying

$$-\Delta v = ae^v \qquad (in\ B_R(0)), \tag{49}$$

$$v = 0 \qquad (on\ \partial\ B_R(0)) \tag{50}$$

and

$$\int_{B_R(0)} ae^v dH^2 = \Sigma \ \ (< 8\pi). \tag{51}$$

Thus we have for $q = ae^v$ that

$$-\Delta \log q = q,\ q = q(r)\ (r = \mid x \mid),\ q_r < 0 (0 < r < R), \tag{52}$$

$$q = a \qquad (on\ \partial B_R(0)) \tag{53}$$

and

$$\int_\Omega p dH^2 = \int_{B_R(0)} q dH^2. \tag{54}$$

Now Lemma in §2 implies

$$\max_{\overline{\Omega}}\{\lambda e^u\} = \max_{\overline{\Omega}} p \leq q(0) = ae^{v(0)}$$

$$= a \exp\{-2\log(1 - \frac{\Sigma}{8\pi})\} = \max_{\partial\Omega}\{\lambda e^u\}\{1 - \frac{\Sigma}{8\pi}\}^{-2}$$

or equivalently,

$$\max_{\overline{\Omega}} u \leq \max_{\partial\Omega} u - 2\log(1 - \frac{\Sigma}{8\pi}).$$

Proposition 2 (Mean value theorem). *If $u \in C^0(\overline{B})$ is spherically subharmonic in a ball $B \subset R^2$ and satisfies*

$$\Sigma = \int_B \lambda e^u dH^2 < 8\pi, \tag{55}$$

then

$$u(0) \leq \frac{1}{H^1(\partial B)} \int_{\partial B} u dH^1 - 2\log(1 - \frac{\Sigma}{8\pi}). \tag{56}$$

Proof: We make use of the argument of Bandle[1]. Namely, for each harmonic function h in B and each subdomain $\omega \subset B$ the inequality

$$\{\int_{\partial\omega} e^{h/2} dH^1\}^2 \geq 4\pi \int_\omega e^h dH^2 \tag{57}$$

holds.

In fact, there exists a holomorphic function $g = g(z)$ in B such that

$$\mid g'(z) \mid^2 = e^h \qquad (in\ B). \tag{58}$$

Hence

$$\int_{\partial\omega} e^{h/2} dH^1 = \int_{\partial\omega} \mid g' \mid dH^1 = H^1(\partial\{g(\omega)\})$$

and

$$\int_\omega h dH^2 = \int_\omega \mid g' \mid^2 dH^2 = H^2(g(\omega)).$$

Therefore, (57) is nothing but the usual isoperimetric inequality for the flat Riemann surface $g(\omega)$.

Now we take the harmonic lifting of u denoted by h, that is,

$$-\Delta h = 0 \qquad (in\ B) \tag{59}$$

with

$$h = u \qquad (on\ \partial B). \tag{60}$$

Then the positive function $p = \lambda e^{u-h}$ satisfies

$$-\Delta \log p \leq e^h p \qquad (in\ \Omega) \tag{61}$$

and

$$p = \lambda \qquad (on\ \partial\Omega). \tag{62}$$

We put

$$\mu(t) = \int_{\{p>t\}} e^h dH^2. \tag{63}$$

Co-area formula reads as

$$-\mu'(t) = \int_{\{p=t\}} \frac{e^h}{|\nabla p|} dH^1 \qquad (a.e.\ t \in (\lambda, p_{\max})). \tag{64}$$

Hence the right continuous decreasing function

$$D(t) = \int_{\{p>t\}} pe^h dH^2 = \int_t^\infty rd(-\mu(r)) \quad for\ t \geq \lambda \tag{65}$$

satisfies

$$\int_{\{p=t\}} |\nabla p| \, dH^1 \leq \frac{D(t)}{(\log t)'} \qquad (a.e.\ t \in (\lambda, p_{\max})) \tag{66}$$

and

$$\int_{\{p=t\}} \frac{e^h}{|\nabla p|} dH^1 = -\frac{D'(t)}{t} \qquad (a.e.\ t \in (\lambda, p_{\max})). \tag{67}$$

Therefore,

$$\begin{aligned} \{\int_{\{p=t\}} e^{h/2} dH^1\}^2 &\leq \int_{\{p=t\}} |\nabla p| \, dH^1 \int_{\{p=t\}} \frac{e^h}{|\nabla p|} dH^1 \\ &= -D(t)D'(t) \ \ (a.e. t > \lambda), \end{aligned} \tag{68}$$

where the isoperimetric inequality (57) applies to deduce

$$\mu(t) = \int_{\{p>t\}} e^h dH^2 = \int_t^\infty \frac{1}{r} d(-D(r))$$

$$\leq \frac{1}{4\pi}\{\int_{\{p=t\}} e^{h/2} dH^1\}^2 \leq -\frac{1}{8\pi}\{D(t)^2\}' \quad (a.e.\ t > \lambda)$$

by (68). We now utilize an integration by parts to obtain

$$\frac{D(t)}{t} - \int_t^\infty \frac{D(r)}{r^2} dr \le -\frac{1}{8\pi}\{D(t)^2\}' \quad (a.e. t > \lambda). \tag{69}$$

As in the proof of Proposition 1 we can take a radial function $q = q(r)$ $(r = | x |)$ on a ball $B_R(0)$ satisfying

$$-\Delta \log q = q, \quad q \ge 0, \quad q_r < 0 \quad (0 < r < R), \tag{70}$$

with

$$q = \lambda \quad (on\ \partial B_R(0)) \tag{71}$$

and

$$\int_{B_R(0)} q dH^2 = \Sigma \ \ (= \int_B \lambda e^u dH^2) \in [0, 8\pi). \tag{72}$$

Recalling the proof of Lemma in §2, we see that

$$\overline{D}(t) = \int_{\{q>t\}} q dH^2 \tag{73}$$

satisfies the equality in (69). Noting

$$\overline{D}(\lambda) = \int_{B_R(0)} q dH^2 = \Sigma \tag{74}$$

and

$$D(\lambda) \le \int_B p dH^2 = \int_B \lambda e^u dH^2 = \Sigma, \tag{75}$$

we realize that the relation

$$\max_{\overline{B}} p \le q(0) = \lambda \exp\{-2\log(1 - \frac{\Sigma}{8\pi})\}$$

follows from the same argument as in the proof of Lemma.

In other words,

$$\max_{\overline{B}} \{u - h\} \le -2\log(1 - \frac{\Sigma}{8\pi}). \tag{76}$$

Since

$$h(0) = \frac{1}{H^1(\partial B)} \int_{\partial B} u dH^1$$

by the mean value theorem for harmonic functions, we obtain (56).

Remark. Some variants for (46) and (76) are seen in [1] for real analytic u.

Corollary 1. *Under the same assumption, the inequality*

$$\lambda e^{u(0)} \le \frac{\Sigma}{H^2(B)}\{1 - \frac{\Sigma}{8\pi}\}^{-2} \tag{77}$$

holds.

Proof: For each concentric ball $B_r \subset B$ we have

$$u(0) \leq \frac{1}{H^1(\partial B_r)} \int_{\partial B_r} u dH^1 - 2\log(1 - \frac{\Sigma}{8\pi})$$

and hence

$$u(0) \leq \frac{1}{H^2(B)} \int_B u dH^2 - 2\log(1 - \frac{\Sigma}{8\pi}).$$

Now, Jensen's inequality gives

$$\lambda e^{u(0)} \leq \{1 - \frac{\Sigma}{8\pi}\}^{-2} \lambda \exp\{\frac{1}{H^2(B)} \int_B u dH^2\} \tag{78}$$

$$\leq \{1 - \frac{\Sigma}{8\pi}\}^{-2} \frac{1}{H^2(B)} \int_B \lambda e^u dH^2 = \frac{\Sigma}{H^2(B)} \{1 - \frac{\Sigma}{8\pi}\}^{-2}.$$

Corollary 2 (Harnack's inequality). *If $u \in C^0(\overline{B})$ is spherically subharmonic, super harmonic, and nonnegative in a ball $B = B_R(0)$, i.e.,*

$$0 \leq -\Delta u \leq \lambda e^u, \quad u \geq 0 \qquad (in\ B = B_R(0) \subset R^2) \tag{79}$$

then the inequality

$$u(0) \leq \frac{R + |x|}{R - |x|} u(x) - 2\log(1 - \frac{\Sigma}{8\pi})_+ \tag{80}$$

holds for any $x \in B = B_R(0)$, where $\Sigma = \int_B \lambda e^u dH^2$.

Proof: Since $u \geq 0$ is superharmonic, we have from Possion's formula that

$$u(re^{i\theta}) \geq \frac{1}{2\pi} \int_0^{2\pi} \frac{R^2 - r^2}{r^2 - 2Rr\cos(\theta - \phi) + R^2} u(Re^{i\phi}) d\phi \quad (0 < r < R) \tag{81}$$

$$\geq \frac{R - r}{R + r} \frac{1}{2\pi} \int_0^{2\pi} u(Re^{i\phi}) d\phi.$$

Combining (81) with (56), we obtain (80).

4 Applications

The inequalities derived in the previous section can be utilized to deduce the finiteness of blow-up set for solutions of semilinear equations.

We recall that the nonnegative functions $\{u_k\}$ on $\overline{\Omega}$ are said to make blow-up if $\|u_k\|_{L^\infty(\Omega)} \to +\infty$ as $k \to +\infty$. The blow-up set S is defined as

$$S = \{x_0 \in \overline{\Omega} \mid there\ exists\ some\ \ \{x_k\} \subset \Omega\ \ such\ that \tag{82}$$

$$x_k \to x_0\ \ and\ \ u_k(x_k) \to +\infty\}.$$

We call $S_I = S \cap \Omega$ the interior blow-up set.

Theorem 1 (Harnack principle). *The interior blow-up set S_I of functions $\{u_k\}_{k=1}^{\infty} \subset C^2(\Omega) \cap C^0(\overline{\Omega})$ satisfying*

$$-\Delta u_k \leq e^{u_k} \qquad (in\ \Omega \subset R^2) \tag{83}$$

is finite, provided that

$$\Sigma_k \equiv \int_{\Omega} e^{u_k} dH^2 \in O(1). \tag{84}$$

More precisely,

$$\# S_I \leq \liminf_{k \to \infty} [\Sigma_k / 8\pi] \tag{85}$$

holds.

Proof: Take $x_0 \in S_I$ and sufficiently small $r > 0$ so that $B_r(x_0) \subset \Omega$. Then from Corollary 1 to Proposition 2 we have

$$\liminf_{k} \int_{B_r(x_0)} e^{u_k} dH^2 \geq 8\pi. \tag{86}$$

This relation holds for each $x_0 \in S_I$ and hence (85) follows.

This theorem applies to the semilinear parabolic equation

$$\frac{\partial u}{\partial t} - \Delta u = e^u \quad (in\ Q = \Omega \times (0,T)) \tag{87}$$

with

$$u = 0 \qquad (on\ \partial\Omega), \tag{88}$$

where $\Omega \subset R^2$ is a bounded domain with smooth boundary $\partial\Omega$. We suppose that the initial value $u_0 \in C^2(\overline{\Omega})$ satisfies

$$-\Delta u_0 \leq e^{u_0}, \quad u_0 \geq 0 \qquad (in\ \Omega). \tag{89}$$

Then it holds that the nonnegative function $u = u(x,t)$ is nondecreasing in t for each $x \in \overline{\Omega}$ and hence

$$-\Delta u(\cdot,t) \leq e^{u(\cdot,t)} \qquad (in\ \Omega) \tag{90}$$

for any $t \in [0,T)$. We suppose that $\{u(\cdot,t)\}_{0 \leq t < T}$ make blowing-up at $t = T < +\infty$, i.e.,

$$\lim_{t \uparrow T} \|u(\cdot,t)\|_{L^\infty(\Omega)} = +\infty.$$

We set

$$\Sigma(t) = \int_{\Omega} e^{u(\cdot,t)} dH^2, \tag{91}$$

which is a nondecreasing function of $t \in [0,T)$. Furthermore, S denotes the blow-up set of $\{u(\cdot,t)\}_{0 \leq t < T}$.

Corollary. *Under those situations stated above, the finiteness of S follows from $\Sigma(T) < \infty$. More precisely, we have $S \subset \Omega$ and*

$$\# S \leq [\Sigma(T)/8\pi]. \tag{92}$$

The limiting function $u_T = u(\cdot, T)$ must be of the form

$$u_T(x) = \sum_{j=1}^{m} a_j \log \frac{1}{|x - x_j|} + smooth function \tag{93}$$

for some constants $a_j < 2$ with $\sum_{j=1}^{m} a_j \le \Sigma(T)/2\pi$ where $S = \{x_j\}_{j=1}^{m} \subset \Omega$.

On the other hand if $\int_0^T \Sigma(t)dt = +\infty$ then $S_I = \Omega$.

Proof: From the Gidas-Ni-Nirenberg theory [5], there exist some $\alpha > 0$ and $s_0 > 0$ such that $\xi \in R^2, |\xi| = 1$ and $\xi \cdot n(x) \ge \alpha$ at $x \in \partial\Omega$ imply

$$\frac{d}{ds} u(x + s\xi, t) < 0 \quad (-s_0 < s < 0, 0 \le t < T)$$

for any $t \in [0, T)$, where $n = n(x)$ denotes the outer unit normal vector. Then De Figueiredo-Lions-Nussbaum's argument [3] assures us of the boundary estimate. Namely, in the case $\|u\|_{L^1_{loc}(\Omega)} \in O(1)$ there exists an $\overline{\Omega}$-neighbourhood of $\partial\Omega$ denoted by ω such that

$$\|u\|_{L^\infty(\omega)} \in O(1).$$

In the case that $\Sigma(T) \equiv \int_\Omega e^{u_T} dH^2 < +\infty$ we have $\|u_T\|_{L^1(\Omega)} < +\infty$ so that $S = S_I \subset \Omega$. The estimate (92) follows from Theorem 1. More precisely, the relation

$$\int_{B_r(x_0)} (-\Delta u_T) dH^2 \le \Sigma(T) \tag{94}$$

holds for any $x_0 \in S$ and sufficiently small $r > 0$. Therefore, we have

$$-\Delta u_T = \sum_{j=1}^{m} b_j \delta(x - x_j) + smooth function$$

on Ω with some constants b_j satisfying $\sum_{j=1}^{m} b_j \le \Sigma(T)$. Hence the equality (93) follows.

To study the case $\int_0^T \Sigma(t)dt = +\infty$ we take the first eigenfunction $\phi_1 > 0$ of $-\Delta_D$, the differential operator $-\Delta$ in Ω under the Dirichlet boundary condition, and set

$$\Sigma_1(t) = \int_\Omega e^{u(\cdot,t)} \phi_1 dH^2 \ \ and \ \ J(t) = \int_\Omega u(\cdot, t)\phi_1 dH^2. \tag{95}$$

Then the parabolic equation (87) with (88) implies

$$J'(t) + \mu_1 J = \Sigma_1, \tag{96}$$

that is

$$J(t) = e^{-t\mu_1} J(0) + \int_0^t e^{-(t-\tau)\mu_1} \Sigma_1(\tau) d\tau, \tag{97}$$

μ_1 being the first eigenvalue of $-\Delta_D$. Therefore,

$$(a) \qquad \int_0^T \Sigma(t)dt < +\infty$$

implies

$$(b) \qquad \int_0^T \Sigma_1(t)dt < +\infty,$$

which is equivalent to

$$(c) \qquad J(T) < +\infty.$$

However, if (c) holds then $\|u\|_{L^1_{loc(\Omega)}} \in O(1)$ so that the boundary estimate follows. In particular $(a),(b)$ and (c) are equivalent so that $\int_0^T \Sigma(t)dt = +\infty$ implies $J(T) = +\infty$.

Let $G = G(x,y;t) \geq 0$ be the Green function for $\frac{\partial}{\partial t} - \Delta_D$. Then for each $x \in \Omega$ there exists a constant $\gamma_x > 0$ such that

$$G(x,y;t) \geq \gamma_x e^{-t\mu_1}\phi_1(y) \quad (0 \leq t < \infty, y \in \overline{\Omega}). \tag{98}$$

Hence

$$\begin{aligned} u(x,t) &= \int_\Omega G(x,y;t)u_0(y)dy + \int_0^t d\tau \int_\Omega G(x,y;t-\tau)e^{u(y,\tau)}dy \\ &\geq \gamma_x \int_0^t e^{-(t-\tau)\mu_1}\Sigma_1(\tau)d\tau = \gamma_x\{J(t) - e^{-t\mu_1}J(0)\}. \end{aligned}$$

Therefore, $J(T) = +\infty$ implies $u_T(x) = +\infty$ for any $x \in \Omega$ so that $S_I = \Omega$ holds.

Remark. Suppose that $\Omega = B$ is a ball, $u_0 = u_0(|\ x\ |)$ is radial, $u_{0r} < 0$, and $u_{0rr}(0) < 0$. Then it is known that $S = \{0\}$ (Friedman-McLeod[3]) and

$$\log 2 + \log\frac{1}{|\ x\ |^2} \leq u_T(x) \leq \log\frac{1}{|\ x\ |^2} + \log|\log|\ x\ || + constant$$

(Bebernes-Bressan-Lacey[2]), Itoh[6]). In view of this, the condition $\Sigma(T) < \infty$ seems to be slightly stronger as a criterion for the finiteness of blow-up points.

Theorem 2 (Harnack principle). *The interior blow-up set S_I of functions $\{u_k\}_{k=1}^\infty \subset C^2(\Omega) \cap C^0(\overline{\Omega})$ satisfying*

$$0 \leq -\Delta u_k \leq \lambda_k e^{u_k}, \quad u_k \geq 0 \quad (in\ \Omega \subset R^2) \tag{99}$$

for some positive constants $\{\lambda_k\}$ either has an interior point or is finite, provided that

$$\Sigma_k = \int_\Omega \lambda_k e^{u_k} dH^2 \in O(1). \tag{100}$$

More precisely, if S_I has no interior point we have

$$\# S_I \leq \liminf_k [\Sigma_k / 8\pi]. \tag{101}$$

Proof: Take $x_0 \in S_I$ and sufficiently small $r > 0$ so that $B_r(x_0) \subset \Omega$. Then, Corollary 2 to Proposition 2 implies that either

$$\liminf_{k\to\infty} \int_{B_r(x_0)} \lambda_k e^{u_k} dH^2 \geq 8\pi \tag{102}$$

or $B_r(x_0) \subset S_I$. When S_I has no interior point, the relation (102) holds for each $x_0 \in S_I$ and hence (101) follows .

This theorem applies to the semilinear elliptic eigenvalue problem

$$-\Delta u = \lambda f(u),\ u > 0 \qquad (in\ \Omega \subset R^2) \tag{103}$$

with

$$u = 0 \qquad (on\ \partial\Omega), \tag{104}$$

where λ is a positive constant and Ω is a bounded domain with smooth boundary $\partial\Omega$. The C^1 function f is supposed to satisfy

$$\varepsilon e^s - C \leq f(s) \leq \varepsilon^{-1} e^s + C \qquad (s \geq 0) \tag{105}$$

for some positive constants ε and C. We take a sequence $\{\lambda_k\}$ converging to zero. Let S be the blow-up set of $\{u_k\}$, where u_k solves (103) with (104) for $\lambda = \lambda_k$.

Corollary *Under those situations stated above, either S is finite or $S_I = \Omega$. In the former case the limiting function (singular limit) $u_0 = u_0(x)$ must be of the form*

$$u_0(x) = \sum_{j=1}^{m} a_j G(x, x_j) \tag{106}$$

with some positive constants $\{a_j\}$, where $a_j \geq 8\pi\varepsilon$, $S = \{x_j\}_{j=1}^m$, and $G = G(x,y)$ denotes the Green function for $-\Delta_D$

Proof: Setting

$$\Sigma_k = \int_\Omega \lambda_k f(u_k) dH^2, \tag{107}$$

we see that $\Sigma_k \in O(1)$ if and only if $\tilde{\Sigma}_k \in O(1)$ where

$$\tilde{\Sigma}_k = \int_\Omega \lambda_k e^{u_k} dH^2 \tag{108}$$

by (105).

Introducing the first eigenfunction $\phi_1 > 0$ of $-\Delta_D$, we can prove that $\Sigma_k \longrightarrow +\infty$ implies $S_I = \Omega$ in the same way as in the proof of Corollary to Theorem 1.

Conversely, $\Sigma_k \in O(1)$ implies

$$J_k \equiv \int_\Omega u_k \phi_1 dH^2 = \frac{1}{\mu_1} \int_\Omega (-\Delta u_k) \phi_1 dH^2 \tag{109}$$

$$= \frac{1}{\mu_1} \int_\Omega \lambda_k f(u_k) \phi_1 dH^2 \in O(1),$$

and hence the nonexistence of the interior point of S_I. Furthermore, (109) deduces $\|u_k\|_{L^\infty_{loc}(\Omega)} \in O(1)$ and hence follows the boundary estimate $\|u_k\|_{L^\infty(\omega)} \in O(1)$ as in the parabolic case, where ω is an $\overline{\Omega}$-neighbourhood of $\partial\Omega$. Finally, the finiteness of $S = S_I \subset \Omega$ follows from $\tilde{\Sigma}_k \in O(1)$ and Theorem 2.

The singular limit $u_0 = u_0(x)$ is harmonic in $\overline{\Omega}\backslash S$, is equal to 0 on $\partial\Omega$, and satisfies for sufficiently small $r > 0$ that

$$\int_{B_r(x_j)} (-\Delta u_0) dH^2 \geq \varepsilon \liminf_k \int_{B_r(x_j)} \lambda_k e^{u_k} dH^2 \geq 8\pi\varepsilon \tag{110}$$

where $S = \{x_j\}_{j=1}^m$. Hence (106) follows.

References

[1] Bandle, C., On a differential inequality and its applications to geometry, Math. Z. 147 (1976) 253-261.

[2] Bebernes, J., Bressan, A., Lacey, A., Total blow-up versus single point blow-up, J. Differential Equations 73 (1988) 30-44.

[3] De Figueiredo, D.G., Lions, P.L., Nussbaum, R.D., A priori estimates and existence of positive solutions of semilinear elliptic equations, J. Math. Pure et. Appl. 61 (19982) 41-63.

[4] Friedman, A., McLeod, B., Blow-up of positive solutions of semilinear heat equations, Indiana Univ. Math. J. 34 (1985) 425-447.

[5] Gidas, B., Ni, W.M., Nirenberg, L., Symmetry and related properties via the maximum principle, Comm. Math. Phys., 68 (1979) 209-243.

[6] Itoh, T., Blow-up of solutions for semilinear parabolic equations, In; Suzuki, T. (ed.), Solutions for Nonlinear Elliptic Equations, Kokyuroku RIMS Kyoto Univ. 679 pp. 127-139, 1989.

[7] Liouville, J., Sur l'équation aux différences partielles $\partial^2 \log \lambda / \partial u \partial v \pm \lambda / 2a^2 = 0$, J. de Math., 18 (1853) 71-72.

[8] Nagasaki, K., Suzuki, T., Asymptotic analysis for two dimensional eigenvalue problems with exponentially-dominated nonlinearities, to appear in Asymptotic Analysis.

[9] Suzuki, T., Two dimensional Emden-Fowler equation with the exponential nonlinearity, to appear in; Lloyd, N.G., Ni, W.M., Peletier, L.A., Serrin, J. (eds.), Nonlinear Diffusion Equations and their Equilibrium States, Wales 1989, Birkhäuser.

KDV, BO AND FRIENDS IN WEIGHTED SOBOLEV SPACES

RAFAEL JOSÉ IÓRIO, JR.

Instituto de Matemática Pura e Aplicada
Estrada Dona Castorina 110
cep 22460 - Rio de Janeiro - Brazil

§1. Introduction

The purpose of this paper is to discuss some aspects of the relationship between differentiability and spacial decay of the real-valued solutions of the Cauchy problem for certain nonlinear evolution equations. We will concentrate on the equations of Korteweg-de Vries (KDV) and Benjamin-Ono (BO) but we will also consider the less known, albeit very interesting, equation of Smith (S). Thus, we will study some properties of the following problems,

$$(KDV) \qquad \partial_t u = -\partial_x(u^2 + \partial_x^2 u), \quad u(0) = \phi \tag{1.1}$$

$$(BO) \qquad \partial_t u = -\partial_x(u^2 + 2\sigma\partial_x u), \quad u(0) = \phi \tag{1.2}$$

$$(S) \qquad \partial_t u = -\partial_x(u^2 + 2Lu), \quad u(0) = \phi \tag{1.3}$$

where σ denotes the Hilbert transform

$$(\sigma f)(x) = p \cdot v \cdot \frac{1}{\pi} \int_{\mathbf{R}} \frac{f(y)}{y - x} dy \tag{1.4}$$

and L is the operator given by

$$Lf = \left(\sqrt{-\partial_x^2 + 1} - 1\right) f \tag{1.5}$$

Before proceeding it is convenient to recall that if $f \in L^2(\mathbf{R})$ then ([**BN**],[**T**]),

$$(\sigma f)^\wedge(\xi) = ih(\xi)\hat{f}(\xi) \quad , \quad \xi \ a.e. \tag{1.6}$$

$$h(\xi) = \begin{cases} 1 & , \quad \xi > 0 \\ -1 & , \quad \xi < 0 \end{cases} \tag{1.7}$$

where $\hat{f} = \mathcal{F}f$ denote the Fourier transform of f i.e.,

$$\hat{f} = \mathcal{F}f = \ell \cdot i \cdot m \cdot (2\pi)^{-1/2} \int_{\mathbf{R}} f(x) e^{-i\xi x} dx. \tag{1.8}$$

Actually, we will use $^\wedge$ and $\mathcal{F}$ to represent the Fourier transform in whatever context it occurs. The inverse transform will, as usual, be denoted by $\check{f} = \mathcal{F}^{-1} f$. With this notation the operator L of (1.5) can be written as

$$Lf = \mathcal{F}^{-1} p \mathcal{F} f = (p\hat{f})^\vee \tag{1.9}$$

where p denotes the operator of multiplication by $p(\xi) = \left(\sqrt{\xi^2+1} - 1\right)$.

In order to motivate what follows, it is worthwhile to describe the question we wish to address in a very simple form. Let $P(\xi)$, $\xi \in \mathbf{R}$, be a real-valued continuously differentiable function, write $D = i^{-1}\partial_x$ and note that the unique solution of

$$\begin{cases} u \in C(\mathbf{R}; L^2(\mathbf{R})) \\ \partial_t u = iP(D)u, \quad u(0) = \phi \in L^2(\mathbf{R}) \end{cases} \tag{1.10}$$

where $P(D) = \mathcal{F}^{-1} P(\xi) \mathcal{F}$, is given by $u(t) = (\exp(iP(D)t))\phi = (\exp(iP(\xi)t)\hat{\phi})^\vee$. Assume now that $x\phi$ also belongs to $L^2(\mathbf{R})$. We would like to know if the same is true for $xu(t)$, $t \in \mathbf{R}$. Taking the Fourier transform of $xu(t)$ we obtain

$$\begin{cases} (xu(t))^\wedge = i\partial_\xi(\exp(iP(\xi)t)\hat{\phi}(\xi)) = \\ = -P'(\xi)\exp(iP(\xi)t)\hat{\phi}(\xi) + \exp(iP(\xi)t)(i\partial_\xi\hat{\phi}) \end{cases} \tag{1.11}$$

Since $x\phi \in L^2(\mathbf{R})$ it follows from (1.11) that $xu(t) \in L^2(\mathbf{R})$ for all $t \in \mathbf{R}$ if and only if $P'(\xi)\hat{\phi} \in L^2(\mathbf{R})$. This means in particular that the requirement $xu(t) \in L^2(\mathbf{R})$ for all $t \in \mathbf{R}$ imposes an additional "smoothness" condition on the initial data ϕ, namely $P'(D)\phi \in L^2(\mathbf{R})$. Moreover, in this case both $xu(t)$ and $P'(D)u(t)$ belong to $C(\mathbf{R}; L^2(\mathbf{R}))$. What we want to find out is if this and other similar differentiability-decay relations are preserved under non-linear perturbations of the differential equation in (1.10). In what follows, as already mentioned, we will specialize in the cases of KdV, BO and S. It should be stressed however that the ideas and techniques described below can be used in many other situations such as the intermediate long wave (ILW) equation and the non-linear Schrödinger (NLS) equation (see section 6).

Since in the case of KdV $P'(\xi) = 3\xi^2$ it is easy to see that the smoothness condition imposed by the requirement $xu(t) \in L^2(\mathbf{R})$ is that $\partial_x^2\phi \in L^2(\mathbf{R})$. Similarly in the case of BO and S the initial condition must satisfy $\partial_x\phi \in L^2(\mathbf{R})$. It is therefore convenient to describe first the theory of these equations in the Sobolev spaces $H^s(\mathbf{R})$. This will be done in section 2. In the next three sections we take up the study of our model equations in appropriate weighted Sobolev spaces. In the final section we describe very briefly some related results and a few open problems. It should be noted that although part of what follows has already appeared elsewhere, (see [**BSm**], [**I1**], [**K1**] and the references therein), most of the proofs presented here are new and / or simpler than those previously published. There are of course several new results such as theorems (4.7), (5.1) and (5.2).

Before turning to these labours, however, it is convenient to introduce some notation, recall a few facts and establish a very useful technical lemma which will be used many times in what follows. By $H^s(\mathbf{R})$, $s \in \mathbf{R}$, we denote the set of all $f \in \mathcal{S}'(\mathbf{R})$ such that $(1+\xi^2)^{s/2}\hat{f} \in L^2(\mathbf{R})$. They are Hilbert spaces with respect to the inner product

$$(f|g)_s = \int_{\mathbf{R}} (1+\xi^2)^s \hat{f}(\xi)\overline{\hat{g}(\xi)}d\xi \tag{1.12}$$

The corresponding norms, or any of the equivalent ones, will be denoted by $\|\cdot\|_s$. No confusion will arise from this. If $s \geq s'$ then $H^s(\mathbf{R}) \subseteq H^{s'}(\mathbf{R})$ where the inclusion is continuous and dense. Morever, if $s > 1/2$, $H^s(\mathbf{R})$ is continuously and densely embedded in $C_\infty(\mathbf{R})$, the set of all continuous functions that tend to zero at infinity. In this case $H^s(\mathbf{R})$ is a Banach algebra under the usual multiplication of functions. It should also be noted that if $s = k \in \mathbf{N} = \{0, 1, 2, \cdots\}$ then $f \in H^k(\mathbf{R})$ if and only if $\partial_x^{(j)} f \in L^2(\mathbf{R})$, $0 \leq j \leq k$. With these comments in mind we have,

LEMMA (1.1). *Let $P(\xi)$, $\xi \in \mathbf{R}$ be a continuous real-valued function, $\mu \geq 0$, $\lambda \geq 0$ and $s \in \mathbf{R}$. Define*

$$E_\mu(t) = \exp(-Q_\mu t) \quad , \quad t \geq 0 \tag{1.13}$$

where $Q_\mu = -\mu\partial_x^2 - iP(D)$. Then

$$\|E_\mu(t)\phi\|_{s+\lambda} \leq K_\lambda \left(1 + \left(\frac{1}{2\mu t}\right)^\lambda\right)^{1/2} \|\phi\|_s \tag{1.14}$$

for all $t, \mu \in (0,\infty)$, $\lambda \geq 0$ and $\phi \in H^s(\mathbf{R})$. The map $t \in (0,\infty) \to E_\mu(t)\phi$ is continuous with respect to the topology of $H^{s+\lambda}(\mathbf{R})$. Moreover E_μ defines a C^0 semigroup in $H^s(\mathbf{R})$, $s \in \mathbf{R}$, which can be extended to an unitary group if $\mu = 0$ and $t \mapsto E_\mu(t)$, $\mu \geq 0$, is the unique solution of

$$\begin{cases} u \in C([0,\infty); H^s(\mathbf{R})) \\ \partial_t u = \mu\partial_x^2 u + iP(D)u \quad , \quad u(0) = \phi \end{cases} \tag{1.15}$$

SKETCH OF PROOF: We have

$$\begin{cases} \|E_\mu(t)\phi\|_{s+\lambda}^2 = \int_{\mathbf{R}} (1+\xi^2)^{s+\lambda} |\exp(-\mu t\xi^2)\hat{\phi}(\xi)|^2 d\xi \leq \\ \leq [\sup_\xi (1+\xi^2)^\lambda \exp(-2\mu t\xi^2)] \, \|\phi\|_s^2 \end{cases} \tag{1.16}$$

It is easy to check that the supremum in (1.16) is bounded by $C_\lambda(1 + \sup_\xi \alpha_\lambda(\xi))$ where C_λ depends only on λ, and $\alpha_\lambda(\xi) = \xi^{2\lambda}\exp(-2\mu t\xi^2) \leq \lambda^\lambda e^{-\lambda}(2\mu t)^{-\lambda}$, $\xi \in \mathbf{R}$. This implies (1.14). Continuity follows from a similar argument. The remaining statements are trivial. ∎

Next, let $L^2_s(\mathbf{R})$, $s \in \mathbf{R}$ be the collection of all measurable functions $f\colon \mathbf{R} \to \mathbf{C}$ such that

$$\|f\|^2_{L^2_s} = \int_{\mathbf{R}} (1+x^2)^s \, |f(x)|^2 \, dx < \infty \tag{1.17}$$

It is clear that $L^2_s(\mathbf{R}) = (H^s(\mathbf{R}))^\wedge$ so that they are all Hilbert spaces with respect to obvious inner product which we denote by $(\,\cdot \mid \cdot\,)_{L^2_s}$. We are now in position to introduce the weighted Sobolev spaces with which we will work most of the time, namely, $\mathcal{F}_{s,r}(\mathbf{R}) = H^s(\mathbf{R}) \cap L^2_r(\mathbf{R})$. They become Hilbert spaces when provided with the inner product $(f|g)_{s,r} = (f|g)_s + (f|g)_{L^2_r}$. The corresponding norms will be represented by $\|\cdot\|_{s,r}$ and we will sometimes write $\|\cdot\|_{s,0} = \|\cdot\|_s$ and $\|\cdot\|_{0,r} = \|\cdot\|_{L^2_r}$. Note that $\mathcal{F}_{r,s}(\mathbf{R})$, $s > 1/2$, $r \in \mathbf{R}$ is a Banach algebra with respect to pointwise multiplication. Indeed,

$$\begin{cases} \|fg\|^2_{s,r} = \|fg\|^2_{s,0} + \|fg\|^2_{0,r} \leq \\ \leq C_s \|f\|^2_{s,0} \|g\|^2_{s,0} + C'_r \|f\|^2_{L^\infty} \|g\|^2_{0,r} \leq \tilde{C}_{s,r} \|f\|^2_{s,r} \|g\|^2_{s,r} \end{cases} \tag{1.18}$$

Finally, since we work in $\mathbf{R}$ throughout the paper, we will write simply H^s, L^2_s and $\mathcal{F}_{s,r}$ for the spaces defined above. Symbols like $C_s, K_\lambda, \tilde{C}_{r,s}$ and so on will denote constants, the precise values of which are devoid of interest.

§2. Well posedness in Sobolev spaces

In this section we will sketch briefly the theory of KdV, BO and S in the usual Sobolev spaces. We start with

THEOREM 2.1. *Let $P(\xi)$, $\xi \in \mathbf{R}$, be a continuous real-valued odd function and $\phi \in H^s$, $s > 3/2$. Assume that $P(D) \in B(H^s, H^{s-k})$ for some $k \in \mathbf{Z}^+ = \{1, 2, \cdots\}$. Then, there exists a $T = T(s, \|\phi\|_s)$ and a unique $u_0 \in C([0,T]; H^s)$ such that $u_0(0) = \phi$, $\partial_t u_0 \in C([0,T]; H^{s-k})$ and*

$$\partial_t u_0 = iP(D)u_0 + \alpha u_0 \partial_x u_0 \tag{2.1}$$

where $\alpha \in \mathbf{R}$ is a constant.

SKETCH OF PROOF: In order to solve (2.1) we will use the technique known as parabolic regularization. Let $\mu > 0$ be fixed and look for $u = u_\mu$ satisfying the continuity properties of the theorem and such that

$$\partial_t u = \mu \partial_x^2 u + iP(D)u + \alpha u \partial_x u, \quad u(0) = \phi. \tag{2.2}$$

A simple argument involving lemma (1.1) (with $\lambda = 1$), shows that (2.2) is equivalent to the integral equation

$$u(t) = E_\mu(t)\phi + \alpha \int_0^t E_\mu(t-t')u(t')\partial_x u(t')dt' \tag{2.3}$$

in $C([0,T],H^s)$, $s>1/2$. Using lemma (1.1) (with $\lambda=1$), once again it is easy to show that there exists a sufficiently small $\tilde{T}=\tilde{T}(\mu,\|\phi\|_s)$ such that the map

$$(Af)(t)=E_\mu(t)\phi+\alpha\int_0^t E_\mu(t-t')f(t')\partial_x f(t')dt' \tag{2.4}$$

has a unique fixed point in the complete metric space $(\mathcal{X}_s(\tilde{T}),d_s)$ defined by $\mathcal{X}_s(\tilde{T})=\{f\in C([0,\tilde{T}];H^s)|\,\|f(t)-E_\mu(t)\phi\|_s\le\|\phi\|_s,\ t\in[0,\tilde{T}]\}$ and $d_s(f,g)=\sup_{[0,\tilde{T}]}\|f(t)-g(t)\|_s$. Standard arguments ([**KF**],[**H**]) show that this solution is actually unique in $C([0,\tilde{T}];H^s)$. Before proceeding, it is important to note that $u(t)$ is smooth for all $t>0$. Indeed, a straightforward bootstrapping argument involving (2.3) and lemma (1.1) (with fixed $\lambda\in(1,2)$) shows that $u\in C((0,\tilde{T}];H^\infty)$ where $H^\infty=\bigcap_{s\in\mathbf{R}}H^s$ is endowed with its natural Frechet space topology ([**RS**], vol I). We now turn to the limit of $u=u_\mu$ as $\mu\downarrow 0$. First of all we must show that u_μ, $\mu>0$ can be extended to an interval independent of μ. Since $u_\mu(t)$ is smooth if $t>0$ we have $\partial_t\|u\|_s^2=2(u|\partial_t u)_s$. Integration by parts and the fact that $P(\xi)$ is odd imply at once that $\partial_t\|u\|_s^2\le 2|\alpha|\,|(u|u\partial_x u)_s|$. But then lemma (A.5) of the appendix of [**K1**] shows that $\partial_t\|u\|_s^2\le 2|\alpha|\,C_s\|u\|_s^3$ if $s>3/2$. It follows that $\|u\|_s^2\le\rho(t)$ with

$$\partial_t\rho=2|\alpha|\,C_s\rho^{3/2},\quad \rho(0)=\|\phi\|_s^2 \tag{2.5}$$

Thus $\rho(t)^{1/2}=\rho(0)^{1/2}(1-|\alpha|\,C_s\rho(0)^{1/2}t)^{-1}$ and $u_\mu(t)$ can be extended to any interval $0\le t\le T=T(s,\|\phi\|_s)$ where $T\in(0,(|\alpha|\,C_s\|\phi\|_s)^{-1})$. Now choose any such interval and write $u=u_\mu$, $v=u_\nu$, $\mu,\nu>0$. It is easy to check that

$$\partial_t\|u-v\|_0^2\le 4M^2|\mu-\nu|+M\tilde{C}_s|\alpha|\,\|u-v\|_0^2 \tag{2.6}$$

where $\tilde{C}_s$ depends only on s and $M=\sup_{[0,T]}\rho(t)$. Gronwall's inequality then implies that $u_0=\lim_{\mu\downarrow 0}u_\mu$ exists in L^2. Thus $t\in[0,T]\to u_\mu(t)$ is continuous and uniformly bounded in L^2. It follows that u_μ converges weakly to u_0 in H^s uniformly over $[0,T]$. In particular $u_0(t)$ is weakly continuous and uniformly bounded by the function $\rho(t)^{1/2}$. Combining these remarks with the weak continuity of the map $t\in[0,T]\mapsto iP(D)u_0+\alpha u_0\partial_x u_0\in H^{s-k}$ we obtain,

$$u_0(t)-u_0(\tau)=\int_\tau^t(iP(D)u_0(t')+\alpha u_0(t')\partial_x u_0(t'))dt' \tag{2.7}$$

in H^{s-k}, where the R.H.S. is a Bochner integral. Thus $u_0\in AC([0,T];H^{s-k})\cap L^\infty([0,T];H^s)$ and satisfies (2.1) almost everywhere in $[0,T]$. Here $AC(I;X)$ denotes the collection of all absolutely continuous functions from the interval I into the Banach space X. Next we claim that there is only one such function. Indeed, if u_0 and v_0 satisfy the above conditions, a calculation similar to that leading to (2.6) implies $\partial_t\|u_0-v_0\|_0^2\le\tilde{C}_s M|\alpha|\,\|u_0-v_0\|_0^2$ and uniqueness follows from Gronwall's inequality.

Since $\|u_0(t)\|_s \leq \rho(t)^{1/2}$ it follows at once that $\limsup_{t\to 0+} \|u_0(t)\|_s = \|\phi\|_s$ so continuity at $t = 0$ is a consequence from weak-continuity. If $\tau \in (0,T]$ we obtain right continuity from continuity at zero and uniqueness. Left continuity follows from the invariance of (2.1) under the transformation $(t,x) \mapsto (\tau - t, -x)$. ∎

Global existence is far more difficult. In principle one needs additional information to obtain the global estimates required to extend the solutions to the half-line $[0,\infty)$. In the case of KdV and BO the idea is to exploit their Hamiltonian structure ([L]). Roughly speaking this means that both equations can be written in the form

$$\partial_t u = J\Phi'(u) \quad , \quad J = -J^* \tag{2.8}$$

where Φ' is the Gateaux (i.e. directional) derivative of some real valued function Φ (the Hamiltonian of the system), and that there is a "sufficiently large" collection of "conservation laws of order $k \in \mathbf{N}$"(i.e., conserved quantities containing $\|\partial_x^k u\|_0^2$). Smith's equation, as pointed out in [**ABFS**], can be treated as a perturbation of BO (note that the symbols of the linear parts of both equations have the same behavior as $|\xi| \to \infty$). In what follows we will try to explain how to obtain the necessary estimates in the context of parabolic regularization, without going into too much technical detail. Further information on the structure of the equations, different methods, generalizations and extensions can be found in [**ABFS**], [**BSc**], [**BSm**], [**C**], [**I1**], [**K2**] and the references therein; see also theorem (4.5) of [**K1**] where the generalized KdV equation is dealt with by means of a convenient identity which reduces to necessary conservation law in the usual case. In order to fix ideas we will first focus our attention on KdV. It is not difficult to show that the functionals

$$\Phi_0(u) = \|u\|_0^2 \quad , \quad \Phi_1(u) = \int_{\mathbf{R}} \left[\frac{(\partial_x u)^2}{2} - \frac{u^3}{3}\right] dx \tag{2.9}$$

$$\Phi_2(u) = \int_{\mathbf{R}} [(\partial_x^2 u)^2 + au(\partial_x u)^2 + bu^4] dx \tag{2.10}$$

where $a, b \in \mathbf{R}$ can be easily calculated, are formally conserved by the KdV flow. Moreover, Φ_1 is the associated Hamiltonian. A little more precisely KdV can be written as in (2.8) with $\Phi = \Phi_1$, $J = \partial_x$ and if $u(t)$ is a sufficiently smooth solution then $\partial_t \Phi_j(u(t)) = 0$, $j = 0,1,2$. Note that this is equivalent to $(\Phi_j'(f)|J\Phi_k'(f))_0 = 0$ for $k, j = 0,1,2$ and all sufficiently smooth functions f (see [**L**] where it is also shown that there is actually an infinite number of conserved quantities). In order to circumvent the smoothness requirements we again use parabolic regularization. There is a price to pay however: the quantities $\Phi_j(u)$ are no longer conserved so that one must estimate how they change in time.

LEMMA (2.2). *Let $\phi \in H^2$ and $u_\mu \in C([0,T]; H^2)$, $\mu > 0$, be the solution of*

$$\partial_t u = \mu \partial_x^2 u - \partial_x \Phi_1'(u) \quad , \quad u(0) = \phi \tag{2.11}$$

Then there are continuous functions $F_k\colon [0,\infty)^3 \to [0,\infty)$ which are monotone non-decreasing with respect to the third variable such that

$$\|u_\mu(t)\|_k^2 \le F_k(t,\mu;\|\phi\|_k), \quad k = 0,1,2 \tag{2.12}$$

SKECTH OF PROOF: Let $\mu > 0$ be fixed. To simplify the notation we write $u = u_\mu$ in what follows. It is easy to check that $\partial_t \|u(t)\|_0^2 \le 0$ so that $\|u(t)\|_0^2 \le \|\phi\|_0^2$. We will indicate the proof in the case $k = 1$. The remaining estimate can be obtained in the same way. Since u is smooth we have $(\Phi_k'(u)|J\Phi_\ell'(u)) = 0$, $k,\ell = 0,1,2$. Combining this fact with Hölder's inequality and Sobolev's lemma ([**RS**], vol II), we obtain

$$\left\{\begin{aligned} &\partial_t\Phi_1(u(t)) = (\Phi_1'(u(t))|\partial_t u(t))_0 = \mu(\Phi_1'(u(t))|\partial_x^2 u(t))_0 = \\ &= -\mu(u(t)^2 + \partial_x^2 u(t)|\partial_x^2 u(t))_0 \le \mu[\gamma\left\|\partial_x^2 u(t)\right\|_0 - \left\|\partial_x^2 u(t)\right\|_0^2] \end{aligned}\right. \tag{2.13}$$

where $\gamma = C\|\phi\|_1 \|u(t)\|_1$ and C is a positive constant. Since $f(y) = \gamma y - y^2$ attains its maximum at $y = 2^{-1}\gamma$, we have

$$\Phi_1(u(t)) - \Phi_1(\phi) \le \frac{\mu}{2}\|\phi\|_1^2 C^2 \int_0^t \|u(t')\|_1^2\, dt' \tag{2.14}$$

Combining (2.14) with the definition of Φ_1 and some routine estimates, we get

$$\left\{\begin{aligned} &\|u(t)\|_1^2 \le f(\|\phi\|_1) + k_1\|\phi\|_1^2\left(\|u(t)\|_1^2 + \tfrac{\mu}{2}\textstyle\int_0^t \|u(t')\|_1^2\, dt'\right) \\ &f(\|\phi\|_1) = \|\phi\|_1^2 + k_2\|\phi\|_1^3 \end{aligned}\right. \tag{2.15}$$

where $k_1, k_2 \ge 0$ are constants. It is not difficult to show that (2.15) implies the usual setting for the application of Gronwall's inequality and result follows. ∎

We are now in position to prove

THEOREM (2.3). *Let $\phi \in H^s$, $s \ge 2$ and $\mu \ge 0$ be fixed. Then there exists a unique $u_\mu \in C([0,\infty);H^s)$ such that $u_\mu(0) = \phi$, $\partial_t u_\mu \in C([0,\infty);H^{s-3})$ and (2.11) (which coinscides with KdV when $\mu = 0$) is satisfied.*

PROOF: Global existence in H^2 follows from lemmas (1.1) and (2.2). Next, lemma (A.5) of the appendix of [**K1**] shows that $\partial_t \|u_\mu(t)\|_s^2 \le K_s \|u_\mu(t)\|_2 \|u_\mu(t)\|_s^2$ if $\mu > 0$, so global existence follows in this case. Another application of the limiting process described in theorem (2.1) concludes the proof. ∎

Next we turn to BO which can be written in Hamiltonian form taking $J = -\partial_x$ and

$$\Phi(u) = \Phi_3(u) = \int_{\mathbf{R}} \left[\frac{u^3}{3} + u\sigma\partial_x u\right] dx \tag{2.16}$$

Arguments similar to those employed in the case of KdV, involving the L^2 norm and the conserved quantities ([**C**]),

$$\Phi_4(u) = \int_{\mathbf{R}} \left\{ \frac{u^4}{4} + \frac{3}{2}u^2\sigma\partial_x u + 2(\partial_x u)^2 \right\} dx \tag{2.17}$$

$$\begin{cases} \Phi_6(u) = \int_{\mathbf{R}} \left\{ \frac{u^6}{6} + \left[\frac{5}{4}u^4\sigma\partial_x u + \frac{5}{3}u^3\sigma(u\partial_x u)\right] \right\} dx + \\ +\frac{5}{2}\int_{\mathbf{R}} \left\{ 5u^2(\partial_x u)^2 + u^2(\sigma\partial_x u)^2 + 2u(\sigma\partial_x u)(\sigma u\partial_x u) \right\} dx + \\ -10\int_{\mathbf{R}} \left\{ (\partial_x u)^2\sigma\partial_x u + 2u\partial_x^2 u(\sigma\partial_x u) \right\} dx + 8\int_{\mathbf{R}} (\partial_x^2 u)^2 dx \end{cases} \tag{2.18}$$

lead to the following result ([**I1**]).

THEOREM (2.4). *Lemma (2.2) and theorem (2.3) hold for BO (with $C([0,\infty); H^{s-3})$ replaced by $C([0,\infty); H^{s-2})$).*

Next we turn to S which, following [**ABFS**], will be treated as a pertubation of BO. We have,

THEOREM (2.5). *Let $\phi \in H^s$, $s \geq 2$. Then there exists a unique $u \in C([0,\infty); H^s) \cap C^1([0,\infty); H^{s-2})$ satisfying (1.3).*

SKETCH OF PROOF: Let $\phi \in H^s$, $\mu \geq 0$ and consider

$$\partial_t u = \mu\partial_x^2 u + 2A\partial_x u - \partial_x\Phi_3'(u) \quad , \quad u(0) = \phi \tag{2.19}$$

where $A = (\sigma\partial_x - L) \in B(H^s)$, $s \in \mathbf{R}$. The structure of the proof is the same as that of theorems (2.3) and (2.4). First one establishes global existence in H^2 with $\mu > 0$, using $\Phi_4(u)$ and $\Phi_6(u)$, and extends this for all $s > 2$ using lemma (A.5) of [**K1**]. An application of the limiting procedure described in theorem (2.1) concludes the proof. The *a priori* estimates needed for the H^s theory can be obtained as follows. Let $T > 0$ be fixed (but arbitrary) and $u = u_\mu \in C([0,T]; H^2)$, $\mu > 0$, be the solution of (2.19). It is easy to check that $u \in C((0,T]; H^\infty)$ and $\|u(t)\|_0 \leq \|\phi\|_0$. Next, using $\Phi_4(u)$, Abdelouhab et al. (section 6 of [**ABFS**]) have shown that there exists a $C = C(T, \|\phi\|_1)$ such that $\|u(t)\|_1^2 \leq C$ for all $t \in [0,T]$. In order to handle $\|u(t)\|_2$ one must estimate

$$\partial_t\Phi_6(u(t)) = \mu(\Phi_6'(u(t))|\partial_x^2 u(t))_0 + 2(\Phi_6'(u(t))|\partial_x A u(t))_0 \tag{2.20}$$

The first inner product on the R.H.S. of (2.20) was estimated in lemma (B.3) of [**I1**], while the second can easily be shown to satisfy,

$$\begin{cases} |(\Phi_6'(u(t))|\partial_x A u(t))_0| \leq \alpha(\|\phi\|_2) + \beta(\|\phi\|_2)\left\|\partial_x^2 u\right\|_0 + \\ +\gamma(\|\phi\|_2)\left\|\partial_x^2 u\right\|_0^{3/2} + \theta(\|\phi\|_2)\left\|\partial_x^2 u\right\|_0^2 \end{cases} \tag{2.21}$$

where $\alpha, \beta, \gamma, \theta$ are non-negative non-decreasing functions of their arguments. The desired estimate then follows from (2.20), (2.21), lemma (B.3) of [**I1**], the formula of $\|u(t)\|_0^2$ in terms of $\Phi_6(u(t))$ and Gronwall's inequality. ∎

In order to keep this article reasonably short we will not pursue the question of continuity with respect to the initial data in detail. We would like to point out however that the equations under study can be treated from a unified point of view using Kato's theory of linear equations of "hyperbolic" type. Indeed, if u and v are solutions of (2.1) it is easy to verify that $\omega = u - v$ satisfies the linear equation

$$\partial_t \omega = iP(D)\omega + \frac{\alpha}{2}[(u+v)\partial_x \omega + (\partial_x u + \partial_x v)\omega] \tag{2.22}$$

One is therefore led to the analysis of the Cauchy problem

$$\partial_t \Theta = iP(D)\Theta + a(t)\partial_x \Theta + b(t)\Theta \quad , \quad \Theta(0) = \Theta_0 \tag{2.23}$$

where $a(t)$ and $b(t)$ are continuous functions of t with values in H^s and H^{s-1} respectively. Theorems I and IV of [**K3**] can then be applied (see [**I2**], [**I3**], [**K1**] and [**K4**]) to prove,

THEOREM (2.6). *Let $\phi \in H^s$, $s > 3/2$ and $u \in C([0,T]; H^s)$ be the corresponding solution of (2.1). Assume that $\phi_n \to \phi$ in H^s and let u_n be the solutions of (2.1) satisfying $u_n(0) = \phi_n$. Then given $T' \in (0,T)$, $u_n(t)$ exists in $[0,T']$ for all n sufficiently large and,*

$$\lim_{n\to\infty} \sup_{[0,T']} \|u_n(t) - u(t)\|_s = 0 \tag{2.24}$$

Moreover, if $s \geq 2$ and $(iP(D))$ stands for the linear part of KdV, BO or S, the solutions exist globally and (2.24) holds for all fixed $T' > 0$.

There are of course other possible approaches to the question dealt with in theorem (2.6). We refrain from further discussion and simply refer the reader to [**ABFS**], [**Al**] and [**BSc**].

§3. The Korteweg-de Vries equation in weighted Sobolev spaces

First of all we consider what happens in the case of the associated linear equation, i.e., (1.10) with $P(\xi) = \xi^3$. As pointed out in the introduction, if $\phi \in \mathcal{F}_{2,1}$ then $u(t) = \exp(-t\partial_x^3)$ belongs to $C(\mathbf{R}, \mathcal{F}_{2,1})$. More generally we have,

THEOREM (3.1). *Let $\phi \in \mathcal{F}_{2s,s}$, $s \geq 0$. Then $u(t) = \exp(-t\partial_x^3)\phi \in \mathcal{F}_{2s,s}$ for all $t \in \mathbf{R}$. In fact, $u \in C(\mathbf{R}, \mathcal{F}_{2s,s})$ is the unique solution of*

$$\begin{cases} u \in C(\mathbf{R}, \mathcal{F}_{2s,s}) \\ \partial_t u = -\partial_x^3 u \quad , \quad u(0) = \phi \end{cases} \tag{3.1}$$

where the derivative with respect to time is taken in the topology of H^{s-3}. In particular if $\phi \in \mathcal{S}(\mathbf{R})$ (the Schwartz space) then $u \in C(\mathbf{R}; \mathcal{S}(\mathbf{R}))$, where $\mathcal{S}(\mathbf{R})$ is endowed with its usual Frechet space topology.

Sketch of proof: Assume first that $s = k = 0, 1, 2, \cdots$. Then an easy induction argument shows that

$$(x^k \exp(-t\partial_x^3)\phi)^\wedge = i^k \exp(i\xi^3 t)(3it\xi^2 + \partial_\xi)^k \hat{\phi}(\xi) \tag{3.2}$$

Combining the inverse Fourier transform of (3.2) with theorem (A.7) of [**K1**], it is not difficult to conclude that $u \in C(\mathbf{R}, \mathcal{F}_{2k,k})$ and that $\|u(t)\|_{2k,k} \leq p_k(t) \, \|\phi\|_{2k,k}$ where $p_k(t)$ is a polynomial of degree k with positive coefficients. In particular $t \in \mathbf{R} \mapsto \exp(-t\partial_x^3)$ is a continuous function with values in $B(\mathcal{F}_{2k,k})$. An application of the Riesz-Thorin interpolation theorem concludes the proof. ∎

We are now in position to discuss the main result of this section, namely the following theorem due to T. Kato ([**K1**]),

Theorem (3.2). *Let $\phi \in \mathcal{F}_{2r,r}$, $r = 1, 2, 3 \cdots$. Then there exists a unique $u \in C([0,\infty); \mathcal{F}_{2r,r}) \cap C^1((0,\infty); H^{2r})$ solving (1.1). In particular $u \in C([0,\infty); \mathcal{S}(\mathbf{R}))$ if $\phi \in \mathcal{S}(\mathbf{R})$.*

Sketch of proof: Uniqueness is of course trivial in view of the H^s theory. In order to establish existence we again use parabolic regularization. Combining lemma (1.1) with arguments similar to those described in the preceeding theorem it is possible to show that the operator $E_\mu(t) = \exp(t(\mu\partial_x^2 - \partial_x^3))$ satisfies

$$\|E_\mu(t)\phi\|_{2r+\lambda,r} \leq G(t; \mu, \lambda, r) \, \|\phi\|_{2r,r} \tag{3.3}$$

for all $\phi \in \mathcal{F}_{2r,r}$, $t > 0$, $\lambda \geq 0$, where G is locally integrable with respect to t. Morever the map $t \in (0,\infty) \mapsto E_\mu(t)\phi$ is continuous in the topology of $\mathcal{F}_{2r+\lambda,r}$. These remarks together with Banach's fixed point theorem imply local existence for the regularized problem. In order to prove global existence, let $u = u_\mu \in C([0,T]; \mathcal{F}_{2r,r})$, $\mu > 0$, be the solution of the regularized equation in some interval $[0,T]$, $T > 0$, satisfying $u(0) = \phi$. Since $\mathcal{F}_{2r,r}$ is a Banach algebra it follows that $u^2 \in C([0,T]; \mathcal{F}_{2r,r})$. Next, using the exponential decay of the symbol of $E_\mu(t)$ and the integral equation satisfied by $u(t)$ it is not difficult to check that $\partial_x^k u \in C((0,T]; \mathcal{F}_{2r,r})$, $k = 0, 1, 2, \cdots$. Moreover, theorem (A.7) of [**K1**] implies that $\left\|[\omega, \partial_x^j] f\right\|_{0,0} \leq C(r,j) \, \|f\|_{2r,r}$, $j = 2, 3$, $f \in \mathcal{F}_{2r,r}$, where $\omega(x) = (1 + x^2)^{r/2}$. Taking into account the previous remarks we get,

$$\begin{cases} \partial_t \|u\|_{0,r}^2 = 2(u|\partial_t u)_{0,r} = -2\mu \|\partial_x \omega u\|_{0,0} - 2(\omega u|[\omega, \partial_x^2]u)_{0,0} - \\ 4(\omega u|\omega u \partial_x u)_{0,0} + (\omega u|[\omega, \partial_x^3]u)_{0,0} \leq \\ \leq C \|u\|_{2r,r}^2 + C' \|\partial_x u\|_{L^\infty} \|u\|_{2r,r}^2 \end{cases} \tag{3.4}$$

where we have used the fact that $(\partial_x \omega u|\partial_x^2 \omega u)_{0,0} = 0$. Since $\|\partial_x u\|_{L^\infty} \leq \|u\|_{2,0}$, (3.4) and lemma (2.2) imply that $\partial_t \|u\|_{0,r}^2 \leq \tilde{F}_r(t; \mu, \|\phi\|_{2r,r}) \, \|u\|_{2r,r}^2$ where $\tilde{F}_r \colon [0,\infty)^3 \to$

$[0,\infty)$ is a continuous function. The theorem then follows from the H^s theory and Gronwall's inequality. ∎

Some remarks are now in order. Since $u \in C([0,\infty);\mathcal{F}_{2r,r})$, theorem (A.7) of [**K1**] shows that $\omega^h \partial_x^k u \in C([0,\infty);L^2)$ for all $k = 0,1,\cdots,2r$, $h \in [0,1-k/2r]$ and $\|\omega^h \partial_x^k u(t)\|_{0,0} \le C\|u(t)\|_{2r,r}$. Continuous dependence can be proved by the method indicated at the end of last section in connection with the H^s theory. Furthermore both theorem (3.2) and the continuity with respect to the initial data can be extended for non-integer values of $r \ge 1$ by means of the Tartar-Bona-Scott interpolation theorem ([**BSc**]). Finally, it is not difficult to obtain the preceeding results with $[0,\infty)$ replaced by $(-\infty,0]$: one either repeats the previous arguments with $\mu\partial_x^2$ replaced by $(-\mu\partial_x^2)$ or uses the invariance of KdV under the change of variables $(t,x)\to(-t,-x)$.

§4. BO in weighted Sobolev spaces

In this section we will consider the differentiability – decay relationship for the Benjamin-Ono equation. As we shall see its behavior in weighted spaces is rather different from that of KdV. In fact the situation discussed below is far more complicated than the one considered in the preceeding section. We will concentrate mainly in what happens in spaces of integer order, although a very pleasing result in fractional order spaces will be sketched.

Let $\mu \ge 0$ be fixed and introduce

$$Q_\mu = -((\mu - 2\sigma)\partial_x^2) \tag{4.1}$$

$$F_\mu(t,\xi) = \exp(-t(\mu - 2ih(\xi))\xi^2) \tag{4.2}$$

$$E_\mu(t)f = (F_\mu(t,\ \cdot\ )\hat{f})^\vee \tag{4.3}$$

where $h(\xi)$ is defined in (1.7). It follows at once that if $s \in \mathbf{R}$ and $\phi \in H^s$ the function

$$u(t) = E_\mu(t)\phi = (F_\mu(t,\ \cdot\ )\hat{\phi})^\vee \tag{4.4}$$

is the unique solution of the Cauchy problem for the *linear μ-BO equation*, namely

$$\partial_t u = \mu\partial_x^2 u - 2\sigma\partial_x^2 u \quad , \quad u(0) = \phi \tag{4.5}$$

In fact $t \in [0,\infty) \mapsto E_\mu(t)$ has all the properties stated in lemma (1.1) (take $P(\xi) = -2h(\xi)\xi^2 = -2\xi|\xi|$).

Suppose that $\phi \in L_1^2$. We would like to determine whether or not the solution $u(t)$ of (4.5) remains in this space. The answer is of course no in general. As in the case of KdV differentiability condition must be satisfied in order to insure $u(t) \in L_1^2$ for all $t \in [0,\infty)$. Indeed, from the formulas

$$\partial_\xi \hat{u}(t,\xi) = (\partial_\xi F_\mu(t,\xi))\hat{\phi} + F_\mu(t,\xi)\partial_\xi\hat{\phi} \tag{4.6}$$

$$\partial_\xi F_\mu(t,\xi) = (-2t\xi)(\mu - 2ih(\xi))F_\mu(t,\xi) \tag{4.7}$$

it is easy to conclude that $\partial_\xi \hat{u} \in L^2$ if and only if $\xi\hat{\phi} \in L^2$ or, equivalently, if and only if $\phi \in H^1$. This argument also shows that if $\phi \in H^1 \cap L_1^2 = \mathcal{F}_{1,1}$ then $u \in C([0,\infty);\mathcal{F}_{1,1})$. A similiar result holds for L_2^2. Differentiating (4.7) with respect to ξ we obtain,

$$\begin{cases} \partial_\xi^2 F_\mu(t,\xi) & = -2t(\mu - 2ih(\xi))F_\mu(t,\xi) + \\ & +(-2t\xi)^2(\mu - 2ih(\xi))^2 F_\mu(t,\xi) \end{cases} \tag{4.8}$$

Combining (4.8) with Leibniz rule it follows that if $\phi \in H^2 \cap L_2^2 = \mathcal{F}_{2,2}$ then $u \in C([0,\infty);\mathcal{F}_{2,2})$. Next we turn to L_3^2 which is where the trouble starts. Note that

$$\begin{cases} \partial_\xi^3 F_\mu(t,\xi) = 4it\delta + 3(-2t)^2\xi^2(\mu - 2ih(\xi))^2 F_\mu(t,\xi) + \\ +(-2t)^3\xi^3(\mu - 2ih(\xi))^3 F_\mu(t,\xi) \end{cases} \tag{4.9}$$

and we are left with a δ function that must be eliminated. In view of Leibniz rule this is possible if and only if $\hat{\phi}(0) = 0$. The next derivative contains δ' which requires that $\hat{\phi}(0) = \partial_\xi\hat{\phi}(0) = 0$. With these comments in mind, an easy induction argument implies,

THEOREM (4.1). *Let $r \in \mathbf{R}$ and $\phi \in \mathcal{F}_{r,r}$. Then for each fixed $\mu \geq 0$ we have,*
a) $E_\mu(t)\phi \in C([0,\infty);\mathcal{F}_{r,r})$, $r = 0,1,2$ and satisfies,

$$\|E_\mu(t)\phi\|_{r,r} \leq \Theta_\mu(t)\,\|\phi\|_{r,r} \tag{4.10}$$

where $\Theta_\mu(t)$ is a polynomial of degree r with positive coefficients depending only on r and on μ;
b) if $r \geq 3$ the function $t \mapsto E_\mu(t)\phi$ belongs to $C([0,\infty);\mathcal{F}_{r,r})$ if and only if

$$\partial_\xi^j\hat{\phi}(0) = 0 \quad , \quad j = 0,1,\cdots,r-3 \tag{4.11}$$

In this case an estimate of the form (4.10) also holds.

In view of theorem (4.1) and our previous experience with KdV we are led to the following conjectures,

CONJECTURE (4.2). *Let $\phi \in \mathcal{F}_{2,2}$. Then the Cauchy problem*

$$\partial_t u = \mu\partial_x^2 u - \partial_x(u^2 + 2\sigma\partial_x u) \quad , \quad u(0) = \phi \tag{4.12}$$

has a unique solution $u \in C([0,\infty);\mathcal{F}_{2,2})$.

CONJECTURE (4.3). *Let $r \geq 0$ and*

$$\tilde{\mathcal{F}}_r = \{\psi \in \mathcal{F}_{r,r} | \partial_\xi^j\hat{\psi}(0) = 0 \quad , \quad j = 0,1,\cdots,r-3\} \tag{4.13}$$

Then if $\phi \in \tilde{\mathcal{F}}_r$, (4.12) has a unique solution $u \in C([0,\infty);\tilde{\mathcal{F}}_r)$.

The first of these two statements does hold and is a special case of theorem (4.7). Before sketching its proof however, we will consider the second conjecture which leads to some interesting consequences. It is convenient to start with,

LEMMA (4.4). *Assume that* $u \in C([0,T];\mathcal{F}_{3,3})$, $T > 0$, *is a solution of (4.12) with* $\mu \geq 0$. *Then* $\hat{u}(t,0) = 0$ *for all* $t \in [0,T]$, *that is,* u *necessarily belongs to* $C([0,T];\tilde{\mathcal{F}}_3)$.

PROOF: We will restrict ourselves to the most important case, namely $\mu = 0$. The general case can be handled in essentially the same way with a little extra effort. Multiply BO by x^3 to get,

$$\partial_t(x^3 u) = -2x^3 u\partial_x u - 2x^3\sigma\partial_x^2 u \tag{4.14}$$

Since, by assumption, $x^3 u(t) \in L^2$ we have,

$$\begin{cases} \left\|x^3 u\partial_x u\right\|_0 \leq \left\|\partial_x u\right\|_{L^\infty}\left\|x^3 u\right\|_0 \leq \\ \leq \left\|u\right\|_2\left\|x^3 u\right\|_0 \leq F_2(t;0;\left\|\phi\right\|_2)\left\|x^3 u\right\|_0. \end{cases} \tag{4.15}$$

where $F_2\colon [0,\infty)^3 \to [0,\infty)$ is the function mentioned in lemma (2.2). It follows that $\gamma(t) = (x^3 u\partial_x u)(t) \in L^2, t \in [0,T]$. A similar argument shows that $\gamma \in C([0,T];L^2)$. Taking the Fourier transform of (4.14) we obtain,

$$\partial_t(\partial_\xi^3 \hat{u}) = 2i(\gamma(t))^\wedge + 2i\partial_\xi^3(h(\xi)\xi^2\hat{u}) \tag{4.16}$$

which, in view of the assumptions on u, theorem (A.2) of appendix (A.1) of [**I1**] and the fact that $\xi^k\partial_\xi^j\delta = 0$, $k, j \in \mathbf{Z}^+$, $k \geq j$, can be rewritten in the form

$$\partial_t(\partial_\xi^3 \hat{u}) = 2i(\gamma(t))^\wedge + 2i(\Gamma(t))^\wedge + 8i\delta\hat{u}(t,0) \tag{4.17}$$

where $\Gamma(t)^\wedge = 2i[2h(\xi)\partial_\xi + 2i\partial_\xi(4\xi h(\xi)\partial_\xi + \xi^2 h(\xi)\partial_\xi^2)]\hat{u} \in C([0,T];L^2_{-2})$. Integrating (4.17) with respect to t we conclude that $\delta(\xi)\int_0^t \hat{u}(t',0)dt' \in C([0,T],L^2_{-2}(\mathbf{R}))$. The lemma then follows since the elements of L^2_{-2} are measurable functions. ∎

Now, a proof along the lines described in the previous sections (i.e., parabolic regularization, commutator estimates and so on), establishes the following result.

THEOREM (4.5). *Let* $\phi \in \tilde{\mathcal{F}}_3$. *Then there exists a unique* $u = u_\mu \in C([0,\infty);\tilde{\mathcal{F}}_3)$, $\mu \geq 0$ *which solves (4.12).*

The previous result seems to confirm conjecture (4.3). The situation is not so simple however. Indeed, $\tilde{\mathcal{F}}_3$ is a very natural place in which to look for solutions since, as can easily be checked, the quantity $\hat{u}(t,0) = (2\pi)^{-1/2}\int_{\mathbf{R}} u(t,x)dx$ is conserved by the BO flow. This does not hold in the case of $\partial_\xi\hat{u}(t,0) = -i(2\pi)^{-1/2}\int_{\mathbf{R}} xu(t,x)dx$. In fact $\partial_\xi\hat{u}(t,0)$ can be computed explicit (as already pointed out in [**O**]) and this computation leads to

THEOREM (4.6). *Let* $u = u_\mu \in C([0,T];\mathcal{F}_{4,4})$, $T > 0$, $\mu \geq 0$, *be a solution of the PDE in (4.12). Then* $u(t) = 0$ *for all* $t \in [0,T]$.

PROOF: Once again we consider only the case of BO itself, i.e., $\mu = 0$. The general case can be treated similarly. Let $f(t) = i(2\pi)^{1/2}\partial_\xi\hat{u}(t,0)$. Then differentiating under the integral sign and integrating by parts we obtain,

$$\partial_t f(t) = \int_{\mathbf{R}} x\partial_t u(t,x)dx = \int_{\mathbf{R}}(u^2 + 2\sigma\partial_x u)dx = \|u(t)\|_0^2 \tag{4.18}$$

where we have used the fact that the integral of $(\partial_x \sigma u)$ over $\mathbf{R}$ is zero. Since $\|u(t)\|_0^2$ is a conserved quantity it follows that

$$\partial_\xi \hat{u}(t,0) = \partial_\xi \hat{\phi}(0) - i(2\pi)^{-1/2} \|\phi\|_0^2 \, t \tag{4.19}$$

Proceeding as in lemma (4.4) we conclude that $u \in C([0,T]; \tilde{\mathcal{F}}_4)$. Combining this with (4.19) we obtain $\|\phi\|_0 = 0$ and the proof is complete. ∎

The follwing result takes care of conjecture (4.2). We have,

THEOREM (4.7). *Let $\phi \in Y_\gamma = H^2 \cap L^2_{2\gamma}$, $\gamma \in [0,1]$. Then there exists a unique solution $u = u_\mu \in C([0,\infty); Y_\gamma)$ of (4.12) for each fixed $\mu \geq 0$. Moreover*

$$\begin{cases} \omega_\gamma' \partial_x u \in C([0,\infty)); L^2) \\ \left\|\omega_\gamma' \partial_x u\right\|_{0,0} \leq C(\gamma,\mu) \|u\|_{Y_\gamma} \end{cases} \tag{4.20}$$

where $\omega_\gamma(x) = (1+x^2)^\gamma$ and the map $\phi \mapsto u$ is continuous in the sense described in theorem (2.6) with H^s replaced by Y_γ.

SKETCH OF PROOF: The proof is similar to that of theorem (3.2). One starts out studying the properties of the map $t \in [0,\infty) \mapsto E_\mu(t)\phi$, $\mu > 0$, $\phi \in H^{s+\lambda} \cap L^2_{2\gamma}$, $s > 3/2$, $\lambda \geq 0$, $\gamma \in [0,1]$, in order to obtain the analogue of (3.3) in the present case. Local existence for $\mu > 0$ then follows from Banach's fixed point theorem. The next step is to estimate $\partial_t \|u(t)\|_{Y_\gamma}^2$ and apply Gronwall's inequality. The crucial technical point is to show that $[\omega_\gamma, \sigma\partial_x^2] \in B(Y_\gamma, L^2)$. This can be done as follows. Note that,

$$[\omega_\gamma, \sigma\partial_x^2] = [\omega_\gamma, \sigma]\partial_x^2 - \sigma[\omega_\gamma, \partial_x^2]. \tag{4.21}$$

A routine computation combined with the unitarity of σ and the easily verifiable fact that $\omega_\gamma' \partial_x f \in L^2$ if $f \in Y_\gamma$ (integrate $(\omega_\gamma')^2(\partial_x f)(\overline{\partial_x f})$ by parts), takes care of the second term on the R.H.S. of (4.21). In order to handle the other one, integrate by parts to obtain,

$$[\omega_\gamma, \sigma]\partial_x^2 f = \frac{1}{\pi}\int_{\mathbf{R}} \frac{\omega_\gamma'(y) - \omega_\gamma'(x)}{y-x} \partial_y f \, dy + \sigma(\omega_\gamma' \partial_x f) \tag{4.22}$$

Since $\omega_\gamma' \partial_x f \in L^2$ the desired estimate follows from Calderon's theorem on the first commutator, as generalized by Coiffman, McIntosh and Meyer ([**CMcM**]; see also [**T**]), if $\gamma \in [0,1/2]$. Another integration by parts shows that $[\omega_\gamma, \sigma]\partial_x^2$ can be written as a Hilbert-Schmidt operator if $\gamma \in [1/2,1]$. A limiting argument concludes the proof of global existence. Continuity with respect to the initial data can be proved along the lines suggested in the end of section 2. ∎

It should be noted that the use of Calderon's theorem in the preceeding proof can be avoided: in [**I2**] we present an elementary argument due to J. Hounie ([**Ho**]). However, apart from providing a satisfyingly short proof of the required estimate, this powerful theorem can be used to obtain results in situations which are far more general than the one considered here. For example, the preceeding theorem holds with $L^2_{2\gamma}$ replaced by any $L^2(\mathbf{R}, \rho(x)^2dx)$ where ρ is Lipschtzian and $\rho' \in L^\infty(\mathbf{R})$, ([**I3**]).

Finally we would like to remark that the results of appendix A of [**I1**] imply that the solution constructed in theorem (4.5) satisfies,

$$x\partial_x u, x^2\partial_x u, x\partial_x^2 u \in C([0,\infty); L^2) \tag{4.23}$$

Moreover these three quantities can be estimated by the $\mathcal{F}_{3,3}$ norm of u. The corresponding property in $Y_\gamma = H^2 \cap L^2_{2\gamma}$, i.e., $\omega'_\gamma \partial_x u \in C([0,\infty), Y_\gamma)$ has already been mentioned in theorem (4.7).

§5. Smith's equation in weighted Sobolev spaces

In this section we consider what happens in the case of S, which in a sense lies somewhere in between KdV and BO. A little more precisely, if $P(\xi) = -2\xi(\sqrt{\xi^2+1}-1)$ is the symbol of the linear part of S then $P'(\xi)$ behaves like $(\mp 4\xi)$ as $\xi \to \pm\infty$ so one is led to expect that the appropriate spaces for S are the same as those used in the case of BO, namely $\mathcal{F}_{r,r} = H^r \cap L^2_r$, $r \geq 2$. On the other hand, since $P(\xi)$ is infinitely differentiable, no restrictions on the Fourier transforms should occur. In particular, global existence must hold in $\mathcal{F}_{r,r}$ for all $r \geq 2$. The following results show that these conjectures are indeed true. Let Q_μ, $E_\mu(t)$, $\mu \geq 0$ denote the operators defined in lemma (1.1) in the case of S.

THEOREM (5.1). *Let $\phi \in \mathcal{F}_{s,s}$, $s \geq 0$. Then $u(t) = E_\mu(t)\phi \in C(\mathbf{R}; \mathcal{F}_{s,s})$. Moreover, if $\mu > 0$ then $u(t) = E_\mu(t)\phi \in C((0,\infty); \mathcal{F}_{s+\lambda,s})$ for all $\lambda \geq 0$ and satisfies*

$$\|E_\mu(t)\phi\|_{s+\lambda,s} \leq G(t;\mu,\lambda,s)\,\|\phi\|_{s,s} \tag{5.1}$$

where G is locally integrable with respect to t.

SKETCH OF PROOF: In view of lemma (1.1) it is enough to estimate the L^2_s norm of $E_\mu(t)\phi$. If s is a non-negative integer the result follows from Leibniz' rule combined with

$$\begin{cases} Q_\mu(\xi) = \mu\xi^2 - iP(\xi) \\ \partial_\xi^j(\exp(-tQ_\mu(\xi))) = \exp(-tQ_\mu(\xi))(-tQ'_\mu(\xi) + \partial_\xi)^j 1, \; j \in \mathbf{N} \end{cases} \tag{5.2}$$

and theorem (A.2) of appendix A of [**I1**] (where $\mathcal{F}_{r,r} = \mathcal{F}_r$) which states that if $f \in \mathcal{F}_{r,r}$ then $x^k\partial_x^\ell f \in L^2$ and $\|x^k\partial_x^\ell f\|_{0,0} \leq C(k,\ell)\,\|f\|_{r,r}$, $0 \leq k+\ell \leq s$. The general case follows from the previous one and the Riesz-Thorin interpolation theorem ([**RS**], vol. II). ∎

Before proceeding it is worthwhile to note that if $s = r$ is a non-negative integer then due to theorem (A.1) of [**I1**] we have $x^k\partial_x^\ell E_\mu(t)\phi \in C(\mathbf{R}; L^2)$ and $\|x^k\partial_x^\ell E_\mu(t)\phi\|_{0,0} \leq C(k,\ell)\,\|E_\mu(t)\phi\|_{r,r}$. Next,

THEOREM (5.2). *Let $\phi \in \mathcal{F}_{s,s}$, $s \geq 2$. Then there exists a unique $u = u_\mu \in C([0,\infty);\mathcal{F}_{s,s})$, $\mu \geq 0$, such that $\partial_t u_\mu \in C([0,\infty);H^{s-2})$ and*

$$\partial_t u = \mu\partial_x^2 u - iP(D)u - \partial_x u^2 \quad , \quad u(0) = \phi \tag{5.3}$$

Moreover, the map $\phi \mapsto u$ is continuous in the sense described in theorem (2.6) with H^s replaced by $\mathcal{F}_{s,s}$ and if $s = r$ is a non-negative integer we have,

$$\begin{cases} x^k\partial_x^\ell u \in C([0,\infty);L^2) \quad , \quad 0 \leq k+\ell \leq r \\ \|x^k\partial_x^\ell u\|_{0,0} \leq C(k,\ell;\mu)\,\|u\|_{r,r} \end{cases} \tag{5.4}$$

SKETCH OF PROOF: The first step in the proof is to show that the result holds in case $s \geq 2$ is an even integer. The general case follows from the non-linear interpolation theorem of Tartar, Bona and Scott ([**BSc**]). The key step in the first part is to show that the commutator $[\omega, P(D)]$ belongs to $B(\mathcal{F}_{s,s};L^2)$ where $\omega(x) = (1+x^2)^{s/2}$. This can be done by taking the Fourier transform and applying theorem (A.1) of [**I1**] to obtain the desired estimates. ∎

§6. Some final comments

As remarked in the introduction, many other situation can be covered by the methods described in this work. For example, the results of [**Ts**] on the nonlinear Schrödinger equation (NLS) can be recovered without much difficulty. Many other equations remain to be treated. The intermediate long wave equation (ILW; see [**ABFS**] and the references therein) has yet to be studied in spaces of the type $\mathcal{F}_{s,r}$. Moreover, smoothing properties for KdV, BO and NLS have been obtained in [**P1**] and [**P2**] assuming that the initial data decays sufficiently fast at infinity. Another interesting question is what happens in spaces equipped with non-spherically-symmetric weights. As far as we know, this problem remains open, except for KdV which was treated in [**BS**], [**K1**] (sections 10 and 11) and [**KrF**]. In view of the smoothing effects established in these papers, such a study could prove to be quite profitable. One would also like to know if the "non-existence" result (theorem (4.6)) holds if the solution decays fast enough in only one direction.

AKNOWLEDGEMENTS: We would like to thank Professors S.T. Kuroda, T. Ikebe and K. Yajima for their gracious hospitality during our visit to Japan. We would also like to express our deep gratitude to Professor T. Kato who, through his constant advice and friendship, is the person ultimately responsible for our participation in the conference in his honor.

REFERENCES

[ABFS] L. Abdelouhab, J. Bona, M. Felland and J. Saut, "*Non- local models for nonlinear, dispersive waves*", preprint 1989. To appear in Physica D.

[Al] E. A. Alarcon, "*The Cauchy problem for the generalised Ott-Sudan equation*", Doctoral Thesis, IMPA, 1990.

[BN] P. Butzer and R. Nessel, "*Fourier Analysis and Approximation vol. 1. One Dimensional Theory*" Birkhauser Verlag, (1971).

[BS] J. Bona and J. Saut, "*Singularités dispersives de solutions d'equations de type Korteweg-de Vries*", C.R. Acad. Sc. Paris, t. 303, Série I, no.4, (1986), 101-103.

[BSc] J. Bona and R. Scott, "*Solutions of the Korteweg-de Vries equation in fractional order Sobolev spaces*", Duke Mathematical Journal, vol. 43, no.1, (1976), 87-99.

[BSm] J. Bona and R. Smith, "*The initial value problem for the Korteweg-de Vries equation*", Philos. Trans. Roy. Soc. London, Ser. A 278, (1975), 555-601.

[C] K. M. Case, "*Benjamin-Ono related equations and their solutions*", Proc. Nat. Acad. Sci. U.S.A, vol. 76, no.1,(1979), 1-3.

[CMcM] R. Coiffman, A. McIntosh and Y. Meyer, "*L'integrale de Cauchy sur les courbes lipschtziennes*", Ann. of Math, 116, (1982), 361-387.

[H] Dan Henry "*Geometric theory of semilinear parabolic equations*", Lecture notes in Mathematics 840, Springer Verlag, (1981).

[Ho] J. Hounie, "*Remarks on the Benjamin-Ono equation*", private communication, (1989).

[I1] R. J. Iorio, Jr., "*On the Cauchy problem for the Benjamin-Ono equation*" Commun. in Partial Differential Equations, 11(10), (1986), 1031-1081.

[I2] R. J. Iório, Jr., "*The Benjamin-Ono equation in weighted Sobolev spaces*", preprint IMPA, (1989). Submitted for publication in Journal of Mathematical Analysis and Applications.

[I3] R. J. Iório, Jr., "*On the relationship between differentiability and spacial decay satisfied by the solutions of certain non- linear evolution equations*", in preparation.

[K1] T. Kato, "*On the Cauchy problem for the (generalized) Korteweg-de Vries equations*", Studies in Applied Mathematics, Advances in Mathematics Supplementary Studies, vol. 8, Academic Press, (1983), 93-128.

[K2] T. Kato, "*Weak solutions of infinite-dimensional Hamiltonian systems*", preprint U. C. Berkeley, (1989).

[K3] T. Kato, "*Linear evolution equations of "hyperbolic" type, II*", J. Math. Soc. Japan, vol. 25, no.4, (1973), 648-666.

[K4] T. Kato, "*Quasi-linear equations of evolution with application to partial differential equations*", Spectral Theory and Differential Equations, Lecture Notes in Mathematics, 448, Springer Verlag, (1975), 25-70.

[K5] T. Kato, "*Linear and quasi-linear equations of evolution of hyperbolic type*", Hyperbolicity, C.I.M.E. II CICLO, (1976), 125-191.

[K6] T. Kato, "*Non-linear equations of evolution in Banach spaces*", Proc. Sympos. Pure Math., vol. 45, Part 2, A.M.S., (1986), 9-23.

[KF] T. Kato and H. Fujita, "*On the non-stationary Navier- Stokes system*", Rend. Sem. Mat. Univ. Padova, vol. 32, (1962), 243-260.

[KrF] S. Kruzhkov and A. Faminskii, "*Generalized solutions of the Cauchy problem for the Korteweg-de Vries equation*", Math. USSR Sbornik, vol. 48, no.2, (1984), 391-421.

[L] P. Lax, "*A Hamiltonian approach to the KdV and other equations*", Nonlinear Evolution Equations, Ed. M. G. Crandall, Academic Press, (1978).

[O] H. Ono, "*Algebraic solitary waves in stratified fluids*", J. Phys. Soc. Japan, vol. 39, no.4, (1975), 1082-1091.

[P1] G. Ponce, "*Regularity of solutions to nonlinear dispersive equations*", preprint U. of Chicago, (1987).

[P2] G. Ponce, "*Smoothing properties of solutions to the Benjamin-Ono equation*", preprint, U. of Chicago, (1988).

[RS] M. Reed and B. Simon, "*Methods of Modern Mathematical Physics*", vol. I and II, Academic Press, (1972, 1975).

[T] A. Torchinsky, "*Real variable methods in harmonic analysis*", Academic Press (1986).

[Ts] M. Tsutsumi, "*Weighted Sobolev spaces and rapidly decreasing solutions of some nonlinear dispersive wave equations*", J. Diff. Equations 42, (1981), 260-281.

THE SQUARE ROOT PROBLEM FOR ELLIPTIC OPERATORS

A SURVEY

Alan McIntosh

School of Mathematics, Physics, Computing and Electronics
Macquarie University, N.S.W. 2109
AUSTRALIA

Consider an elliptic sesquilinear form defined on $\mathcal{V} \times \mathcal{V}$ by

$$J[u,v] = \int_\Omega \left\{ \sum a_{jk} \frac{\partial u}{\partial x_k} \overline{\frac{\partial v}{\partial x_j}} + \sum a_k \frac{\partial u}{\partial x_k} \overline{v} + \sum \alpha_j u \overline{\frac{\partial v}{\partial x_j}} + a u \overline{v} \right\} dx$$

where $\mathcal{V}$ is a closed linear subspace of the Sobolev space $H^1(\Omega)$ which contains $C_c^\infty(\Omega)$, Ω is an open subset of $\mathbb{R}^n$, a_{jk}, a_k, α_j, $a \in L_\infty(\Omega)$ and $Re \sum a_{jk}(x)\zeta_k \overline{\zeta_j} \geq \kappa|\zeta|^2$ for all $\zeta = (\zeta_j) \in \mathbb{C}^n$ and some $\kappa > 0$. Let A be the operator in $L_2(\Omega)$ with largest domain $\mathcal{D}(A) \subset \mathcal{V}$ such that $J[u,v] = (Au, v)$ for all $u \in \mathcal{D}(A)$ and all $v \in \mathcal{V}$. Then $A + \lambda I$ is a maximal accretive operator in $L_2(\Omega)$ for some positive number λ and so has a maximal accretive square root $(A + \lambda I)^{\frac{1}{2}}$. The problem of determining whether its domain $\mathcal{D}((A + \lambda I)^{\frac{1}{2}})$ is equal to $\mathcal{V}$, possibly for particular choices of Ω and $\mathcal{V}$, has become known as the square root problem of Kato for elliptic operators. It seems to be a more difficult problem now than when posed by Kato almost 30 years ago.

1. Fractional Powers

A linear operator in a complex Banach space is of *type* ω for some $\omega < \pi$ if, for every positive real number t, $(A+tI)$ is invertible, and $\|t(A+tI)^{-1}\|$ is uniformly bounded in t. Fractional powers A^α of such operators have been defined for $0 < \alpha \leq 1$, and studied by many people including R. Phillips, V. Balakrishnan, K. Yoshida, M. Krasnosel'skii, P. Sobolevskii, T. Kato, H. Tanabe, M. Watanabe, H. Komatsu, A. Yagi and many others. They are uniquely defined operators of type $\alpha\omega$ which satisfy $A^\alpha A^\beta = A^{\alpha+\beta}$. In particular, $(A^{\frac{1}{2}})^2 = A$.

These operators satisfy many formulae which are valid for complex numbers in the cut complex plane $\{z \in \mathbb{C} : z \notin (-\infty,0)\}$, such as

$$A^{\frac{1}{2}}u = \frac{2}{\pi}\int_0^\infty (I + t^2A)^{-1}A\,u\,dt$$

for all u in the domain of A [K,1976], [T,1979]. Many of the earlier papers were concerned with proving the equality of operators defined by different such formulae, but it is now easier to see them in terms of a functional calculus.

Professor Tosio Kato has played an influential role in the development and application of this theory, though of course it represents but a minor part of his remarkable contributions in many areas of mathematics. He has been particularly motivated by problems arising from the study of hyperbolic and parabolic evolution equations associated with an elliptic partial differential operator A.

I shall concentrate on elliptic operators, but shall not discuss the applications to evolution equations or to other problems in this survey. There are also versions of the square root problems which I shall not discuss, concerning, for example, elliptic operators in non-divergence form, and operators in $L_p(\Omega)$. But we cannot do everything, so I apologize to those people whose results I did not include.

I would like to thank the organizers of this very successful conference for inviting me to participate, and for the opportunity it provided of a very enjoyable visit to Japan.

I would also like to take this opportunity to acknowledge my great debt to Tosio Kato for the profound influence which he has had on my mathematical development.

2. Maximal accretive operators

Let us restrict our attention to fractional powers of maximal accretive operators in a complex Hilbert space $\mathcal{H}$. These were first studied in depth by Kato in the important paper "Fractional powers of dissipative operators" [K,1961] which appeared in the Journal of the Mathematical Society of Japan in 1961. (A dissipative operator is the negative of an accretive one.) To say that A is a *maximal accretive operator in* $\mathcal{H}$ means that A is a linear mapping from a dense linear subspace $\mathcal{D}(A)$ of $\mathcal{H}$ to $\mathcal{H}$ with the properties that $Re(Au,u) \geq 0$ for all u in $\mathcal{D}(A)$ and that $(A+tI)$ is surjective for positive numbers t.

In the paper [K,1961], Kato proved that the adjoint A^* of every maximal accretive operator A in $\mathcal{H}$ is also maximal accretive, as are the fractional powers

A^α and $A^{*\alpha}$ for $0 < \alpha \le 1$, and moreover $(A^\alpha)^* = A^{*\alpha}$. He also showed that A^α and $A^{*\alpha}$ have the same domains when $0 < \alpha < \frac{1}{2}$, but that A^α and $A^{*\alpha}$ do not necessarily have the same domains when $\frac{1}{2} < \alpha \le 1$. The case $\alpha = \frac{1}{2}$ was left open.

The following year J. L. Lions showed that the domains $\mathcal{D}(A^\alpha)$ form a complex interpolation family. That is, $\mathcal{D}(A^\alpha) = [\mathcal{H}, \mathcal{D}(A)]_\alpha$ whenever A is a maximal accretive operator in $\mathcal{H}$. He then considered the example when $A = \frac{d}{dx}$ in the Hilbert space $\mathcal{H} = L_2[0,\infty)$, with domain

$$\mathcal{D}(A) = \mathring{H}^1[0,\infty) = \{ u \in H^1[0,\infty) : u(0) = 0 \}$$

where $H^1[0,\infty)$ is the Sobolev space

$$H^1[0,\infty) = \{ u \in L_2[0,\infty) : \frac{du}{dx} \in L_2[0,\infty) \}$$

in which case $A^* = -\frac{d}{dx}$ with $\mathcal{D}(A^*) = H^1[0,\infty)$. Then

$$\mathcal{D}(A^\alpha) = [L_2[0,\infty),\mathring{H}^1[0,\infty)]_\alpha = \mathring{H}^\alpha[0,\infty) \text{ and } \mathcal{D}(A^{*\alpha}) = [L_2[0,\infty),H^1[0,\infty)]_\alpha = H^\alpha[0,\infty)$$

and, when $0 < \alpha < \frac{1}{2}$,

$$\mathcal{D}(A^\alpha) = \mathring{H}^\alpha[0,\infty) = H^\alpha[0,\infty) = \mathcal{D}(A^{*\alpha}) ,$$

in agreement with the result of Kato. But it was already known to Lions that $\mathring{H}^\alpha[0,\infty) \ne H^\alpha[0,\infty)$ when $\alpha \ge \frac{1}{2}$. In this way Lions showed the existence of a maximal accretive operator A for which $\mathcal{D}(A^{\frac{1}{2}}) \ne \mathcal{D}(A^{*\frac{1}{2}})$.

3. Regularly accretive operators

A *sesquilinear form* J in a Hilbert space $\mathcal{H}$ is a complex valued function J defined on $\mathcal{V}_J \times \mathcal{V}_J$, where $\mathcal{V}_J$ is a linear subspace of $\mathcal{H}$, such that J is linear in the first variable and conjugate linear in the second variable. The operator A_J *associated with* J is the operator in $\mathcal{H}$ with largest domain $\mathcal{D}(A_J) \subset \mathcal{V}_J$ such that $J[u,v] = (A_J u, v)$ for all $u \in \mathcal{D}(A_J)$ and all $v \in \mathcal{V}_J$.

We say that J is a *regular* sesquilinear form provided

(i) there exists $\omega \in [0,\frac{1}{2}\pi)$ such that $|Im J[u,u]| \le \tan\omega \, Re J[u,u]$ for all $u \in \mathcal{V}_J$, and

(ii) $\mathcal{V}_J$ is a dense linear subspace of $\mathcal{H}$ which is complete under the norm

$$\|u\|_J = \sqrt{\|u\|^2 + Re J[u,u]} .$$

The operator A_J associated with a regular sesquilinear form J is called a *regularly accretive operator*. Such an operator A_J is a maximal accretive operator with numerical range in the closed sector $S_\omega = \{z \in \mathbb{C} : |\arg(z)| \leq \omega\}$.

When $\omega = 0$ we call J a closed non-negative hermitian form, in which case the associated operator A_J is a non-negative self-adjoint operator.

Kato developed the theory of regular forms and regularly accretive operators in the paper [K,1961] and answered some further questions the following year [K,1962]. In particular he showed that if J is a closed non-negative hermitian form with domain $\mathcal{V}_J \times \mathcal{V}_J$, then $\mathcal{D}(A_J^{\frac{1}{2}}) = \mathcal{V}_J$ and

$$J[u,v] = (A_J^{\frac{1}{2}}u\,,\,A_J^{\frac{1}{2}}v)$$

for all $u,v \in \mathcal{V}_J$.

He also considered a family of regular forms J_t all with the same domain $\mathcal{V}_J \times \mathcal{V}_J$, and with associated operators A_t, such that $J_t[u,v]$ is a holomorphic function of t for all $u,v \in \mathcal{V}_J$, and showed that, provided $0 < \alpha < \frac{1}{2}$, then $\mathcal{D}(A_t^\alpha)$ is independent of t and $A_t^\alpha u$ depends holomorphically on t for all $u \in \mathcal{D}(A_t^\alpha)$.

In the paper [K,1961] Kato made two remarks, which I record here (with some minor notational changes). Note that the *real part of* A_J is the self-adjoint operator H associated with the non-negative hermitian form $ReJ[u,v] = \frac{1}{2}\{\,J[u,v] + \overline{J[v,u]}\,\}$.

REMARK 1. We do not know whether or not $\mathcal{D}(A^{\frac{1}{2}}) = \mathcal{D}(A^{*\frac{1}{2}})$ (*where A is a maximal accretive operator*). This is perhaps not true in general. But the question is open even when A is regularly accretive. In this case it appears reasonable to suppose that both $\mathcal{D}(A^{\frac{1}{2}})$ and $\mathcal{D}(A^{*\frac{1}{2}})$ coincide with $\mathcal{D}(H^{\frac{1}{2}}) = \mathcal{V}_J$, where H is the real part of A and J is the regular sesquilinear form which defines A. But all that we know are $\mathcal{V}_J \supset \mathcal{D}(A) \subset \mathcal{D}(A^{\frac{1}{2}}) \supset \mathcal{D}(P)$ (*where P is the real part of $A^{\frac{1}{2}}$*) and a similar chain of inclusions with A replaced by A^*.

REMARK 2. If $A = H$ is self-adjoint, the question raised above is answered in the affirmative, for we have $\mathcal{V}_J = \mathcal{D}(H^{\frac{1}{2}})$. The question is still open, however, whether or not ["J_t is holomorphic in t" implies "A_t^α is holomorphic in t"] is true with $\alpha = \frac{1}{2}$ when A_t are self-adjoint for real t, although it is true that $\mathcal{D}(A_t^{\frac{1}{2}})$ is independent of t as long as t is real. Thus it must be stated that our knowledge is quite unsatisfactory regarding the case $\alpha = \frac{1}{2}$.

Let me remark on remark 1. We have already seen that "$\mathcal{D}(A^{\frac{1}{2}}) = \mathcal{D}(A^{*\frac{1}{2}})$ ... is ... not true in general". Concerning the case when A_J is a regularly accretive operator, we note that if any two of the sets $\mathcal{D}(A_J^{\frac{1}{2}})$, $\mathcal{D}(A_J^{*\frac{1}{2}})$ and $\mathcal{V}_J$ are equal, then all three are equal, and $J[u,v] = (A_J^{\frac{1}{2}}u, A_J^{*\frac{1}{2}}v)$ for all $u,v \in \mathcal{V}_J$.

Although the questions raised in remarks 1 and 2 remained open for some time, it is actually not too hard to construct counter-examples to them, as we shall now see.

4. Counter-examples

A counter-example to the question raised in remark 1 was presented in [Mc,1972], while a counter-example to the question raised in remark 2 was presented in [Mc,1982]. Let us do things slightly differently here, following closely the treatment of related questions in [McY,1989].

I shall first define some matrices which will be useful in constructing the counter-examples.

For $N \geq 1$, consider $\mathbb{C}^{N+1}$ as a Hilbert space with the usual inner product, and let D_N, B, and Z be operators on $\mathbb{C}^{N+1}$ given by matrices $D_N = \mathrm{diag}(\lambda_j)$, $B = (B_{j,k})$ and $Z = (Z_{j,k})$, where $\lambda_j \geq 2\lambda_{j-1} \geq 2$,

$$B_{j,k} = \begin{cases} \dfrac{i}{\pi(k-j)} & \text{if } j \neq k \\ 0 & \text{if } j = k \end{cases} \qquad Z_{j,k} = \begin{cases} \dfrac{\lambda_j\, i}{(\lambda_k+\lambda_j)\pi(k-j)} & \text{if } j \neq k \\ 0 & \text{if } j = k \end{cases}$$

with j and k ranging from 0 to N.

Then D_N and B are self-adjoint operators with $D_N \geq I$ and $\|B\| \leq 1$,

$$D_N Z + Z D_N = D_N B ,$$

while $\|Zu\| \geq (\frac{1}{7}\log N - 1)\|u_N\|$ for $u_N = (1,1,\ldots,1)$. The inequality $\|B\| \leq 1$ is a consequence of the fact that B is the $N \times N$ Toeplitz matrix corresponding to the function $b(\theta) = \pi^{-1}\theta - 1$ on $0 < \theta < 2\pi$, while the lower bound for $\|Zu\|$ can be easily calculated.

Note that Z is the only solution of the above operator equation.

For $|z| < 1$, define $J_{N,z}$ on $\mathbb{C}^{N+1} \times \mathbb{C}^{N+1}$ by $J_{N,z}[u,v] = ((I+zB)D_N u, D_N v)$, which is a regular sesquilinear form with associated operator $A_z = D_N(I+zB)D_N$. Of course

$A_z^{\frac{1}{2}}A_z^{\frac{1}{2}} = D_N(I + zB)D_N$, so on differentiating both sides with respect to z, setting $z = 0$, and substituting $A_0^{\frac{1}{2}} = D_N$, we obtain

$$D_N\left(\frac{d}{dz}A_z^{\frac{1}{2}}\Big|_{z=0}\right) + \left(\frac{d}{dz}A_z^{\frac{1}{2}}\Big|_{z=0}\right)D_N = D_N B D_N$$

or in other words $\frac{d}{dz}A_z^{\frac{1}{2}}\Big|_{z=0} = ZD_N$. Hence, if $w_N = (D_N)^{-1}u_N = (\lambda_0^{-1}, \lambda_1^{-1}, \lambda_2^{-1}, \ldots, \lambda_N^{-1})$, then

$$\left\| \left(\frac{d}{dz}A_z^{\frac{1}{2}}\Big|_{z=0}\right) w_N \right\| \geq M_N \|D_N w_N\|$$

where $M_N = (\frac{1}{7}\log N - 1) \rightarrow \infty$ as $N \rightarrow \infty$.

I claim now that $\|A_z^{\frac{1}{2}} w_N\| \geq \frac{1}{2} M_N \|D_N w_N\|$ for some values of z satisfying $|z| = \frac{1}{2}$. Suppose to the contrary that $\|A_z^{\frac{1}{2}} w_N\| < \frac{1}{2} M_N \|D_N w_N\|$ whenever $|z| = \frac{1}{2}$. Then

$$\left\| \left(\frac{d}{dz}A_z^{\frac{1}{2}}\Big|_{z=0}\right) w_N \right\| = \frac{1}{2\pi} \left\| \int_{|z|=\frac{1}{2}} \frac{1}{z^2} A_z^{\frac{1}{2}} w_N \, dz \right\| < M_N \|D_N w_N\|$$

which is a contradiction. Hence $\|A_z^{\frac{1}{2}} w_N\| \geq \frac{1}{2} M_N \|D_N w_N\|$ for some values of z satisfying $|z| = \frac{1}{2}$ as claimed. Define J_N to be $J_{N,z}$ for such a value of z.

Now let $\mathcal{H} = \oplus \mathbb{C}^{N+1}$, let $\mathcal{V}_J = \{ u=(u_1,u_2,u_3,\ldots) : \sum \|D_N u_N\|^2 < \infty \}$, and define J_t and J on $\mathcal{V}_J \times \mathcal{V}_J$ by $J_t[u,v] = \sum J_{N,t}[u_N,v_N]$ and $J[u,v] = \sum J_N[u_N,v_N]$. These forms provide counter-examples to remarks 2 and 1 respectively.

Incidentally, it is not difficult to modify these examples to answer some of the more specific questions which have been asked. For example, suppose that we are given a closed unbounded linear transformation D from a Hilbert space $\mathcal{H}$ to another Hilbert space $\mathcal{K}$ with dense domain $\mathcal{D}(D) \subset \mathcal{H}$ and $\varepsilon \in (0,1)$. Then there exists a bounded linear operator B on $\mathcal{K}$ with $\|B\| < \varepsilon$ such that the square root of the operator A_J associated with the regular form

$$J[u,v] = ((I + B)Du, Dv)$$

has $\mathcal{D}(A_J^{\frac{1}{2}}) \not\subset \mathcal{V}_J = \mathcal{D}(D)$.

In order to end this section on a positive note, I shall present two conditions on J derived prior to 1981 which imply that $\mathcal{D}(A_J^{\frac{1}{2}}) = \mathcal{V}_J$. Condition (a) was presented in [L,1962] and (b) in [Mc,1982]. Other conditions can be found in the literature (e.g. [G,1974]).

Theorem. *Let J be a regular sesquilinear form in $\mathcal{H}$ with domain $\mathcal{V}_J \times \mathcal{V}_J$.*

(a) *Suppose that $\mathcal{K}$ is a Hilbert space which is continuously and densely embedded in $\mathcal{H}$ such that $\mathcal{V}_J \subset [\mathcal{K}, \mathcal{H}]_{\frac{1}{2}}$. If $\mathcal{D}(A_J) \cup \mathcal{D}(A_J{}^*) \subset \mathcal{K}$ and $\mathcal{V}_J$ is closed in $[\mathcal{K}, \mathcal{H}]_{\frac{1}{2}}$, then $\mathcal{D}(A_J{}^{\frac{1}{2}}) = \mathcal{V}_J$.*

(b) *If there exists $s \in [0,1)$ and $c > 0$ such that*

$$|ImJ[u,u]| \leq c(ReJ[u,u])^s \|u\|^{2(1-s)}$$

for all $u \in \mathcal{V}_J$, then $\mathcal{D}(A_J{}^{\frac{1}{2}}) = \mathcal{V}_J$.

5. Elliptic forms

Although the counter-examples presented in section 4 answer the questions specifically posed by Kato, they do not tell us anything about the elliptic operators in which he was really interested. So let us turn our attention to second order elliptic operators.

Let Ω denote an open subset of $\mathbb{R}^n$, let $\mathcal{V}_J$ be a closed linear subspace of the Sobolev space $H^1(\Omega)$ which contains $C_c^\infty(\Omega)$, and let J be the *elliptic* sesquilinear form defined on $\mathcal{V}_J \times \mathcal{V}_J$ by

$$J[u,v] = \int_\Omega \left\{ \sum a_{jk} \frac{\partial u}{\partial x_k} \overline{\frac{\partial v}{\partial x_j}} + \sum a_k \frac{\partial u}{\partial x_k} \overline{v} + \sum \alpha_j u \overline{\frac{\partial v}{\partial x_j}} + a u \overline{v} \right\} dx$$

where $a_{jk}, a_k, \alpha_j, a \in L_\infty(\Omega)$ and there exists $\kappa > 0$ such that $Re\sum a_{jk}(x)\zeta_k\overline{\zeta_j} \geq \kappa|\zeta|^2$ for all $\zeta = (\zeta_j) \in \mathbb{C}^n$ and almost all $x \in \Omega$. Then there exists $\lambda \geq 0$ such that $J + \lambda$ is a regular sesquilinear form (where $J + \lambda$ is defined on $\mathcal{V}_J \times \mathcal{V}_J$ by $(J + \lambda)[u,v] = J[u,v] + \lambda(u,v)$). Note that

$$A_{(J+\lambda)}u = A_J u + \lambda u = -\sum \frac{\partial}{\partial x_j} a_{jk} \frac{\partial u}{\partial x_k} + \sum a_k \frac{\partial u}{\partial x_k} - \sum \frac{\partial}{\partial x_j} \alpha_j u + (a + \lambda)u$$

if u belongs to the appropriate domain $\mathcal{D}(A_J)$. If $\mathcal{V}_J = \mathring{H}^1(\Omega)$, then A_J is the Dirichlet operator with domain $\mathcal{D}(A_J) = \{u \in \mathring{H}^1(\Omega) : A_J u \in L_2(\Omega)\}$. Other choices of $\mathcal{V}_J$ give rise to operators A_J which satisfy natural or mixed boundary conditions.

The problem of determining whether $\mathcal{D}((A_J + \lambda I)^{\frac{1}{2}}) = \mathcal{V}_J$ holds for such forms J has become known as "the square root problem of Kato for elliptic operators", "the Kato problem" or "the Kato conjecture".

The results presented at the end of the preceding section were proved in order to answer this conjecture when the coefficients and boundary $b\Omega$ are sufficiently smooth, or when the leading term is hermitian [L,1962], [Mc,1982].

Theorem. *Let J be an elliptic sesquilinear form as defined above, and suppose that one of the following conditions holds.*

(a) $H^1(\Omega) = [H^2(\Omega), L_2(\Omega)]_{\frac{1}{2}}$ *and* $\mathcal{D}(A_J) \cup \mathcal{D}(A_J{}^*) \subset H^2(\Omega)$;

(b) $a_{jk} = \overline{a_{kj}}$.

Then $\mathcal{D}((A_J + \lambda I)^{\frac{1}{2}}) = \mathcal{V}_J$.

Condition (a) is satisfied, for example, if the coefficients are Lipschitz functions, Ω is a bounded open set with $C^{1,1}$ boundary, and $\mathcal{V}_J$ is $\overset{\circ}{H}{}^1(\Omega)$ or $H^1(\Omega)$. Its limitations were spelled out by Lions as follows [L,1962].

> REMARQUE. Il est bon de rappeler que les problèmes aux limites *mêlés n'entrent pas* dans la catégorie précédante. Donc, *par example*, pour un opérateur elliptique A du 2ème ordre, non auto-adjoint, avec condition aux limites de Dirichlet sur une partie de la frontière et condition aux limites de Neumann sur le reste de la frontière, on ignore si $\mathcal{D}(A^{\frac{1}{2}}) = \mathcal{D}(A^{*\frac{1}{2}})$. Même chose d'ailleurs avec le problème de Dirichlet et une *frontière irrégulière*.

We shall return to this remark in section 9.

Most of the material outlined so far was presented at the Lions-Brezis Seminar at the Collège de France in Paris in late 1980. In that lecture I noted that there is a connection between the Kato problem in the very special case when $\Omega = \mathbb{R}$ and $J[u,v] = \int a u \overline{v}$ (where $a \in L_\infty(\mathbb{R})$ and $Re\, a \geq \kappa > 0$) and the Calderón commutator theorem [C,1965] which at that time was regarded as a deep result. In that talk I stated that it may be true that every elliptic form J satisfies $\mathcal{D}((A_J + \lambda I)^{\frac{1}{2}}) = \mathcal{V}_J$, but "it may also be that the proof would need to be at least as deep as the proofs required for" the Calderón commutator theorem [Mc,1982]. This was an understatement.

Not long afterwards I showed how a positive answer to this special case of the Kato problem could be used to prove the L_2 boundedness of the singular Cauchy integral on a Lipschitz curve in the complex plane, a result which had been conjectured by Zygmund and Calderón years before and proved by Calderón for curves with small Lipschitz constants [C,1977]. This excited interest in the Kato problem in Y. Meyer and R. Coifman, and resulted in the solution of this special case of the Kato problem. The multilinear estimates which were needed to solve the Kato problem were modified and used to prove the L_2 boundedness of the singular Cauchy integral [CMcM,1982], so the fact that square roots could be used to prove the L_2 boundedness of the singular Cauchy integral on a Lipschitz curve was not needed. However a similar approach was later used in proving the original T(b) theorem [McM,1985]. Other connections between the two problems were presented by Kenig and Meyer [KM,1983].

6. Multilinear expansions

Let us consider a sesquilinear form J in a Hilbert space $\mathcal{H}$ which is expressed as

$$J[u,v] = (FDu\ , Dv) \qquad\qquad u,v \in \mathcal{V}_J$$

where $\mathcal{V}_J$ is a dense linear subspace of $\mathcal{H}$, D is a closed linear transformation from $\mathcal{H}$ to another Hilbert space $\mathcal{K}$ with domain $\mathcal{D}(D) = \mathcal{V}_J$, and F is a bounded linear operator on $\mathcal{K}$ which satisfies $Re(Fu,u) \geq \kappa\|u\|^2$ for all $u \in \mathcal{K}$ and some $\kappa > 0$.

Then J is a regular sesquilinear form with associated operator $A_J = D^*FD$, where D^* is the adjoint of D. In particular $\mathcal{D}(A_J) = \{u \in \mathcal{V}_J : FDu \in \mathcal{D}(D^*)\}$.

It can be shown that (i) there exists $\mu > 0$ and a bounded linear operator B such that $\|B\| \leq \rho < 1$ and $F = \mu(I - B)$, and (ii) there exists $\mu_1 > 0$ and a bounded linear operator B_1 such that $\|B_1\| < 1$ and $F = \mu_1(I - B_1)^{-1}$. So

$$A_J = \mu D^*(I - B)D = \mu_1 D^*(I - B_1)^{-1}D\ .$$

These statements are almost obvious if F is a normal operator (e.g. a multiplication operator on $L_2(\mathbb{R})$), but the general statements require proving.

Let us prove (i). Choose $\mu = \kappa^{-1}\|F\|^2$ and $\rho = 1 - \kappa^2\|F\|^{-2}$, and let $B = I - \mu^{-1}F$. Then

$$\begin{aligned} \|Bu\|^2 &= \|u\|^2 - \mu^{-1}\{(Fu,u) + (u,Fu)\} + \mu^{-2}\|Fu\|^2 \\ &\leq \|u\|^2 \{ 1 - 2\mu^{-1}\kappa + \mu^{-2}\|F\|^2 \} \\ &= \|u\|^2 \{ 1 - \kappa^2\|F\|^{-2} \} = \rho\|u\|^2 \end{aligned}$$

for all $u \in \mathcal{K}$. Part (ii) was proved by Journé [J,1989].

We now present $A_J^{\frac{1}{2}}u$ as an infinite sum of terms which are multilinear in B and u. To simplify things a little we work with $A = \mu^{-1}A_J = D^*(I-B)D$. Then, for $u \in \mathcal{D}(A) = \mathcal{D}(A_J)$,

$$A^{\frac{1}{2}}u = \frac{2}{\pi}\int_0^\infty (I + t^2A)^{-1}A\,u\,dt$$

$$= \frac{2}{\pi}\int_0^\infty (I + t^2D^*(I-B)D)^{-1}D^*(I-B)Du\,dt$$

$$= \frac{2}{\pi}\int_0^\infty (I + t^2D^*D)^{-1}D^*Du\,dt$$

$$+\frac{2}{\pi}\int_0^\infty (I + t^2D^*D)^{-1}\{t^2D^*BD(I+t^2D^*D)^{-1}D^*D - D^*BD\}u\,dt + \dots$$

$$= |D|u - \frac{2}{\pi}\int_0^\infty Q_tBP_t(Du)\frac{dt}{t} - \sum_{k=1}^{\infty}\frac{2}{\pi}\int_0^\infty Q_t(BT_t)^kBP_t(Du)\frac{dt}{t}$$

where

$$Q_t = tD^*(I + t^2DD^*)^{-1} \supset (I + t^2D^*D)^{-1}tD^*$$

$$T_t = t^2DD^*(I + t^2DD^*)^{-1} \qquad \text{and} \qquad P_t = (I + t^2DD^*)^{-1},$$

provided that the integrals exist (as improper integrals) and can be summed.

This is possible if there exists $\delta < \rho^{-1}$ such that the multilinear estimates

$$(\mathrm{ME_k}) \qquad \left\|\frac{2}{\pi}\int_0^\infty Q_t(BT_t)^kBP_t(Du)\frac{dt}{t}\right\| \le C_\delta\,\delta^{k+1}\|B\|^{k+1}\|Du\|$$

hold for all integers $k \ge 0$, all $u \in \mathcal{D}(A)$, and some $C_\delta \ge 1$. Then

$$\|A_J^{\frac{1}{2}}u\| = \mu^{\frac{1}{2}}\|A^{\frac{1}{2}}u\| \le \mu^{\frac{1}{2}}C_\delta\{1 + \sum_{k=0}^{\infty}(\delta\rho)^{k+1}\}\|Du\| = \mu^{\frac{1}{2}}C_\delta(1-\delta\rho)^{-1}\|Du\|$$

for all $u \in \mathcal{D}(A)$. It follows that $\mathcal{V}_J \subset \mathcal{D}(A_J^{\frac{1}{2}})$. A dual estimate gives the desired equality $\mathcal{V}_J = \mathcal{D}(A_J^{\frac{1}{2}})$.

We actually need uniform bounds on the norms of integrals over compact subintervals of $(0,\infty)$. We also need to take much more care in dealing with unbounded operators. Details can be found in [Mc,1985].

Most of the progress which has been made on the Kato problem since 1981 has involved the use of multilinear estimates. They can also be used to prove the holomorphic dependence of $A_t^{\frac{1}{2}}$ on J_t as asked in Remark 2 of Kato quoted in section 3. We shall not pursue this matter here.

7. Multilinear estimates when $\Omega = \mathbb{R}$

When can estimates be obtained for the multilinear expressions arising from the elliptic differential forms introduced in section 5? Let us first treat the simplest case when $\Omega = \mathbb{R}$ and the lower order terms are absent.

Consider the sesquilinear form in $\mathcal{H} = L_2(\mathbb{R})$ defined by

$$J[u,v] = \int_0^\infty f(x)u'(x)\overline{v'(x)}dx$$

for all $u,v \in \mathcal{V}_J = H^1(\mathbb{R})$, where f is a function in $L_\infty(\mathbb{R})$ which satisfies $Re(f(x)) \geq \kappa > 0$ for almost all $x \in \mathbb{R}$. Then J can be expressed as

$$J[u,v] = (FDu\,,Dv) = \mu((I-B)Du\,,Dv)$$

with $D = \frac{1}{i}\frac{d}{dx}$, and F and B are the multiplication operators on $L_2(\mathbb{R})$ given by $Fw = fw$ and $Bw = bw$, where $b(x) = 1 - \mu^{-1}f(x)$, with μ a positive real number chosen so that $\|B\| = \|b\|_\infty \leq \rho < 1$. Then $D^* = D$ and

$$A = \mu^{-1}A_J = -\mu^{-1}\frac{d}{dx}f(x)\frac{d}{dx} = -\frac{d^2}{dx^2} + \frac{d}{dx}b(x)\frac{d}{dx}$$

with $\mathcal{D}(A) = \{\, u \in H^1(\Omega) : fu' \in H^1(\Omega)\,\}$.

Do the required estimates (ME_k) hold? Let us look at the first term, $k = 0$. We need to show that there exists a constant c such that

$$(ME_0)\qquad \Big\|\frac{2}{\pi}\int_0^\infty Q_tBP_tw\,\frac{dt}{t}\Big\| \leq c\,\|b\|_\infty\|w\|_2$$

for all $w \in L_2(\mathbb{R})$ and $b \in L_\infty(\mathbb{R})$, where $Q_t = tD(I+t^2D^2)^{-1}$ and $P_t = (I+t^2D^2)^{-1}$.

Let us digress. A major result of the 1960's was the commutator theorem of A. P. Calderón [C,1965] which states that

$$\left\| \frac{1}{\pi} \text{ p.v} \int_{-\infty}^{\infty} \frac{g(x)-g(y)}{(x-y)^2} w(y)dy \right\|_{L_2(dx)} \leq \text{ const.} \|g'\|_\infty \|w\|_2$$

for all $w \in L_2(\mathbb{R})$ and all Lipschitz functions g. (The integral is a principal value singular integral.) This can be written as

$$\| \,|D|(gw) - g|D|w \,\| \leq \text{ const.} \|g'\|_\infty \|w\|_2$$

On letting B denote the multiplication operator on $L_2(\mathbb{R})$ given by $Bw = bw$ where $b = g' \in L_\infty(\mathbb{R})$, we can re-express this as

$$\left\| \frac{2}{\pi} \int_0^\infty (Q_t B P_t + P_t B Q_t) w \frac{dt}{t} \right\| \leq \text{ const.} \|b\|_\infty \|w\|_2 ,$$

which is a consequence of (ME_0) (since $P_t B Q_t = (Q_t B P_t)^*$).

We see in retrospect the difficulties faced by anyone who tried to solve the Kato problem for elliptic sesquilinear forms in the 1960's or 1970's. The first term in the expansion of the simplest case ($\Omega = \mathbb{R}$ and no lower order terms) was at least as difficult to estimate as the Calderón commutator integral! Of course, the 1980's have seen a great deal of progress in the estimation of such integrals.

How then do we obtain the bounds (ME_k)? Not without difficulty. It is better to use the expansion which we obtain from the alternative expression $A_J = \mu_1 D(I - B_1)^{-1} D$ which was given at the start of section 6. (It is easy to obtain such an expression in this case.) This expansion is

$$A_J^{\frac{1}{2}} u = \mu_1^{\frac{1}{2}} |D| u + \mu_1^{\frac{1}{2}} \sum_{k=1}^{\infty} \frac{2}{\pi} \int_0^\infty Q_t (B_1 P_t)^k (Du) \frac{dt}{t} .$$

So we need to show that there exists $\delta < \|B_1\|^{-1}$ such that the multilinear estimates

$$\left\| \int_0^\infty Q_t (B_1 P_t)^k w \frac{dt}{t} \right\| \leq c_\delta \, \delta^k \|B_1\|^k \|w\|$$

hold for all integers $k \geq 1$, all $w \in L_2(\mathbb{R})$ and some $c_\delta \geq 1$. Such estimates were first obtained in [CMcM,1982]. Using them we conclude that $\mathcal{D}(A_J^{\frac{1}{2}}) = H^1(\mathbb{R})$ as indicated above. (A treatment of the same material from a more operator-theoretic point of view appears in [CMcM,1981].)

The proof of these estimates relied heavily on the work of harmonic analysts such as Zygmund, Calderón, Carleson, Stein, Fefferman, Meyer, Coifman and others. The machinery was not available when the square root problem was posed by Kato almost 20 years earlier.

I have pointed out already that similar estimates were used in the same papers to prove the L_2 boundedness of the singular Cauchy integral on a Lipschitz curve in the complex plane. Now there are many alternative proofs of this result.

Let us turn our attention back to the estimates (ME_k) which we have successfully circumvented. They can be deduced now that we already know that $\mathcal{D}(A_J^{\frac{1}{2}}) = H^1(\mathbb{R})$ (and have related estimates). They can also be obtained with $\delta > 2$ by similar methods to the alternative estimates. A direct (though involved) proof that the estimates (ME_k) hold for all $\delta > 1$ has only recently been given by Journé [J,1989].

8. Multilinear estimates when $\Omega = \mathbb{R}^n$

Consider the sesquilinear form in $\mathcal{H} = L_2(\mathbb{R}^n)$ defined by

$$J[u,v] = \int_{\mathbb{R}^n} \Sigma\, a_{jk} \frac{\partial u}{\partial x_k} \overline{\frac{\partial v}{\partial x_j}}\, dx$$

for all $u,v \in \mathcal{V}_J = H^1(\mathbb{R}^n)$, where a_{jk} are functions in $L_\infty(\mathbb{R}^n)$ which satisfy $Re\Sigma a_{jk}(x)\zeta_k\overline{\zeta_j} \geq \kappa|\zeta|^2$ for all $\zeta = (\zeta_j) \in \mathbb{C}^n$ and almost all $x \in \mathbb{R}^n$, where $\kappa > 0$. Then J can be expressed as

$$J[u,v] = (FDu\, , Dv) = \mu((I - B)Du\, , Dv)$$

where $D = grad$, and F and B are the multiplication operators on $\oplus L_2(\mathbb{R}^n)$ given by $(Fw)_j = \Sigma a_{jk}w_k$ and $(Bw)_j = \Sigma b_{jk}w_k$, where $b_{jk}(x) = \delta_{jk} - \mu^{-1}a_{jk}(x)$, with μ a positive real number chosen so that $\|B\| \leq \rho < 1$. Then $D^* = -div$ and

$$A = \mu^{-1}A_J = -\mu^{-1}\Sigma \frac{\partial}{\partial x_j} a_{jk}(x) \frac{\partial}{\partial x_k} = -\Sigma \frac{\partial^2}{\partial x_j^2} + \Sigma \frac{\partial}{\partial x_j} b_{jk}(x) \frac{\partial}{\partial x_k}\, .$$

Do the required estimates (ME_k) hold?

That they do hold for some (sufficiently large) value of δ was shown by Coifman, Deng and Meyer [CDM,1983], and also by Fabes, Jerison and Kenig

[FJK,1982]. Hence $\mathcal{D}(A_J^{\frac{1}{2}}) = H^1(\mathbb{R}^n)$ provided $\|B\|$ is sufficiently small. Recently Journé has proved [J,1989] that the estimates (ME_k) hold for all $\delta > 1+2\sqrt{n}$, and hence that $\mathcal{D}(A_J^{\frac{1}{2}}) = H^1(\mathbb{R}^n)$ whenever $\|B\| < (1+2\sqrt{n})^{-1}$.

But do the estimates (ME_k) hold for all $\delta > 1$? This is what is needed if we are to conclude that $\mathcal{D}(A_J^{\frac{1}{2}}) = H^1(\mathbb{R}^n)$ for all of the elliptic forms J specified above. This tantalizing question remains open.

And what about the Kato problem for the more general elliptic sesquilinear forms defined in section 5? This seems to be a harder question now than it did when posed by Kato almost 30 years ago!

9. Some multilinear estimates

When can the multilinear estimates (ME_k) be proved without invoking the full machinery of real analysis? Let us recall the problem.

We are given Hilbert spaces $\mathcal{H}$ and $\mathcal{K}$, a closed densely-defined linear transformation D from $\mathcal{H}$ to $\mathcal{K}$ and a bounded linear operator B on $\mathcal{K}$. Let S denote the non-negative self-adjoint operator in $\mathcal{K}$ given by $S = (DD^*)^{\frac{1}{2}}$ and, for $t > 0$, let $Q_t = tD^*(I + t^2S^2)^{-1}$, $T_t = t^2S^2(I + t^2S^2)^{-1}$ and $P_t = (I + t^2S^2)^{-1}$.

We want to prove that for all $\delta > 1$, there exists C_δ such that the multilinear estimates

$$(ME_k) \qquad \left\| \int_0^\infty Q_t(BT_t)^k BP_t w \frac{dt}{t} \right\| \leq C_\delta\, \delta^{k+1} \|B\|^{k+1} \|w\|$$

hold for all integers $k \geq 0$ and all $w \in \mathcal{R}(D)$, the range of D.

In addition to the above operators, define, for $0 < s < 2$,

$$Q_t^{(s)} = (tS)^s(I + t^2S^2)^{-1} = \psi^{(s)}(tS) ,$$

where $\psi^{(s)}(\tau) = \tau^s(1 + \tau^2)^{-1}$. Note that $D^* = VS$ for some partial isometry V (with $\|V\| = 1$), so $Q_t = VQ_t^{(1)}$.

These operators satisfy the quadratic estimates

$$\left\{\int_0^\infty \|Q_t^{(s)}w\|^2 \frac{dt}{t}\right\}^{\frac{1}{2}} = \left\{\int_0^\infty (\psi^{(s)}(tS)^2 w,w) \frac{dt}{t}\right\}^{\frac{1}{2}} \leq \|w\| \left\{\int_0^\infty \psi^{(s)}(\tau)^2 \frac{d\tau}{\tau}\right\}^{\frac{1}{2}} \leq q_s \|w\|$$

for some numbers q_s, so long as $0 < s < 2$. (There are no such estimates in the limiting cases, $Q_t^{(0)} = P_t$ and $Q_t^{(2)} = T_t$.)

(α) Let us see why the estimates (ME$_k$) hold in the very special case when $SB = BS$. In this case

$$|(\int_0^\infty Q_t(BT_t)^k BP_t w \frac{dt}{t}, v)| = |(\int_0^\infty VQ_t^{(\frac{1}{2})}(BT_t)^k BQ_t^{(\frac{1}{2})} w \frac{dt}{t}, v)|$$

$$\leq \sup \|(BT_t)^k B\| \left\{\int_0^\infty \|Q_t^{(\frac{1}{2})} w\|^2 \frac{dt}{t}\right\}^{\frac{1}{2}} \left\{\int_0^\infty \|Q_t^{(\frac{1}{2})} V^* v\|^2 \frac{dt}{t}\right\}^{\frac{1}{2}}$$

$$\leq (q_{\frac{1}{2}})^2 \|B\|^{k+1} \|w\| \|v\|$$

for all $w, v \in \mathcal{K}$ (using $\|T_t\| \leq 1$). Therefore

$$\| \int_0^\infty Q_t(BT_t)^k BP_t w \frac{dt}{t} \| \leq (q_{\frac{1}{2}})^2 \|B\|^{k+1} \|w\|$$

for all $w \in \mathcal{K}$, so in this very special case the multilinear estimates (ME$_k$) hold with $\delta = 1$.

(β) Let us weaken the commutativity assumption as follows. Let $\mathcal{R}$ denote the closure of $\mathcal{R}(D)$ in $\mathcal{K}$, and suppose that for some $s > 0$ there exists a bounded linear operator B_s on $\mathcal{R}$ which satisfies $S^s Bw = B_s S^s w$ for all $w \in \mathcal{R} \cap \mathcal{D}(S^s)$. Then the multilinear estimates (ME$_k$) still hold.

We can see this by modifying the above proof as follows. When $k = 0$,

$$|(\int_0^\infty Q_t BP_t w \frac{dt}{t}, v)| = |(\int_0^\infty VQ_t^{(1-s)} B_s Q_t^{(s)} w \frac{dt}{t}, v)|$$

$$\leq \|B_s\| \left\{\int_0^\infty \|Q_t^{(s)} w\|^2 \frac{dt}{t}\right\}^{\frac{1}{2}} \left\{\int_0^\infty \|Q_t^{(1-s)} V^* v\|^2 \frac{dt}{t}\right\}^{\frac{1}{2}}$$

$$\leq q_s q_{1-s} \|B_s\| \|w\| \|v\|$$

for all $w \in \mathcal{R}$ and $v \in \mathcal{K}$, while when $k \geq 1$,

$$|(\int_0^\infty Q_t(BT_t)^k BP_t w \tfrac{dt}{t}, v)| = |(\int_0^\infty Q_t(BT_t)^{k-1} BQ_t^{(2-s)} B_s Q_t^{(s)} w \tfrac{dt}{t}, v)|$$

$$\leq \|B\|^k \|B_s\| \left\{\int_0^\infty \|Q_t^{(s)} w\|^2 \frac{dt}{t}\right\}^{\frac{1}{2}} \left\{\int_0^\infty \|Q_t^{(1)} V^* v\|^2 \frac{dt}{t}\right\}^{\frac{1}{2}}$$

$$\leq q_s q_1 \|B\|^k \|B_s\| \ \|w\| \ \|v\|$$

for all $w \in \mathcal{R}$ and $v \in \mathcal{K}$ (using $\|T_t\|$, $\|Q_t^{(2-s)}\| \leq 1$). Therefore the multilinear estimates (ME_k) hold, again with $\delta = 1$.

This is actually quite a useful condition. To illustrate, let us consider the sesquilinear form

$$J[u,v] = \int_0^\infty f(x) u'(x) \overline{v'(x)}\, dx$$

discussed in section 7. In this case $\mathcal{H} = \mathcal{K} = \mathcal{R} = L_2(\mathbb{R})$, $D^* = D = \frac{1}{i}\frac{d}{dx}$, $S = |D| = (D^2)^{\frac{1}{2}}$ (with $\mathcal{D}(S) = \mathcal{D}(D) = H^1(\mathbb{R})$) and B is the multiplication operator given by $Bw = bw$ for $b = 1 - \mu^{-1} f \in L_\infty(\mathbb{R})$. The assumption on B can be re-expressed as

$$\| \, |D|^s (bw) \, \| \leq \text{const.} \| \, |D|^s w \, \|$$

for all $w \in \mathcal{D}(S^s) = H^s(\mathbb{R})$. This is precisely the condition that b is an $\dot{H}^s(\mathbb{R})$ multiplier, or equivalently that f is an $\dot{H}^s(\mathbb{R})$ multiplier. So we deduce that $\mathcal{D}(A_J^{\frac{1}{2}}) = H^1(\mathbb{R})$ whenever f is an $\dot{H}^s(\mathbb{R})$ multiplier for some $s > 0$.

Of course it has already been stated in section 7 that $\mathcal{D}(A_J^{\frac{1}{2}}) = H^1(\mathbb{R})$ without any such extra condition on f, though, as we know, this requires a much deeper proof.

As we have seen, however, we do not know the answer to the Kato problem for the general elliptic sesquilinear forms introduced in section 5. So there is some interest in determining whether we can deduce that $\mathcal{D}(A_{(J+\lambda)}^{\frac{1}{2}}) = \mathcal{V}_J$ under extra conditions on the coefficients.

This is indeed the case. In the paper [Mc,1985] I used this approach to prove that $\mathcal{D}(A_{(J+\lambda)}^{\frac{1}{2}}) = \mathcal{V}_J$ provided that a_{jk}, a_k and α_j are $H^s(\Omega)$ multipliers for some $s > 0$ under very general conditions on Ω and $\mathcal{V}_J$. These conditions are satisfied for example if Ω is a strongly Lipschitz bounded domain and $\mathcal{V}_J = \overset{\circ}{H}{}^1(\Omega)$ or $\mathcal{V}_J = H^1(\Omega)$, and also if Ω is a domain with smooth boundary $b\Omega$ and $\mathcal{V}_J = \{ u \in H^1(\Omega) : u|_\sigma = 0 \}$ where σ is a smooth subregion of $b\Omega$. Such $\mathcal{V}_J$ correspond to operators A_J satisfying the Dirichlet condition on σ and the natural Neumann condition on the remainder of $b\Omega$.

Therefore the questions raised in the remark of Lions quoted in section 5 have positive answers, at least for second order operators whose coefficients a_{jk}, a_k and α_j are $H^s(\Omega)$ multipliers for some $s > 0$. Incidentally, if Ω is a strongly Lipschitz bounded domain, then every Hölder continuous function $a \in C^\alpha(\overline{\Omega})$ is an $H^s(\Omega)$ multiplier for $0 < s < \alpha < 1$, as are the characteristic functions of strongly Lipschitz subdomains $\Omega_0 \subset \Omega$ provided $0 < s < \frac{1}{2}$.

(γ) When we drop the assumption that $S^s B = B_s S^s$ we have

$$\Big|\Big(\int_0^\infty Q_t (BT_t)^k BP_t w \,\frac{dt}{t}, v \Big)\Big| = \Big|\Big(\int_0^\infty V Q_t^{(1)} (BT_t)^k BP_t w \,\frac{dt}{t}, v \Big)\Big|$$

$$\le \left\{ \int_0^\infty \|(Q_t^{(1)})^{\frac{1}{2}} (BT_t)^k BP_t w\|^2 \frac{dt}{t} \right\}^{\frac{1}{2}} \left\{ \int_0^\infty \|(Q_t^{(1)})^{\frac{1}{2}} V^* v\|^2 \frac{dt}{t} \right\}^{\frac{1}{2}}$$

$$\le \text{const.} \left\{ \int_0^\infty \|(Q_t^{(1)})^{\frac{1}{2}} (BT_t)^k BP_t w\|^2 \frac{dt}{t} \right\}^{\frac{1}{2}} \|v\|$$

for all $w \in \mathcal{R}(D)$ and $v \in \mathcal{K}$. So what we need is that for all $\delta > 1$ there exists C_δ such that

$$\left\{ \int_0^\infty \|(Q_t^{(1)})^{\frac{1}{2}} (BT_t)^k BP_t w\|^2 \frac{dt}{t} \right\}^{\frac{1}{2}} \le C_\delta \, \delta^{k+1} \|B\|^{k+1} \|w\|$$

for all integers $k \ge 0$ and all $w \in \mathcal{R}(D)$, the range of D. Of course this is not the case for arbitrary operators D and B, even when $k = 0$.

The results which were discussed in sections 7 and 8 were all obtained by using a considerable amount of real analysis to prove estimates something like these. Related results have been obtained in recent years by G. David, J.-L. Journé, S. Semmes, P. Jones, C. Kenig, as well as R. Coifman, Y. Meyer and many others.

It remains a challenge, however, to prove the multilinear estimates needed to solve the square root problem of Kato for elliptic operators, or to find an alternative approach.

REFERENCES

[C,1965] Calderón, A. P., *Commutators of singular integral operators*, Proc. Nat. Acad. Sci. U.S.A., **53** (1965), 1092-1099.

[C,1977] Calderón, A. P., *Cauchy integrals on Lipschitz curves and related operators*, Proc. Nat. Acad. Sci. U.S.A., **74** (1977), 1324-1327.

[CDM,1983] Coifman, R., Deng, D., Meyer, Y., *Domaine de la racine carrée de certains opérateurs différentiels accrétifs*, Ann. Inst. Fourier (Grenoble), **33** (1983), 123-134.

[CMcM,1981] Coifman, R., McIntosh, A., Meyer, Y., *The Hilbert transform on Lipschitz curves,* Miniconference on Partial Differential Equations, 1981, Proceedings of the Centre for Mathematical Analysis, ANU, Canberra, **1** (1982), 26-69.

[CMcM,1982] Coifman, R., McIntosh, A., Meyer, Y., *L'intégrale de Cauchy définit un opérateur borné sur* L_2 *pour les courbes lipschitziennes*, Ann. of Math. **116** (1982), 361-387.

[FJK,1982] Fabes, E., Jerison, D., Kenig, C., *Multilinear Littlewood-Paley estimates with applications to partial differential equations*, Proc. Nat. Acad. Sci. U.S.A. **79** (1982), 5746-5750.

[G,1974] Greenlee, W. M., *On fractional powers of operators in Hilbert space*, Acta Sci. Math., **36** (1974), 55-61.

[J,1989] Journé, J.-L., *Remarks on Kato's square root problem*, preprint.

[K,1961] Kato, T., *Fractional powers of dissipative operators*, J. Math. Soc. Japan, **13** (1961), 246-274.

[K,1962] Kato, T., *Fractional powers of dissipative operators, II*, J. Math. Soc. Japan, **14** (1962), 242-248.

[K,1976] Kato, T., *"Perturbation Theory for Linear Operators"*, second edition, Springer-Verlag, Berlin/Heidelberg/New York, 1976.

[KM,1983] Kenig, C., Meyer, Y., *Kato's square roots of accretive operators and Cauchy kernels on Lipschitz curves are the same*, Institut Mittag-Leffler Technical Report No4, 1983.

[L,1962] Lions, J. L., *Espaces d'interpolation et domaines de puissances fractionnaires d'opérateurs*, J. Math. Soc. Japan, **14** (1962), 233-241.

[Mc,1972] McIntosh, Alan, *On the comparability of $A^{\frac{1}{2}}$ and $A^{*\frac{1}{2}}$*, Proc. Amer. Math. Soc., **32** (1972), 430-434.

[Mc,1982] McIntosh, Alan, *On representing closed accretive sesquilinear forms as $(A^{\frac{1}{2}}u, A^{*\frac{1}{2}}v)$*, Collège de France Seminar, Volume III, edited by H. Brezis and J. L. Lions, Pitman Advanced Publishing Program, Research Notes in Math., **70** (1982), 252-267

[Mc,1985] McIntosh, Alan, *Square roots of elliptic operators*, J. Functional Analysis, **61** (1985), 307-327.

[McM,1985] McIntosh, A., Meyer, Y., *Algèbres d'opérateurs définis par des intégrales singulières*, Comptes Rendus Acad. Sci., Paris, Sér.I, Math., **301** (1985), 395-397.

[McY,1989] McIntosh, A., Yagi, A, *Operators of type ω without a bounded H_∞ functional calculus*, Miniconference on Operators in Analysis, 1989, Proceedings of the Centre for Mathematical Analysis, ANU, Canberra, **24** (1989), 159-172.

[T,1979] Tanabe, H., *Equations of Evolution*, Pitman, London/San Francisco/ Melbourne, 1979.

THE INITIAL VALUE PROBLEM FOR A CLASS OF NONLINEAR DISPERSIVE EQUATIONS

CARLOS E. KENIG

UNIVERSITY OF CHICAGO, USA

GUSTAVO PONCE

PENNSYLVANIA STATE UNIVERSITY, USA

AND

LUIS VEGA

UNIVERSITY OF CHICAGO, USA

Abstract. We consider the initial value problem for a (generalized) equation which arises in the study of propagation of unidirectional nonlinear, dispersive waves. The aim is to study the local and global well-posedness of this problem in classical Sobolev spaces H^s. For the associated linear problem sharp local and global smoothing effects are proven. It is shown how to use these effects to establish well-posedness result for the nonlinear problem.

§1. Introduction.

This paper is concerned with the initial value problem (IVP) for nonlinear dispersive equations of the form

$$
(1.1) \qquad \begin{cases} \partial_t u - D^\alpha \partial_x u + \partial_x \left(u^{k+1}/(k+1)\right) = 0 & x, t \in \mathbb{R} \\ u(x,0) = u_0(x) \end{cases}
$$

where $D = (-\partial_x^2)^{1/2}$, $k \in \mathbb{Z}$, $k \geq 1$, *and* $\alpha \geq 1$ real.

These are model equations for the unidirectional propagation of small-amplitude, nonlinear, dispersive waves. For $(\alpha, k) = (2,1)$ the equation in (1.1) reduces to the celebrated Korteweg-de Vries (KdV) equation. When $(\alpha, k) = (2,2)$ we obtain the modified Korteweg-de Vries equation (mKdV), and for $(\alpha, k) = (1,1)$ the Benjamin-Ono equation.

These three equations possess infinitely many conservation laws, and are integrable by the inverse scattering method (KdV, mKdV), and by its analogue in the case of the BO equation.

In general, solutions of (1.1) satisfy at least three conservation laws, namely

$$
\Phi_1(u) = \int u dx, \quad \Phi_2(u) = \int u^2 dx,
$$

and

$$
\Phi_3(u) = \int ((D^{\alpha/2}u)^2 - c_k u^{k+2}) dx.
$$

Our aim in this paper is to study local and global well-posedness of the IVP (1.1) in classical Sobolev spaces $H^s(\mathbb{R})$. The problem (1.1) is said to be local (resp. global) well-posed in H^s if it generates a continuous local (resp. global) flow in H^s (i.e. existence, uniqueness, persistence, and continuous dependence on the initial data hold).

Well-posedness of the IVP (1.1) can be considered in other function spaces, for example: weighted Sobolev spaces $H^s \cap L^s(\omega(x)dx)$.

In general, global well-posedness in H^s depends on the available local theory, and on the conservation laws $\Phi_k(\cdot)$ (k = 2,3). Therefore, we will be mainly interested in local results.

In this direction we have the following theorem due to T. Kato [14][15] (see also [25]).

THEOREM A. *Let $s > 3/2$ (real).*

For any $u_0 \in H^s(\mathbb{R})$ there exists a unique solution u to (1.1) in the class

$$C([-T,T] : H^s)$$

with T depending on k, and $\| u_0 \|_{s,2}$.

Moreover, for any $T' < T$ there exists a neighborhood Ω of u_0 in H^s such that the map $\tilde{u}_0 \longmapsto \tilde{u}(t)$ is continuous from Ω into $C([-T',T'] : H^s)$. ■

The idea of the proof of this theorem can be reduced to the following "sharp energy estimate" (see lemma 4.1 for its proof),

$$\frac{d}{dt} \| u(t) \|_{s,2} \leq c_{s,k} \| u(t) \|_{\infty}^{k-1} \| \partial_x u(t) \|_{\infty} \| u(t) \|_{s,2} \tag{1.2}$$

for any $s \geq 0$. Hence, if $s > 3/2$ one sees that

$$\frac{d}{dt} \| u(t) \|_{s,2} \leq c_{s,k} \| u(t) \|_{s,2}^{k+1},$$

and consequently

$$\| u(t) \|_{s,2} \leq c_{s,k} \frac{\| u_0 \|_{s,2}}{\sqrt[k]{1 - k \| u_0 \|_{s,2}^{k} \cdot t}}$$

for any $t \in [0, (k \| u_0 \|_{s,2})^{-k}]$.

Therefore for $T^* = c \| u_0 \|_{s,2}^{-k}$ with $c = c(s,k)$

$$\sup_{[0,T^*]} \| u(t) \|_{s,2} \leq 2 \| u_0 \|_{s,2}$$

which provides an *a priori* estimate of the solution in the time interval $[0, T^*]$ with T^* having a lower bound depending only on $\| u_0 \|_{s,k}$, k, and s.

The above argument shows that the proof of Theorem A has almost nothing to do with the dispersive structure of the equation in (1.1). Notice that all the estimates above are independent of α. In fact, the same proof applies to the generalized inviscid Burger's equation

$$\partial_t u + \partial_x \left(\frac{u^{k+1}}{k+1} \right) = 0.$$

In this case the assumption $s > 3/2$ on the Sobolev exponent is known to be sharp. The same is true for any quasilinear symmetric hyperbolic first order system in one space dimension (see [13]).

Recently, several papers have been devoted to studying special properties of solutions of the IVP (1.1). More precisely, different kinds of smoothing effects have been established in solution of the associated linear problem

$$(1.3)\qquad \begin{cases} \partial_t v - D^\alpha \partial_x v = 0 & x, t \in \mathbb{R} \\ v(x,0) = v_0(x) \end{cases}$$

and several techniques have been introduced to prove these effects in solutions of the IVP (1.1).

To explain this approach let us assume that *a priori* we know that the solution of $u(\cdot)$ of (1.1) satisfies that

$$(1.4)\qquad \int_0^T \| \partial_x u(\theta) \|_\infty \, d\theta < M.$$

Inserting this estimate in (1.2) it follows that for any $s > 1/2$

$$\sup_{[0,T']} \| u(t) \|_{s,2} \leq 2 \| u_0 \|_{s,2}$$

with $T' < T$ depending only on k, s, and $\| u_0 \|_{s,2}$.

Thus, roughly speaking one should try to use the regularizing effects mentioned above to obtain (1.4) before (or simultaneously with) the energy estimate (1.2).

To study these smoothing processes it is convenient to consider first the associated linear problem (1.3). In this case the solution $v(x,t)$ is given by the unitary group (in H^s) $\{W^\alpha(t)\}_{-\infty}^{\infty}$, i.e.

$$v(x,t) = W^\alpha(t) v_0 = e^{itD^\alpha \partial_x} v_0 = S_t^\alpha * v_0$$

where

$$S_t^\alpha(x) = c \int e^{i(x\cdot\xi + t|\xi|^\alpha \xi)} d\xi.$$

Since $\{W^\alpha(t)\}_{-\infty}^{\infty}$ is a group in H^s smoothing effects in H^s have to be excluded (i.e. if $v_0 \notin H^s$ then $W^\alpha(t) v_0 \notin H^s$ for any t).

However, the following result of Strichartz type [29] shows that the solution $v(t)$ gain $(\alpha - 1)/4-$ derivatives in $L^\infty(\mathbb{R})$ a.e. in t

$$(1.5)\qquad \left(\int_{-\infty}^{\infty} \| D^{(\alpha-1)/4} W^\alpha(t) v_0 \|_\infty^4 \, dt \right)^{1/4} \leq c \| v_0 \|_2 \, .$$

The proof of (1.5) (in a more general form) will be given in section 2. Notice that for the case $\alpha = 1$ (linear BO equation) we obtain a similar result to that proven in [10] for the one dimensional Schrödinger group $\{e^{it\partial_x^2}\}_{-\infty}^{\infty}$ (for further comments and related results we refer to section 2).

In [15], T. Kato has shown that solutions $u's$ of the KdV satisfy the following local smoothing effect

$$\int_{-T}^{T}\int_{-R}^{R} |\partial_x u(x,t)|^2 dx dt \leq c(T; R; \| u_0 \|_2). \tag{1.6}$$

The estimate (1.6) was used in [15] to establish the existence of a global weak solution for the KdV equation with data $u_0 \in L^2(\mathbb{R})$. The same method of proof used by T. Kato works for the modified KdV and for the case $(\alpha, k) = (2,3)$ in (1.1). However, a similar result for $\alpha = 2$ and $k \geq 4$ remains open (even for the local case).

The local smoothing effect for the group $\{W^\alpha(t)\}_{-\infty}^{\infty}$ is quite nice and simple to prove. It reads

$$\int_{-\infty}^{\infty} |D^{\alpha/2}(W^\alpha(t)v_0)(x)|^2 dt = c_\alpha \| v_0 \|_2^2 \tag{1.7}$$

for any $x \in \mathbb{R}$.

Notice that the estimate (1.7) involves the $L^\infty(\mathbb{R} : L^2(\mathbb{R} : dt))$-norm instead of the $L^2_{loc}(\mathbb{R}^2)$-norm in (1.6). Also we may remark that the estimates (1.6), (1.7) cannot hold for solution of hyperbolic equations. This is not the case of the smoothing effect described in (1.5), which was initially established in solution of the wave equation, and Klein-Gordon equation (see [29] and references therein).

In [8], P. Constantin, and J.C. Saut have shown that the smoothing effect described in (1.6) is a common property of linear dispersive equations, (where roughly speaking the gain of derivatives is equal to $(m-1)/2$, with m denoting the order). This local smoothing effect also appears implicitly in the works of P. Sjölin [27] and L. Vega [32] concerning the pointwise behavior of $e^{it\Delta}u_0$ as t tends to zero (see also [20]). The proof of the estimate (1.7) and its relation with almost everywhere convergence results will be given in section 3.

Once global and local smoothing effects have been established in the associated linear group, the aim is to extend them to solutions of the IVP (1.1) and to use these to prove well-posedness.

For this purpose one should consider the integral equation

$$u(t) = W^\alpha(t)u_0 + \int_0^t W^\alpha(t-\tau)(u^k \partial_x u)(\tau) d\tau. \tag{1.8}$$

In general estimates involving the integral equation (1.8) can present the so-called loss of derivatives. More precisely, if we use (1.8) to obtain an estimate of the norm in (1.4) the term in the integral sign may involve a derivative of order larger than one. To overcome this difficulty one has to rely on the smoothing effects commented above. In section 4 we shall discuss the techniques introduced in [12] [17] and explain their limitations in application to the IVP (1.1).

Finally, in section 5 we state and explain some of our results concerning the well-posedness of the problem (1.1). In particular, the techniques developed here allows us to prove that the KdV equation is globally well posed in H^s for any $s \geq 1$. This settles the question left open by J.C. Saut, and R. Temam [26] (Remark 2.1) and by T. Kato [15] section 3. Previous results were restricted to $s \geq 2$ (see [2] [3] [26]).

NOTATION

— The norm in $L^p(\mathbb{R})$, $1 \leq p \leq \infty$, will be denoted by $\|\cdot\|_p$.
— $J^s = (1-\Delta)^{s/2}$ and $D^s = (-\Delta)^{s/2}$ denote the Bessel and the Riesz potential of order $-s$ respectively.
— $L^p_s = J^{-s}L^p$ whose norm will be denoted by $\|\cdot\|_{s,p} = \| J^s \cdot \|_p$. When $p=2$ we will write H^s instead of L^2_s. $H^\infty = \cap_{s>0} H^s$.
— $S(\mathbb{R})$ denotes the Schwartz space.
— $[A;B] \equiv A\cdot B - B\cdot A$ where A, B are operators. Thus $[J^s; f]g = J^s(fg) - fJ^s g$ in which f is regarded as a multiplication operator.
— $\chi : \mathbb{R} \to \mathbb{R}$ denotes a nondecreasing C^∞-function such that $\chi' \equiv 1$ on [0,1] and support of $\chi' \subseteq (-1,2)$. For any $j \in \mathbb{Z}$, $\chi_j(\cdot) = \chi(\cdot) = \chi(\cdot - j)$.

§2. Global Smoothing Effects.

In this section we shall prove a general estimate which contains (1.5) as a particular case. Also it will be shown how an intermediate step in this proof can be used to improve results concerning the asymptotic behavior of small solutions to the IVP (1.1).

Consider the associated linear problem

$$\begin{cases} \partial_t v - D^\alpha \partial_x v = 0 & x, t \in \mathbb{R} \\ v(x,0) = v_0(x) \end{cases} \tag{2.1}$$

whose solution can be written as

$$v(x,t) = W^\alpha(t)v_0 = S_t^\alpha * v_0$$

where

$$S_t^\alpha(x) = c\int e^{i(x\xi + t|\xi|^\alpha \xi)}d\xi.$$

Our first result is concerned with the time behavior of the derivatives of order $\beta \in [0, (\alpha-1)/2]$ of the oscillatory integral $S_t(x)$.

LEMMA 2.1. *If $\beta \in [0,(\alpha-1)/2]$ and*

$$D^\beta S_t^\alpha(x) = c\int e^{i(x\xi+t|\xi|^\alpha\xi)}|\xi|^\beta d\xi$$

then

$$\| D^\beta S_t(\cdot) \|_\infty \leq c_{\beta,\alpha}|t|^{-(\beta+1)/(\alpha+1)}. \blacksquare \tag{2.2}$$

PROOF: The proof is based on the classical Van der Corput Lemma, and was given in detail in [24]. Therefore it will be omitted here. ■

As a consequence of Lemma 2.1 and its proof we have

COROLLARY 2.2. *For any* $(\theta, \beta) \in [0,1] \times [0, (\alpha-1)/2]$

$$\| D^{\theta\beta} S_t^{\alpha} * v_0 \|_{2/(1-\theta)} \le c|t|^{-\theta(\beta+1)/(\alpha+1)} \| v_0 \|_{2/(1+\theta)} \,. \quad \blacksquare \tag{2.3}$$

PROOF: First we introduce the analytic family of operators $(D^z S_t^{\alpha}) * v_0 = D^z W^{\alpha}(t) v_0$, with $z \in [0, (\alpha-1)/2] \times \mathbb{R}$.

From the argument used in the proof of the previous lemma it follows that

$$\| D^{\beta+i\gamma} W^{\alpha}(t) v_0 \|_{\infty} \le c(1+|z|)^c |t|^{-(\beta+1)/(\alpha+1)} \| v_0 \|_1$$

where $z = \beta + i\gamma$, and $c = c(\alpha, \beta)$.

Since $\{W^{\alpha}(t)\}_{-\infty}^{\infty}$ is a unitary group it follows that $\| D^{i\gamma} W^{\alpha}(t) v_0 \|_2 = \| D^{i\gamma} v_0 \|_2 = \| v_0 \|_2$.

By interpolation (see [28] Chapter V) we obtain (2.3). ■

The above estimates show that for general $v_0 \in L^1(\mathbb{R})$ the solution of IVP (2.1) satisfies the decay estimates

$$\| W^{\alpha}(t) v_0 \|_{\infty} \le c|t|^{-1/(\alpha+1)} \| v_0 \|_1 ,$$

and

$$\| D^{(\alpha-1)/2} W^{\alpha}(t) v_0 \|_{\infty} \le c|t|^{-1/2} \| v_0 \|_1 \,.$$

The last inequality tells us that the $(\alpha-1)/2$-derivative of the solution decays faster than the solution itself or any of its derivatives of order less than $(\alpha-1)/2$.

This fact was used in [24] to obtain lower bounds on the degree of the nonlinear perturbation in the IVP (1.1) with $\alpha = 2$ which guarantees that small solutions of (1.1) behaves asymptotically like the solutions of the associated linear problem. Further improvement in this direction has recently been given in [5] (again for the case $\alpha = 2$). For general $\alpha \ge 1$ we refer to [18] section 6.

Next we use the above estimates to obtain the global smoothing effect of Strichartz [29] type commented in the introduction.

LEMMA 2.3. *For any* $(\theta, \beta) \in [0,1] \times [0, (\alpha-1)/2]$

$$\| D^{\theta\beta/2} W^{\alpha}(t) v_0 \|_{L^q(\mathbb{R}:L^p)} = \left(\int_{-\infty}^{\infty} \| D^{\theta\beta/2} W^{\alpha}(t) v_0 \|_p^q \, dt \right)^{1/q} \le c \| v_0 \|_2 , \tag{2.4}$$

$$\| \int_{-\infty}^{\infty} D^{\theta\beta} W^{\alpha}(t-\tau) g(\cdot, \tau) d\tau \|_{L^q(\mathbb{R}:L^p)} \le c \| g \|_{L^{q'}(\mathbb{R}:L^{p'})} , \tag{2.5}$$

and

$$\| \int_0^t D^{\theta\beta} W^{\alpha}(t-\tau) g(\cdot, \tau) d\tau \|_{L^q([-T,T]:L^p)} \le c \| g \|_{L^{q'}([-T,T]:L^{p'})} \tag{2.6}$$

where $(q,p) = (2(\alpha+1)/\theta(\beta+1), 2/(1-\theta))$, *and* $\frac{1}{p} + \frac{1}{p'} = \frac{1}{q} + \frac{1}{q'} = 1$. ■

REMARKS: By fixing $\beta = (\alpha - 1)/2$ it follows from (2.4) that for any $\theta \in [0,1]$ and any $v_0 \in L^2(\mathbb{R})$

$$D^{\theta(\alpha-1)/4} W^\alpha(t) v_0 \in L^{2/(1-\theta)}(\mathbb{R}) \quad \text{a.e. in t.}$$

When $(\theta, \beta) = (1, (\alpha-1)/2)$, (2.4) reduces to the estimate (1.5) mentioned in the introduction.

We may remark that this lemma generalized those in [17] [18].

The proof combines the arguments used in [31] [21] [22] [10].

PROOF: By duality

$$\int\int D^{\theta\beta/2} W^\alpha(t) v_0(\cdot) g(x,t)\,dx\,dt = \int v_0(x) \left(\int D^{\theta\beta/2} W^\alpha(t) g(\cdot,t)\,dt \right) dx.$$

Thus (2.4) is equivalent to

$$\| \int D^{\theta\beta/2} W^\alpha(t) g(\cdot,t)\,dt \|_2 \leq c \| g \|_{L^{q'}(\mathbb{R}:L^{p'})} . \tag{2.7}$$

But P. Tomas [31] argument shows that

$$\begin{aligned} &\| \int D^{\theta\beta/2} W^\alpha(t) g(\cdot,t)\,dt \|_2^2 \\ &= \int \left(\int D^{\theta\beta/2} W^\alpha(t) g(\cdot,t)\,dt \right) \left(\overline{\int D^{\theta\beta/2} W^\alpha(\tau) g(\cdot,\tau)\,d\tau} \right) dx \\ &= \int\int g(x,t) \left(\int D^{\theta\beta} W^\alpha(t-\tau) \overline{g(\cdot,\tau)}\,d\tau \right) dx\,dt. \end{aligned}$$

Hence (2.5), (2.7) are equivalent, and consequently (2.4), (2.5) are too.

To prove (2.5) we follow the method introduced in [21] [22]. Thus, from Minkowski's integral inequality, the decay estimate (2.3), and fractional integration one finds that

$$\begin{aligned} &\| \int D^{\theta\beta} W^\alpha(t-\tau) g(\cdot,\tau)\,dt \|_{L^q(\mathbb{R}:L^p)} \\ &\leq \| \int \| D^{\theta\beta} W^\alpha(t-\tau) g(\cdot,\tau) \|_p \,d\tau \|_{L^q(\mathbb{R}:dt)} \\ &\leq c \| \int |t-\tau|^{-\theta(\beta+1)/(\alpha+1)} \| g(\cdot,\tau) \|_{p'} \,d\tau \|_{L^q} \\ &\leq c \| g \|_{L^{q'}(\mathbb{R}:L^{p'})} \end{aligned}$$

with $\frac{1}{q} = 1 - \frac{1}{q'} = \frac{1}{q'} - \left(1 - \frac{\theta(\beta+1)}{\alpha+1}\right)$; i.e. $q = 2(\alpha+1)/\theta(\beta+1)$. ■

§3. Local Smoothing Effects.

In the previous section we study global smoothing effects in $L^p_s(\mathbb{R})$ with $p > 2$. As was earlier mentioned similar results cannot hold when $p = 2$. However, a smoothing effect in H^s_{loc} will not contradict this principle. In this section we deal with the L^s-theory which is simpler.

Our main result in this section is the following

LEMMA 3.1. *If $v(x,t) = W^\alpha(t)v_0 = S_t^\alpha * v_0$ denotes the solution of the linear IVP (2.1) then*

$$\int_{-\infty}^{\infty} |D^{\alpha/2} v(x,t)|^2 dt = c_\alpha \parallel v_0 \parallel_2^2 \tag{3.1}$$

for any $x \in \mathbb{R}$. ■

REMARKS: Since the $W^\alpha(t)\cdot$'s are linear operators (3.1) can be written in the equivalent form

$$\int_{-\infty}^{\infty} |v(x,t)|^2 dt = c_\alpha \int_{-\infty}^{\infty} \frac{|\hat{v}_0(\xi)|^2}{|\xi|^\alpha} d\xi$$

whose value may be infinity.

In particular, for $v_0 > 0$, $v_0 \in L^1 \cap L^2$ we have that

$$\int_{-\infty}^{\infty} |W^\alpha(t) v_0|^2 dt = +\infty.$$

In the non-linear case for similar data ($v_0 > 0$, $v_0 \in L^1 \cap L^2$) this may not happen. Example: when $(\alpha, k) = (2,1)$ the solitary wave solution (with speed c) $u(x,t) = c_0 sech^2\left(\frac{1}{2}\sqrt{c}(x-ct)\right)$ satisfies

$$\int_{-\infty}^{\infty} |u(x,t)|^2 dt = \int_{-\infty}^{\infty} |u_0(x)|^2 dx < \infty.$$

PROOF: It is straightforward application of the Plancherel Theorem.

Thus changing variables $\eta = \xi \cdot |\xi|^\alpha$ $(\xi = \phi(\eta))$, using Plancherel, and returning to the original variable ξ one obtains that

$$\begin{aligned}
\int_{-\infty}^{\infty} |v(x,t)|^2 dt &\equiv \int_{-\infty}^{\infty} |\int e^{it|\xi|^\alpha \xi} e^{ix\xi} \hat{v}_0(\xi) d\xi|^2 dt \\
&= \int_{-\infty}^{\infty} |\int e^{it\eta} e^{ix\phi(\eta)} \hat{v}_0(\phi(\eta))\phi'(\eta) d\eta|^2 dt \\
&= c \int_{-\infty}^{\infty} |e^{ix\phi(\eta)} \hat{v}_0(\phi(\eta))\phi'(\eta)|^2 d\eta \\
&= c \int_{-\infty}^{\infty} |\hat{v}_0(\xi)|^2 |\phi'(\eta)| d\xi = c \int_{-\infty}^{\infty} \frac{|\hat{v}_0(\xi)|^2}{|\xi|^\alpha} d\xi.
\end{aligned}$$

As was mentioned in the introduction this local smoothing effect was first proved by T. Kato [15] (section 6) in the form given in (1.6) for solutions of the KdV equation. The same type of estimate appears implicitly in the work of P. Sjölin [27] and L. Vega [32] (see also [8], [20]) concerning the following problem proposed by L. Carleson [4]:

For which s can one guarantee that for $u_0 \in H^s(\mathbb{R}^n)$

$$\lim_{t \to 0} e^{it\Delta} u_0(x) = u_0(x) \quad \text{a.e. in } x?$$

In [4] it was shown that when $n = 1$, $s = 1/4$ is sufficient. It turns out that in this case ($n = 1$) the condition $s = 1/4$ is also necessary (see [9] [19]).

For $n \geq 2$ the best results known are: $s \geq n/4$, and $s > 1/2$ (see [27] [32] and reference therein).

This problem can be reduced to the proof of the following estimate for the Maximal Function : $\sup_{[0,1]} |e^{it\Delta} \cdot |(x)$,

$$\left(\int_{|x|\leq M} \sup_{0\leq t\leq 1} |e^{it\Delta} u_0|^2 dx \right)^{1/2} \leq c_M \parallel u_0 \parallel_{s,2} \tag{3.2}$$

The local smoothing effect proven in [8] [27] [32] for the free Schrödinger group is

$$\left(\int_{|x|\leq M} \int_{-T}^{T} |D^{1/2} e^{it\Delta} u_0|^2 dt dx \right)^{1/2} \leq c(M,T) \parallel u_0 \parallel_2 \tag{3.3}$$

which is similar to the estimate (1.6) commented in the introduction.

Using (3.3) we can sketch the proof of (3.2) for the case $s > 1/2$. By Sobolev Theorem it suffices to show that

$$\left(\int_{|x|\leq M} \int_{-1}^{2} |D_t^{1/2^+} e^{it\Delta} u_0|^2 dt dx \right)^{1/2} \leq c_M \parallel u_0 \parallel_{s,2} \tag{3.4}$$

where a^+ denotes any real number larger than a.

From the equation we have that $1/2^+$-derivatives in time are equivalent to 1^+-derivatives in space. Thus from (3.3) we obtain (3.4) whenever $s > 1/2$.

§4. Nonlinear Estimates.

This section is concerned with estimates for solutions of the IVP (1.1). We shall compare these with those deduced in sections 2, and 3 for solutions of the associated linear problem, and use them to establish well-posedness of the IVP (1.1) with $\alpha = 2$, and $k \geq 4$ in H^s for $s > 3/4$.

We begin by proving the energy estimate (1.2)

LEMMA 4.1. *If $u \in C([0,T] : H^\infty)$ is a solution of the IVP (1.1) then for any $s > 0$*

$$\frac{d}{dt} \parallel u(t) \parallel_{s,2} \leq c_{s,k} \parallel u(t) \parallel_\infty^{k-1} \parallel \partial_x u(t) \parallel_\infty \parallel u(t) \parallel_{s,2} \;. \; \blacksquare \tag{4.1}$$

To prove (4.1) we need the following commutator estimates proved by T. Kato and G. Ponce [16] as a consequence of the multilinear estimates of R.R. Coifman and Y. Meyer in [6].

LEMMA 4.2. *If $f, g \in S(\mathbb{R}^n)$, $s > 0$, and $1 < p < \infty$ then*

$$\parallel [J^s; f]g \parallel_p \leq c\{\parallel \nabla f \parallel_{p_1} \parallel g \parallel_{s-1,p_2} + \parallel f \parallel_{s,p_3} \parallel g \parallel_{p_4}\}, \tag{4.2}$$

and

$$\parallel fg \parallel_{s,p} \leq c\{\parallel f \parallel_{p_1} \parallel g \parallel_{s,p_2} + \parallel f \parallel_{s,p_3} \parallel g \parallel_{p_4}\} \tag{4.3}$$

where $p_2, p_3 \in (1, \infty)$ such that $\frac{1}{p} = \frac{1}{p_1} + \frac{1}{p_2} = \frac{1}{p_3} + \frac{1}{p_4}$. $\blacksquare$

PROOF: For $p_1 = p_4 = \infty$ the proof was given in [16] (Appendix). The general case follows easily by combining the argument used in [16] with the version of the Coifman-Meyer result found in [7] (page 22). ■

PROOF OF LEMMA 4.1: Since $u(t)$ satisfies the identity

$$\partial_t J^s u - D^\alpha \partial_x J^s u + u^k \cdot J^s \partial_x u + [J^s; u^k]\partial_x u = 0,$$

integration by parts and (4.2), (4.3) show that

$$\begin{aligned}\frac{d}{dt} \| u(t) \|_{s,2}^2 &= \int \partial_x(u^k)(J^s u)^2 - 2\int [J^s; u^k]\partial_x u \cdot J^s u \\ &\leq c \| u(t) \|_\infty^{k-1} \| \partial_x u(t) \|_\infty \| u(t) \|_{s,2}^2,\end{aligned}$$

which completes the proof. ■

Next we shall deduce two *a priori* estimates for smooth solutions of the IVP (1.1). For this purpose we restrict ourselves to the case $\alpha = 2$. In this case the operator modelling the dispersive effects in (1.1) is local. The proof of these estimates for solutions of the IVP (1.1) with general $\alpha \geq 1$ seems to be unknown (specially for (4.8)). However, in the case $\alpha = 1$ several techniques have been introduced to obtain the expected result (see [11] [23] [30])

First we shall prove the following version of Kato's local smoothing effect for solutions of (1.1) with $\alpha = 2$ and general k.

LEMMA 4.3. *If $u \in C([0,T] : H^\infty)$ is a solution of the IVP (1.1) with $\alpha = 2$ then for any $s > 0$*

(4.4)

$$\begin{aligned}&\sup_j \int_0^T \int |J^{s+1}(u\chi_j')|^2 dx dt \\ &\leq c\{1 + T + \int_0^T (\| \partial_x u \|_\infty \| u \|_\infty^{k-1} + \| u \|_\infty^k)(t) dt\} \sup_{[0,T]} \| u(t) \|_{s,2}^2\end{aligned}$$

where the constant c depends on k, s, and $\chi(\cdot)$. ■

PROOF: Integration by parts shows that

$$\begin{aligned}(4.5) \qquad &\frac{1}{2}\frac{d}{dt}\int (J^s u)^2 \chi_j dx + \frac{3}{2}\int (J^s \partial_x u)^2 \chi_j' dx \\ &- \frac{1}{2}\int (J^s u)^2 \chi_j''' dx + \int J^s(u^k \cdot \partial_x u) J^s u \cdot \chi_j = 0.\end{aligned}$$

Since the last term in (4.5) can be rewritten as

(4.6)

$$\begin{aligned}\int J^s(u^k \cdot \partial_x u) J^s u \chi_j &= \int u^k J^s \partial_x u \cdot J^s u \chi_j + \int [J^s; u^k]\partial_x u J^s u \chi_j \\ &= -\frac{1}{2}\int \partial_x(u^k \chi_j)(J^s u)^2 + \int [J^s; u^k]\partial_x u \cdot J^s u \chi_j,\end{aligned}$$

using (4.2) its absolute value can be bounded by

$$c(\| \partial_x u \|_\infty \| u \|_\infty^{k-1} + \| u \|_\infty^k) \| J^s u \|_2^2 .$$

Thus, integrating in the time interval $[0,T]$ the identity (4.5) we find from (4.6) that

(4.7)
$$\int_0^T \int (J^s \partial_x u)^2 \chi_j' dx dt$$
$$\leq c\{1 + T + \int_0^T (\| \partial_x u \|_\infty \| u \|_\infty^{k-1} + \| u \|_\infty^k)(t) dt\} \sup_{[0,T]} \| J^s u(t) \|_2^2 .$$

On the other hand

$$\| J^{s+1}(u\chi_j') \|_2 \leq \| J^s(u\chi_j') \|_2 + \| J^s(\partial_x u \chi_j') \|_2 + \| J^s(u\chi_j'') \|_2,$$

and

$$J^s(\partial_x u \chi_j') = J^s \partial_x u \cdot \chi_j' + [J^s; \chi_j'] \partial_x u.$$

Therefore from (4.2), (4.3), and the Sobolev Embedding Theorem it follows that

$$\| u\chi_j' \|_{s+1,2} \leq c\{\| J^s \partial_x u \cdot \chi_j' \|_2 + \| u \|_{s,2}\}$$

which combined with (4.7) leads to (4.4). ■

Similar to the splitting argument introduced by J. Ginibre and Y. Tsutsumi in [12] we have

LEMMA 4.4. *If $u \in C([0,T] : H^\infty)$ is a solution of the IVP (1.1) then for any $s > 0$*

(4.8)
$$\sum_j \sup_{[0,T]} \int |J^{s-1}(u \cdot \chi_j')|^2 dx$$
$$\leq c\{1 + T + \int_0^T \| u(t) \|_\infty^{2k} dt\} \sup_{[0,T]} \| u(t) \|_{s,2}^2$$

where the constant c depends only on k, s, and $\chi(\cdot)$. ■

PROOF: From the equation (1.1) with $\alpha = 2$ one has that

$$\partial_t(u\chi_j') + \partial_x^3(u\chi_j') - 3\partial_x(\partial_x u \chi_j'') - u\chi_j^{(4)} + u^k \partial_x u \chi_j' = 0.$$

Thus integration by parts leads to

(4.9)
$$\frac{d}{dt} \| u\chi_j' \|_{s,2}^2 \leq c \int (|J^{s-1}(\partial_x u \chi_j'')| \cdot |J^{s-1}\partial_x(u\chi_j')|$$
$$+ |J^{s-1}(u\chi_j^{(4)})| \cdot |J^{s-1}(u\chi_j')| + |J^{s-1}(u^k \partial_x u \chi_j')| \cdot |J^{s-1}(u\chi_j')|) dx.$$

Now integrating (4.9) in the time interval $[0, T]$, taking supremum in this interval, summing on j, using (4.2) (4.3) and the following localized version of the H^s-norm with $s \geq 0$: there exists $c_s > 0$ such that

$$c_s^{-1} \| f \|_{s,2} \leq \left(\sum_{-\infty}^{\infty} \| f \cdot \chi_j' \|_{s,2}^2 \right)^{1/2} \leq c_s \| f \|_{s,2} \tag{4.10}$$

for any $f \in H^s(\mathbb{R})$, we obtain (4.8). ■

The proof of (4.10) when s is a positive integer is immediate. The general case follows by interpolation (see [1] Chapter 5). Also we may remark that the right hand side of (4.10) still holds with $\chi_j^{(\ell)}(\cdot), \ell = 2, \ldots$, instead of $\chi_j'(\cdot)$.

Combining lemmas 2.3, 4.1-4.4 we shall obtain an *a priori* estimate of the $L^4([0,T] : L^\infty(\mathbb{R}))$-norm of $\partial_x u$ with T, and the bound depending only on $\| u_0 \|_{s,2}$, and $s = 3/4^+$ for $k \geq 4$.

This estimate basically proves that the IVP (1.1) with $\alpha = 2$ and $k \geq 4$ is locally well posed in H^s for any $s > 3/4$ (in Theorem A in the introduction this was proved for $s > 3/2$).

From (2.4), (4.2), (4.10) one sees that

(4.11)

$$\begin{aligned}
\| \partial_x u(t) \|_{L^4([0,T]:L^\infty)} &\leq c \| u_0 \|_{3/4,2} + c \int_0^T \| D^{7/4} u^{k+1}(t) \|_2 \, dt \\
&\leq c \| u_0 \|_{3/4,2} + c T^{1/2} \{ \int_0^T \| u^{k+1}(t) \|_{7/4}^2 \, dt \}^{1/2} \\
&\leq c \| u_0 \|_{3/4,2} + c T^{1/2} \{ \sum_j \int_0^T \int | J^{7/4}((u \cdot \chi_j')^{k+1}) |^2 dx dt \}^{1/2} \\
&\leq c \| u_0 \|_{3/4,2} + c T^{1/2} \{ \sup_j \int_0^T \int | J^{7/4}(u \chi_j') |^2 dx dt \}^{1/2} \cdot \\
&\cdot \{ \sum_j \sup_{[0,T]} \| u \chi_j' \|_\infty^{2k} \}^{1/2}.
\end{aligned}$$

Thus it remains to bound the two factors in the last term above. For the first we use (4.4), (4.1), and for the second we use the Three Lines Theorem to interpolate between the following spaces:

$$\ell^\infty L^\infty H^s \quad \text{(estimate (4.1))},$$

and

$$\ell^2 L^\infty H^{s-1} \quad \text{(estimates (4.8) and (4.1))}$$

to obtain an estimate for

$$\ell^{2k} L^\infty H^{1/2^+}.$$

In this step we have used that $k \geq 4$.

Inserting the above estimates in (4.11) we obtain for sufficiently small T the desired *a priori* estimate of $\int_0^T \| \partial_x u(\theta) \|_\infty^4 \, d\theta$.

This essentially shows that the IVP (1.1) is locally well posed in H^s for $s > 3/4$ for $\alpha = 2$ and $k \geq 4$.

The same method proves local well-posedness for $\alpha = 2$ in H^s with $s > 9/8$ if $k = 1$, $s > 11/12$ if $k = 2$, and $s > 13/16$ if $k = 3$ (see [17]).

To understand the restriction of this approach and where some improvements can be obtained we compare the results of Lemmas 4.3, and 4.4 with those obtained in section 3 for the associated linear problem.

As was already mentioned (4.4) is an $L^2_{loc}(\mathbb{R}^2)$ estimate in comparison with the linear result (3.1) which is an $L^\infty((\mathbb{R}) : L^2(\mathbb{R} : dt))$ estimate. The main point here is that the L^∞-norm is global in space.

Next we compare (4.8) with that corresponding to the linear case. Thus estimating the solution of the associated linear problem in the norm in the left hand side of (4.8), and using a result of L. Vega [33] it follows that

$$(4.12) \qquad \sum_j \sup_{[0,T]} \int |J^{s-1}(W^2(t)u_0 \cdot \chi_j'(\cdot))|^2 dx \leq c(1+T)^2 \cdot \| u_0 \|^2_{\ell,2}$$

for any $\ell > s - 1/4$ (see(4.10)).

Thus the loss of derivatives in the right hand side is $3/4^+$ instead of 1 in (4.8). More importantly, for the last inequality the splitting argument is not necessary (see estimate (5.4) below).

§5. Final Results.

In this section we shall state and discuss our results concerning the well- posedness of the IVP (1.1) in H^s. To simplify the exposition we restrict ourselves to the case $\alpha = 2$. However, we may remark that our approach applies to any $\alpha \geq 1$, which is not the case of the techniques discussed in the previous section.

THEOREM 5.1. *Let $s > 3/4$ (real).*

For any $u_0 \in H^s(\mathbb{R})$ there exists a unique solution $u(\cdot)$ of the IVP (1.1) with $\alpha = 2$ satisfying

$$(5.1) \qquad u \in C([-T,T] : H^s) \cap L^4([-T,T] : L^\infty_{s+1/4}),$$

$$(5.2) \qquad \sup_x \int_{-T}^{T} |D^{s+1}u(x,t)|^2 dt < c,$$

and

$$(5.3) \qquad \int \sup_{[-T,T]} |u(x,t)|^2 dx < c$$

where $c = c(\| u_0 \|_{s,2}; k)$ and T has a lower bound depending only on k and on $\| u_0 \|_{3/4^+,2}$.

Moreover, for any $T' < T$ there exists a neighborhood Ω of u_0 in H^s such that the map $\tilde{u}_0 \longmapsto \tilde{u}(t)$ from Ω in $C([-T', T'] : H^s)$ is continuous. ■

As a consequence of this theorem and the third conservation law $\Phi_3(\cdot)$ (see introduction) we have,

THEOREM 5.2. *Let $s \geq 1$ (real).*

For any $u_0 \in H^s(\mathbb{R})$ with $\| u_0 \|_{1,2} \leq \nu_k$ the solution of the IVP (1.1) provided by Theorem 5.1 can be extended to any time interval $[-T_0, T_0]$ in which it remains in the same class.

The ceiling $\nu_k = \infty$ for $k = 1, 2, 3$. ■

Notice that the estimates (5.1)-(5.2) are similar to those deduced in sections 2 and 3 for solutions of the associated linear problem. Thus, Theorem 5.1 shows that the solution of the nonlinear problem has (locally) the same regularity as that of the associated linear problem. In particular, we avoid the use of the splitting argument (lemma 4.4) of Ginibre-Tsutsumi by combining the sharp form of the local smoothing effect (3.1) with the following stronger version of (4.12) found in [33]:

$$\int \sup_{[-T,T]} |W^2(t)u_0|^2 dx \ < c(T) \cdot \| u_0 \|_{\ell,2}^2 \tag{5.4}$$

for any $\ell > 3/4$. Notice that this inequality is similar to that in (3.2) (involving the maximal function), and the one in (5.3).

The proof of Theorems 5.1 and 5.2 and their extension to the case $\alpha \geq 1$ will appear somewhere else. However, it should be remarked that the proof of Theorem 5.1 can be obtained by using the contraction principle. In this case, one just needs to combine the integral equation (1.8) with the estimates (2.4), (3.1), (4.1)-(4.3), and (5.4).

As was commented in the introduction Theorem 5.2 answers the question left open by J.C. Saut and R. Temam [26] and by T. Kato [15]. Previous results in the case of local well-posedness for $\alpha = 2$ were restricted to $s > 9/8$ for $k = 1$, $s > 11/12$ for $k = 2$, and $s > 13/16$ for $k = 4$ (see [17]). Theorem 5.1 for $k \geq 4$ was essentially proved in section 4.

In the case of global well-posedness previous results for the case $(\alpha, k) = (2, 1)$ were restricted to $s \geq 2$ (see [2] [3] [26]).

For $k = 2, 3$ a slightly weaker version of Theorem 5.2 was proved in [17]. For $k \geq 4$ a similar version (which in particular contains global well-posedness) follows by combining the results in section 4 with $\Phi_3(\cdot)$ (third conservation law).

References

[1] Bergh, J., and Löfstöm, J., *Interpolation Spaces*, Springer-Verlag, Berlin and New York (1970).

[2] Bona, J. L., and Scott, R., *Solutions of the Korteweg-de Vries equation in fractional order Sobolev spaces*, Duke Math. J. **43** (1976), 87–99.

[3] Bona, J. L., and Smith, R., *The initial value problem for the Korteweg-de Vries equation*, Philos. Trans. Roy. Soc. London **Ser A 278** (1975), 555–601.

[4] Carleson, L., *Some analytical problems related to statistical mechanics, Euclidean Harmonic Analysis*, Lecture Notes in Math., Springer-Verlag, Berlin and New York, **779** (1979), 9–45.

[5] Christ, F. M., and Weinstein, M. I., *Dispersion of small amplitude solutions of the generalized Korteweg-de Vries equation*, preprint.

[6] Coifman, R. R., and Meyer, Y., *Au delá des opérateurs pseudodifféntiels*, Asterisque **57**, Société Mathématique de France (1978).

[7] Coifman, R. R., and Meyer, Y., *Nonlinear harmonic analysis, operator theory and P.D.E.*, Beijing Lectures in Harmonic Analysis, Princeton University Press (1986), 3–45.

[8] Constantin, P. and Saut, J. C., *Local smoothing properties of dispersive equations*, Journal Amer. Math. Soc. **1** (1988), 413–446.

[9] Dahlberg, B. and Kenig, C. E., *A note on the almost everywhere behavior of solutions to the Schrödinger equation, Harmonic Analysis*, Lecture Notes in Math., Springer-Verlag, Berlin and New York, **908** (1982), 205–208.

[10] Ginibre, J. and Velo, G., *Scattering theory in the energy space for a class of nonlinear Schrödinger equation*, J. Math. Pures et Appl. **64** (1985), 363–401.

[11] Ginibre, J. and Velo, G., *Commutator expansions and smoothing properties of generalized Benjamin-Ono equations*, Ann. IHP (Phys. Theor.) **51** (1989), 221–229.

[12] Ginibre, J., and Tsutsumi, Y., *Uniqueness for the generalized Korteweg-de Vries equations*, SIAM J. Math. Anal. **20** (1989), 1388–1425.

[13] Kato, T., *Quasilinear equations of evolutions, with applications to partial differential equations*, Lecture Notes in Math., Springer-Verlag, Berlin and New York, **448** (1975), 27–50.

[14] Kato, T., *On the Korteweg-de Vries equation*, Manuscripta Math. **29** (1979), 89–99.

[15] Kato, T., *On the Cauchy problem for the (generalized) Korteweg-de Vries equation*, Advances in Mathematics Supplementary Studies, Studies in Applied Math. **8** (1983), 93–128.

[16] Kato, T., and Ponce, G., *Commutator estimates and the Euler and Navier-Stokes equations*, Comm. Pure Appl. Math. **41** (1988), 891–907.

[17] Kenig, C. E., Ponce, G., and Vega, L., *On the (generalized) Korteweg-de Vries equation*, Duke Math. J. **59** (1989), 585–610.

[18] Kenig, C. E., Ponce, G., and Vega, L., *Oscillatory integrals and regularity of dispersive equations*, preprint.

[19] Kenig, C.E., and Ruiz, A., *A strong type (2,2) estimate for the maximal function associated to the Schrödinger equation*, Trans. Amer. Math. Soc. **280** (1983), 239–246.

[20] Kruzhkov, S. N., and Framinskii, A. V., *Generalized solutions of the Cauchy problem for the Korteweg-de Vries equation*, Math. U.S.S.R. Sbornik **48** (1984), 93–138.

[21] Marshall, B., *Mixed norm estimates for the Klein-Gordon equation*, Proceedings of a Conference in Harmonic Analysis, Chicago (1981), 638–649.

[22] Pecher, H., *Nonlinear small data scattering for the wave and Klein-Gordon equation*, Math. Z. **185**(1985), 261–270.

[23] Ponce, G., *Smoothing properties of solutions of the Benjamin-Ono equation*, Lecture Notes Pure Appl. Math. **122** (C. Sadosky Ed) Marcel Dekker, Inc. (1990), 667–679.

[24] Ponce, G. and Vega, L., *Nonlinear small data scattering for the generalized Korteweg-de Vries equation*, to appear in J. Funct. Anal.

[25] Saut, J.C., *Sur quelque généralisations de l'equation de Korteweg-de Vrie*, J. Math. Pure Appl. **58** (1979), 21–61.

[26] Saut, J. C., and Temam, R.*Remarks on the Korteweg-de Vries equation*, Israel J. Math. **24** (1976), 78–87.

[27] Sjölin, P., *Regularity of solutions to the Schrödinger equation*, Duke Math. J. **55** (1987), 699–715.

[28] Stein, E. M., and Weiss, G., *Introduction to Fourier Analysis in Euclidean Spaces*, Princeton University Press (1971).

[29] Strichartz, R. S., *Restriction of Fourier transform to quadratic surfaces and decay of solutions of wave equations*, Duke Math. J. **44** (1977), 705–714.

[30] Tom, M. M., *Smoothing properties of some weak solutions of the Benjamin-Ono equation*, to appear in Diff. and Int. Eqs.

[31] Tomas, P., *A restriction theorem for the Fourier transform*, Bull. A.M.S. **81** (1975), 477–478.

[32] Vega, L., *Schrödinger equations: pointwise convergence to the initial data*, Proc. Amer. Math. Soc. **102** (1988), 874–878.

[33] Vega, L., *Doctoral Thesis, Universidad Autonoma*, Madrid, Spain(1988).

On Schrödinger operators with magnetic fields

By

Akira IWATSUKA

Department of Mathematics, Kyoto University
Sakyo-ku, Kyoto 606, JAPAN

1 Introduction

We shall consider the Schrödinger operators with magnetic fields:

$$L = -\sum_{j=1}^{n}(\partial_j - ib_j)^2 + V, \tag{1.1}$$

where $\partial_j = \partial/\partial x_j$, $i = \sqrt{-1}$, and b_j and V are the operators of multiplication by real-valued functions on $\boldsymbol{R}^n$, $b_j(x)$ and $V(x)$, respectively. V and $\boldsymbol{b} = (b_1, \ldots, b_n)$ are called a scalar potential and a (magnetic) vector potential, respectively, and the corresponding magnetic field is the skew symmetric matrix-valued function (or distribution) $B \equiv \operatorname{curl} \boldsymbol{b}$ with (j, k) components

$$B_{jk} = \partial_j b_k - \partial_k b_j \quad \text{for } j,\, k = 1, \ldots, n. \tag{1.2}$$

The operator L is the time-independent Schrödinger Hamiltonian of a particle moving under the influence of the electric potential V and the magnetic field B. First of all, there is an important physical fact that the "physics" of the system depends only on B. This corresponds to gauge invariance of the Schrödinger operators: all $L = L(\boldsymbol{b})$ with common $B = \operatorname{curl} \boldsymbol{b}$ are unitarily equivalent to each other. It should be noted that there is a fact called the Aharonov-Bohm effect to the contrary: if the region to which the motion of the particle is confined is not the whole space $\boldsymbol{R}^n$ (more precisely, if it is not simply-connected), gauge invariance is not necessarily true; we shall present some related result later (see Section 3). Gauge invariance holds, however, for the operators $L = L(\boldsymbol{b})$ considered in the whole space $\boldsymbol{R}^n$: if two different vector potentials $\boldsymbol{b}$ and $\boldsymbol{b}'$ in $\boldsymbol{R}^n$ give the same magnetic field $\operatorname{curl} \boldsymbol{b} = \operatorname{curl} \boldsymbol{b}'$, then there exists a real-valued function g in $\boldsymbol{R}^n$ satisfying $\nabla g = \boldsymbol{b}' - \boldsymbol{b}$, which gives $e^{ig}(\partial_j - ib_j)e^{-ig} = \partial_j - ib'_j$ and thus $L(\boldsymbol{b}') = e^{ig}L(\boldsymbol{b})e^{-ig}$. Though this reasoning is rather formal unless $\boldsymbol{b}$ and $\boldsymbol{b}'$ are smooth, this can be justified also in the case of singular potentials as is shown in [17]. Note also that, in the case of the space dimension $n = 1$, the vector potential is always gauged away, i.e., L is unitarily equivalent to $-\dfrac{d^2}{dx^2} + V(x)$.

In the present lecture, we would like to discuss four topics concerning the operator L:

(i) Essential self-adjointness.

(ii) The Aharonov-Bohm effect on the ground state energy.

(iii) 2-dimensional systems.

(iv) Purely discrete spectrum.

These are only a few aspects of Schrödinger operators with magnetic fields. We refer the reader to the reviews [9, 2] for a more extensive survey and a more complete list of references.

2 Self-adjoint realization of L

To realize properly the formal operator L as a self-adjoint operator in $L^2(\boldsymbol{R}^n)$ is a fundamental mathematical problem from the beginning of quantum mechanics. One way is to consider quadratic forms and the other is to extend an operator defined at first in a good subspace of $L^2(\boldsymbol{R}^n)$, say, $C_0^\infty(\boldsymbol{R}^n)$ (= the set of all C^∞ functions in $\boldsymbol{R}^n$ with compact support).

As for the form extension, consider the assumption

(H2.1) $\boldsymbol{b} \in L^2_{\mathrm{loc}}(\boldsymbol{R}^n)^n$ and $V \in L^1_{\mathrm{loc}}(\boldsymbol{R}^n)$, $V \geq 0$,

and the maximal form h associated with the formal operator L. Namely, define a quadratic form h in $L^2(\boldsymbol{R}^n)$ with the domain of definition

$$Q(h) \equiv \{ \phi \in L^2(\boldsymbol{R}^n) \mid (\nabla - i\boldsymbol{b})\phi \in (L^2(\boldsymbol{R}^n))^n,\ V^{1/2}\phi \in L^2(\boldsymbol{R}^n) \}$$

by

$$h(\phi, \psi) \equiv \sum_{j=1}^{n}((\partial_j - ib_j)\phi, (\partial_j - ib_j)\psi) + (V^{1/2}\phi, V^{1/2}\psi)) \tag{2.1}$$

for ϕ, $\psi \in Q(h)$, where differentiation is understood in the distribution sense. Note that $b_j\phi$, $V^{1/2}\phi$ are well defined as distribution since they belong to L^1_{loc} for $\phi \in L^2(\boldsymbol{R}^n)$. Then h is densely defined since $Q(h) \supset C_0^\infty(\boldsymbol{R}^n)$. Moreover it is easy to see that h is closed. Therefore we have a unique nonnegative self-adjoint operator H associated with this non-negative form h. It is known that h is identical to the minimal form h_{min} = the form closure of h restricted to $C_0^\infty(\boldsymbol{R}^n)$:

Theorem 2.1 (Kato[16], Simon[23]) *Suppose that (H2.1) holds. Then $C_0^\infty(\boldsymbol{R}^n)$ is a form core for H. Namely $C_0^\infty(\boldsymbol{R}^n)$ is dense in $Q(h)$ with respect to the norm $\|\phi\|_h \equiv (h(\phi,\phi) + (\phi,\phi))^{1/2}$.*

As for the proof, we only mention that this theorem and its proof are closely related to the following operator inequality:

$$|e^{-tH}f| \leq e^{-t(-\Delta+V)}|f| \quad \text{a.e. pointwise in } \boldsymbol{R}^n$$

for all $f \in L^2(\boldsymbol{R}^n)$ and $t > 0$. This inequality holds under the assumption (H2.1) and is equivalent to

$$|(H+\lambda)^{-1}f| \leq (-\Delta + V + \lambda)^{-1}|f| \quad \text{a.e. pointwise in } \boldsymbol{R}^n \tag{2.2}$$

for all $f \in L^2(\boldsymbol{R}^n)$ and $\lambda > 0$. These operator inequalities are operator versions of Kato's inequality (see, e.g., [9, 6, 24, 19]):

$$-\Delta|u| \leq -\operatorname{Re}\{(\operatorname{sgn}\overline{u})(\sum_j(\partial_j - ib_j)^2u)\}$$

in the distribution sense, where $\operatorname{sgn}\overline{u} = \overline{u}/|u|$ for $u \neq 0$, 0 for $u = 0$. These inequalities are very important and various results are obtainable by applying them (see, e.g., [9]).

Next let us consider the operator $\dot{L} = L$ restricted to $C_0^\infty(\boldsymbol{R}^n)$. Then the problem is to investigate conditions under which $\dot{L}$ determines a unique self-adjoint operator in the Hilbert space $L^2(\boldsymbol{R}^n)$, i.e., $\dot{L}$ is essentially self-adjoint. The operator L is singular in two respects: first, the potentials may have local singularities; second, the region of the space is the whole $\boldsymbol{R}^n$. In the case where $V \geq 0$, we have a result of [19]:

Theorem 2.2 ([19]) *Suppose that the assumption*

(H2.2) $\boldsymbol{b} \in L^4_{\text{loc}}(\boldsymbol{R}^n)^n$, $\operatorname{div}\boldsymbol{b} \in L^2_{\text{loc}}(\boldsymbol{R}^n)$, *and* $0 \leq V \in L^2_{\text{loc}}(\boldsymbol{R}^n)$

holds. Then $\dot{L}$ is essentially self-adjoint.

This result is decisive as to the assumption on the local singularity of the potentials in the sense that (H2.2) is minimal to assure that $\dot{L}$ defines an operator from $C_0^\infty(\boldsymbol{R}^n)$ to $L^2(\boldsymbol{R}^n)$, for we have

$$L = -\Delta + 2i\boldsymbol{b}\cdot\nabla + i\operatorname{div}\boldsymbol{b} + |\boldsymbol{b}|^2 + V.$$

As to the assumption on the behavior at infinity of the potentials, the result given by [10] is fundamental ([10] allows some local singularities of the potentials; for simplicity, we assume they are C^∞):

Theorem 2.3 ([10]) *Suppose that $\boldsymbol{b}$ and V are C^∞ and there exists a continuous non-decreasing function $Q(r) > 0$ for $r \in [0, \infty)$ such that*

$$V(x) \geq -Q(|x|) \quad \text{for } x \in \boldsymbol{R}^n, \tag{2.3}$$

$$\int^\infty Q(r)^{-1/2}dr = \infty, \tag{2.4}$$

where $|x| = \sqrt{x_1^2 + \cdots + x_n^2}$, $x = (x_1, \ldots, x_n)$. Then $\dot{L}$ is essentially self-adjoint.

Roughly speaking, this sufficient condition for the essential self-adjointness of $\dot{L}$ is almost necessary as well in the one dimensional case, if one requires a suitable condition on the decay rate at infinity of V' and V'' (see [21, Th.X.9]). Note that the condition (2.3) concerns the growth rate at infinity of the negative part $V_- \equiv \max(0, -V)$ of V. As for the magnetic potential, no conditions other than its local regularity are required.

Recently we have shown ([14]) that the condition on the growth rate at infinity of V_- can be relaxed in the presence of the magnetic field B (rather than the magnetic potential because of gauge invariance). Define the magnitude of the skew-symmetric matrix $B(x)$ by

$$|B(x)| = \Big\{\sum_{j<k} B_{jk}(x)^2\Big\}^{1/2}.$$

Then we have

Theorem 2.4 *Suppose that* $\boldsymbol{b}$ *and* V *are* C^∞ *and there exists a continuous non-decreasing function* $Q(r) > 0$ *for* $r \in [0, \infty)$ *such that*

$$V(x) + |B(x)| \ \geq \ -Q(|x|) \quad \textit{for } x \in \boldsymbol{R}^n, \tag{2.5}$$

$$\left\{\frac{|\partial^\alpha B_{jk}(x)|}{|B(x)|+1}\right\}^{3-|\alpha|} \ \leq \ Q(|x|) \tag{2.6}$$

$$\textit{for } x \in \boldsymbol{R}^n,\ j,\ k = 1,\ldots,n,\ |\alpha| = 1,\ 2,$$

$$\int^\infty Q(r)^{-1/2}dr \ = \ \infty, \tag{2.7}$$

where α *are multi-indices* $(\alpha_1,\ldots,\alpha_n)$, $\partial^\alpha = \partial_1^{\alpha_1}\cdots\partial_n^{\alpha_n}$. *Then* $\dot{L}$ *is essentially self-adjoint.*

In view of (2.5) in the above theorem, $V_-(x)$ is allowed to grow as fast as $|B(x)| + Q(|x|)$, where Q is a function satisfying (2.7), which is the same as in Theorem 2.3 and by which Q should not grow faster than r^δ ($\delta > 2$). Hence, in the case where $|B(x)|$ grows sufficiently fast, say, at a rate comparable to $|x|^\delta$ ($\delta > 2$), Theorem 2.4 gives a wider class of potentials assuring the essential self-adjointness of $\dot{L}$ than those given in Theorem 2.3.

A quantum mechanical interpretation of essential self-adjointness is that the uniqueness of a self-adjoint realization means the uniqueness of the dynamics of the quantum mechanical particle. If the particle reaches infinity in a finite time, some boundary condition at infinity should be imposed so as to determine a reflection law, in which case $\dot{L}$ is *not* essentially self-adjoint. Thus Theorem 2.4 can be interpreted as follows: the magnetic field can prevent the particle from going to infinity in a finite time even though the scalar potential is highly repulsive so that the particle would go to infinity in a finite time if the magnetic field were absent.

The proof of Theorem 2.4 can be reduced to a lemma concerning a second-order elliptic operator

$$T = \sum_{j,k=1}^{n} (i\partial_j + f_j)G_{jk}(i\partial_k + f_k) + W, \tag{2.8}$$

where G_{jk}, f_j, W are operators of multiplication by functions on $\boldsymbol{R}^n$, $G_{jk}(x)$, $f_j(x)$, $W(x)$, respectively. We assume

(H2.3) $G_{jk}(x)$ are C^∞ complex-valued functions on $\boldsymbol{R}^n$, $f_j(x)$, $W(x)$ are C^∞ real-valued functions on $\boldsymbol{R}^n$ and $G_{jk}(x) = \overline{G_{kj}(x)}$ for $x \in \boldsymbol{R}^n$ and $j,\ k = 1,\ldots,n$.

We further assume the ellipticity condition on T:

(H2.4) The symmetric matrix $(\alpha_{jk}(x))$ is positive-definite at each point $x \in \boldsymbol{R}^n$ where $\alpha_{jk}(x) = \operatorname{Re} G_{jk}(x)$ (Re means the real part).

We define $\alpha^*(r)$ for $r > 0$ by

$$\alpha^*(r) = \max_{|x|=r}\{\text{the greatest eigenvalue of } (\alpha_{jk}(x))\}. \tag{2.9}$$

Then we have

Lemma 2.5 *Let the assumptions (H2.3) and (H2.4) hold. Suppose that the Hermitian matrix* $(G_{jk}(x))$ *is nonnegative-definite, i.e.,*

$$(G_{jk}(x)) \geq 0 \quad \text{for } x \in \mathbf{R}^n. \tag{2.10}$$

Moreover, suppose that there exist a continuous non-decreasing function $M(r) > 0$ *for* $r \in [0, \infty)$ *such that*

$$W(x) \geq -M(|x|) \quad \text{for } x \in \mathbf{R}^n, \tag{2.11}$$

$$\int^{\infty} \{\alpha^*(r)M(r)\}^{-1/2} dr = \infty, \tag{2.12}$$

Then T *restricted to* $C_0^\infty(\mathbf{R}^n)$ *is essentially self-adjoint in* $L^2(\mathbf{R}^n)$.

This lemma is quite similar to [10, Theorem 1] except that [10] treats real-valued G_{jk}, and this lemma can be proven in the same manner as [10] with little modification.

Proof of Theorem 2.4 (assuming Lemma 2.5). Let

$$G_{jk} \equiv \delta_{jk} + i\beta_{jk}, \tag{2.13}$$

where δ_{jk} are the Kronecker delta and β_{jk} are C^∞ real-valued functions on $\mathbf{R}^n$ such that

$$\beta_{jk} + \beta_{kj} = 0 \quad \text{for } j,\, k = 1, \ldots, n, \tag{2.14}$$

which will be specified later. Let $f_j = b_j + e_j$ and $\Pi_j = i\partial_j + b_j$. Then we have

$$\begin{aligned} T &= \sum_{j,k} (\Pi_j + e_j) G_{jk} (\Pi_k + e_k) + W \\ &= \sum_{j,k} \Pi_j G_{jk} \Pi_k + \sum_{j,k} e_j G_{jk} \Pi_k + \sum_{j,k} \Pi_j G_{jk} e_k \\ &\quad + \sum_{j,k} e_j G_{jk} e_k + W, \end{aligned} \tag{2.15}$$

Further we have

$$\sum_{j,k} \Pi_j G_{jk} e_k = \sum_{j,k} G_{jk} e_k \Pi_j + i \sum_{j,k} \partial_j (G_{jk} e_k). \tag{2.16}$$

and we have by (2.13)

$$\sum_{j,k} \Pi_j G_{jk} \Pi_k = \sum_j \Pi_j^2 + i \sum_{j,k} \beta_{jk} \Pi_j \Pi_k - \sum_{j,k} (\partial_j \beta_{jk}) \Pi_k, \tag{2.17}$$

$$\sum_{j,k} e_j G_{jk} e_k = \sum_j e_j^2 + i \sum_{j,k} e_j \beta_{jk} e_k = \sum_j e_j^2, \tag{2.18}$$

where we used the equality $\sum_{j,k} e_j \beta_{jk} e_k = 0$ obtainable from (2.14) in the last step. Therefore we have by (2.15), (2.16), (2.17) and (2.18)

$$\begin{aligned} T &= \sum_j \Pi_j^2 + i \sum_{j,k} \beta_{jk} \Pi_j \Pi_k - \sum_{j,k} (\partial_j \beta_{jk}) \Pi_k + \sum_{j,k} e_j G_{jk} \Pi_k \\ &\quad + \sum_{j,k} G_{jk} e_k \Pi_j + i \sum_{j,k} \partial_j (G_{jk} e_k) + \sum_j e_j^2 + W \\ &= \sum_j \Pi_j^2 + i \sum_{j<k} \beta_{jk} [\Pi_j, \Pi_k] - \sum_{j,k} (\partial_j \beta_{jk}) \Pi_k \\ &\quad + 2 \sum_{j,k} e_j (\operatorname{Re} G_{jk}) \Pi_j + i \sum_{j,k} \partial_j (G_{jk} e_k) + \sum_j e_j^2 + W, \end{aligned}$$

where $[A, B]$ is the commutator $AB - BA$ and we used (2.14). Put $e_j = \frac{1}{2}\sum_k(\partial_k\beta_{kj})$ to cancel the first-order term with respect to Π_j in the last member. Then, since $[\Pi_j, \Pi_k] = iB_{jk}$, we have

$$T = \sum_j \Pi_j^2 - \sum_{j<k} \beta_{jk}B_{jk} + R + W,$$

where, by (2.13),

$$\begin{aligned} R &= i\sum_j \partial_j e_j - \sum_{j,k} \partial_j(\beta_{jk}e_k) + \sum_j e_j^2 \\ &= -\frac{1}{2}\sum_{j,k,m}\left\{(\partial_j\beta_{jk})(\partial_m\beta_{mk}) + \beta_{jk}(\partial_j\partial_m\beta_{mk})\right\} \\ &\qquad + \frac{1}{4}\sum_j(\sum_k \partial_j\beta_{jk})^2, \end{aligned} \tag{2.19}$$

since $\sum_j \partial_j e_j = \frac{1}{2}\sum_{j,k}(\partial_j\partial_k\beta_{kj}) = 0$ by (2.14). Let $\psi \in C^\infty[0, \infty)$ such that $\psi(\lambda) \le 1/\lambda$ for all $\lambda > 0$ and

$$\psi(\lambda) = \begin{cases} 1 & 0 \le \lambda \le 1/2 \\ 1/\lambda & \lambda \ge 2. \end{cases}$$

By putting $\beta_{jk} = B_{jk}\psi(|B|)$, we have $T = L$ with $W = V + |B|^2\psi(|B|) - R$. It is not difficult to verify by using (2.5), (2.6) and (2.19) that $W(x) \ge -\text{const.}Q(|x|) \equiv -M(|x|)$, which assures (2.11). Since $\alpha_{jk} = \delta_{jk}$ and thus $\alpha^*(r) \equiv 1$, (2.12) is satisfied. Moreover we have (2.10) because $\left(\sum_{jk}\beta_{jk}^2\right)^{1/2} = |B|\psi(|B|) \le 1$. Thus we have Theorem 2.4 by Lemma 2.5. □

3 The Aharonov-Bohm effect on the ground state energy

In this section, we are going to present a result concerning the Aharonov-Bohm effect on the ground state energy, which is essentially due to Lavine and O'Carroll [18] and formulated and proven in a rigorous form by Helffer [7].

Let Ω be a region of $\boldsymbol{R}^n$ and V and $b_j \in C^\infty(\overline{\Omega})$. Suppose that V is bounded from below. We fix the region Ω and the electric potential V. Define the form $h = h(\boldsymbol{b})$ with domain of definition $Q(h) = C_0^\infty(\Omega)$ by

$$h(\phi, \psi) \equiv \sum_{j=1}^n((\partial_j - ib_j)\phi, (\partial_j - ib_j)\psi) + (V\phi, \psi)$$

for ϕ, $\psi \in Q(h)$. Let $L(\boldsymbol{b})$ denote the self-adjoint operator in $L^2(\Omega)$ associated with the closure $\overline{h} = \overline{h}(\boldsymbol{b})$ of $h(\boldsymbol{b})$, which is bounded from below.

It is known from Kato's inequality that

$$\inf\sigma(L(\boldsymbol{b})) \ge \inf\sigma(L(\boldsymbol{0})) \tag{3.1}$$

for all $\boldsymbol{b} \in C^\infty(\overline{\Omega}, \boldsymbol{R}^n)$, where $L(\boldsymbol{0}) = -\Delta + V$. Moreover we have

Theorem 3.1 ([7]) *Suppose that* $\inf\sigma(L(\boldsymbol{b})) \equiv \lambda(\boldsymbol{b})$ *and* $\inf\sigma(L(\boldsymbol{0})) \equiv \lambda(\boldsymbol{0})$ *are both eigenvalues of* $L(\boldsymbol{b})$ *and* $L(\boldsymbol{0})$*, respectively. Then the following three conditions are equivalent to each other:*

(i) $\lambda(\boldsymbol{b}) = \lambda(\boldsymbol{0})$.
(ii) $\operatorname{curl} \boldsymbol{b} \equiv 0$ *in* Ω *and for all closed curve* γ *in* Ω $\int_\gamma \boldsymbol{b} \cdot dx \in 2\pi \boldsymbol{Z}$.
(iii) $L(\boldsymbol{b})$ *and* $L(\boldsymbol{0})$ *are unitarily equivalent through gauge transformation, i.e., there exists a complex-valued function* Ψ *in* Ω *with* $|\Psi| \equiv 1$ *such that* $L(\boldsymbol{b}) = \Psi^{-1}L(\boldsymbol{0})\Psi$.

Proof. (iii) $\Rightarrow$ (i): Obvious.

(ii) $\Rightarrow$ (iii): Define $\Psi(Q) \equiv e^{i\int_P^Q \mathbf{b}\cdot dx}$ for $Q \in \Omega$ with some fixed $P \in \Omega$. Then Ψ is well-defined by (ii). Moreover we have $\Psi^{-1}(\partial_j - ib_j)\Psi = \partial_j$ and thus $\Psi^{-1}L(\boldsymbol{b})\Psi = L(\boldsymbol{0})$.

(i) $\Rightarrow$ (ii): We have by partial integration

$$\int |(\nabla - i\boldsymbol{b} - \boldsymbol{f})\phi|^2 = \int |(\nabla - i\boldsymbol{b})\phi|^2 + \int (|\boldsymbol{f}|^2 + \nabla \cdot \boldsymbol{f})|\phi|^2 \tag{3.2}$$

for all $\boldsymbol{f} \in C^\infty(\Omega, \boldsymbol{R}^n)$ and $\phi \in C_0^\infty(\Omega)$, since

$$-((\nabla - i\boldsymbol{b})\phi, \boldsymbol{f}\phi) - (\boldsymbol{f}\phi, (\nabla - i\boldsymbol{b})\phi) = -\int \boldsymbol{f}\{(\nabla\phi)\overline{\phi} + \phi(\nabla\overline{\phi})\} = \int \boldsymbol{f}\nabla|\phi|^2.$$

Let $\lambda = \lambda(\boldsymbol{0})$ and u a ground state of $L(\boldsymbol{0})$: $-\Delta u + (V - \lambda)u = 0$. Then it is known that $u \in L^2(\Omega) \cap C^\infty(\Omega)$ and $u > 0$ (see, e.g., [22, Section XIII.12]). Put $\boldsymbol{f} = (\nabla u)/u$ in (3.2). Then we have

$$|\boldsymbol{f}|^2 + \nabla \cdot \boldsymbol{f} = (\Delta u)/u = V - \lambda,$$

which gives by (3.2)

$$\int \left|\left(\nabla - i\boldsymbol{b} - \frac{\nabla u}{u}\right)\phi\right|^2 = h(\phi, \phi) - \lambda\|\phi\|^2$$

for all $\phi \in C_0^\infty(\Omega)$, and hence,

$$\int u^2 \left|(\nabla - i\boldsymbol{b})\left(\frac{\phi}{u}\right)\right|^2 = \overline{h}(\phi, \phi) - \lambda\|\phi\|^2 \tag{3.3}$$

for all $\phi \in Q(\overline{h})$. Let ϕ_0 be a ground state of $L(\boldsymbol{b})$ with $\|\phi_0\| = 1$ and let $\Psi \equiv \phi_0/u \in C^\infty(\Omega)$. Then $\nabla\Psi = i\boldsymbol{b}\Psi$ by (3.3) and the assumption that $\lambda(\boldsymbol{b}) = \lambda(\boldsymbol{0})$. Therefore $\nabla|\Psi|^2 = \Psi\nabla\overline{\Psi} + \overline{\Psi}\nabla\Psi \equiv 0$. Hence $|\Psi| = const \equiv c \neq 0$ (because, if $c = 0$, ϕ_0 would be 0). Define a locally continuous (possibly multivalued) function g in Ω by $c^{-1}\Psi = e^{ig}$. Then, by $\nabla\Psi/\Psi = i\nabla g$, we have $\boldsymbol{b} = \nabla g$, which implies $\operatorname{curl} \boldsymbol{b} \equiv 0$. Moreover $\int_\gamma \boldsymbol{b} \cdot dx = \int_\gamma dg \in 2\pi\boldsymbol{Z}$. □

Remark. The equivalence in Theorem 3.1 fails if both of $\lambda(\boldsymbol{b})$ and $\lambda(\boldsymbol{0})$ are not eigenvalues: For example, let $\Omega = \boldsymbol{R}^n$ and $V \equiv 0$. Then it is known (see [17]) that $\sigma(L(\boldsymbol{b})) = \sigma_{ess}(L(\boldsymbol{b})) = [0, \infty)$ for all $\boldsymbol{b}$ satisfying $B = \operatorname{curl} \boldsymbol{b} \to 0$ at infinity. In this case we have $\lambda(\boldsymbol{b}) = \lambda(\boldsymbol{0}) = 0$, so that (i) of Theorem 3.1 holds, but (ii) does not hold unless B identically vanishes.

This theorem shows an aspect of the Aharonov-Bohm effect: Let us consider the case where $\Omega = \boldsymbol{R}^2 \setminus \{|x| \leq \delta\}$ with some $\delta > 0$ and $V \in C^\infty(\boldsymbol{R}^2)$ and $V \to \infty$ at infinity, which assures that $\inf \sigma(L(\boldsymbol{b}))$ is a bound state energy for all $\boldsymbol{b}$. Then Theorem 3.1 shows that $\lambda(\boldsymbol{b}) = \lambda(\boldsymbol{0})$ if and only if $\operatorname{curl} \boldsymbol{b} = 0$ and $\int_{|x|=\delta'} \boldsymbol{b} \cdot dx \in 2\pi\boldsymbol{Z}$ for $\delta' > \delta$. In particular,

$\lambda(\boldsymbol{b}) \neq \lambda(\boldsymbol{0})$ when $\int_{|x|=\delta'} \boldsymbol{b} \cdot dx \notin 2\pi \boldsymbol{Z}$ for some $\delta' > \delta$ even if the magnetic field vanishes identically in Ω. Therefore, a particle does feel the presence of a certain *magnetic field* contained in a small disc even if the motion of the particle is limited to Ω, which is disjoint from the disc, which means that it is not the magnetic field B but the vector potential that determines the "physics".

We remark that [7] gives an asymptotic formula for the difference $\lambda(\boldsymbol{b}) - \lambda(\boldsymbol{0})$ in the semi-classical limit, under a similar situation with some additional conditions on V.

4 Two-dimensional systems

In the case of the space dimension $n = 2$, the magnetic field $B(x)$ is considered as a *scalar* function $\partial_1 b_2(x) - \partial_2 b_1(x)$ on $\boldsymbol{R}^2$ and the operator L describes the motion of an electron confined in a 2-dimensional plane under the influence of a magnetic field perpendicular to the plane whose intensity $B(x)$ may depend on the point x.

Let us first consider the uniform magnetic field, i.e., $B(x) \equiv B_0$ for some constant $B_0 > 0$. If one chooses a suitable gauge, the operator L has the form

$$
\begin{aligned}
L_0 &= L_0(B_0) = \Pi_1^2 + \Pi_2^2, \\
\Pi_1 &= i\partial_1 - B_0 x_2/2, \quad \Pi_2 = i\partial_2 + B_0 x_1/2.
\end{aligned}
\tag{4.1}
$$

We have the commutation relation $[\Pi_1, \Pi_2] = iB_0$, which is C.C.R. (= the canonical commutation relation) and L_0 can be analysed in a manner similar to the case of the harmonic oscillator using the creation and annihilation operators: $A = \Pi_1 + i\Pi_2$ is the annihilation operator and $A^* = \Pi_1 - i\Pi_2$ is the creation operator. They satisfy the relations

$$
\begin{cases}
L_0 A &= A(L_0 - 2B_0) \\
L_0 A^* &= A^*(L_0 + 2B_0)
\end{cases}
$$

It is known by using these relations that we have

Theorem 4.1 (see, e.g., [1]) $\sigma(L_0) = \{\ (2n-1)B_0 \mid n :$ *integer* $> 0\}$ *and each point of* $\sigma(L_0)$ *is an eigenvalue with infinite degeneracy. Moreover* A^* *gives a 1-to-1 and onto correspondence from the eigenspace of* L_0 *with eigenvalue* $(2n-1)B_0$ *to that with eigenvalue* $(2n+1)B_0$ *for each* $n \geq 1$.

$(2n-1)B_0$ are called the Landau levels. The fact that L_0 has a complete set of eigenfunctions corresponds to the fact that classical orbits of charged particles in uniform magnetic fields are bounded in the x_1 and x_2 directions.

L has a similar spectral property if the magnetic field $B(x)$ tends to some non-zero constant B_0 at infinity.

Theorem 4.2 ([11]) *Suppose that* $B(x) \to B_0$ *as* $|x| \to \infty$ *with some constant* $B_0 > 0$. *Then* $\sigma_{ess}(L) = \{\ (2n-1)B_0 \mid n$ *is integer* $> 0\ \}$.

By this theorem, L also has a complete set of eigenfunctions, with eigenvalues which, with their multiplicities taken account of, have the accumulation points equal to $\sigma_{ess}(L)$.

On the other hand, there are examples of the magnetic field $B(x)$ such that L is absolutely continuous:

Theorem 4.3 ([12]) *Suppose that $B(x)$ depends only on the variable x_1, i.e., $B(x) = B(x_1)$ and there exist constants $M_\pm$ such that $0 < M_- \leq B(x_1) \leq M_+ < \infty$ for all $x_1 \in \boldsymbol{R}$. Then L is absolutely continuous if either of the following (B1) or (B2) holds:*

(B1) $\limsup_{x_1 \to -\infty} B(x_1) < \liminf_{x_1 \to \infty} B(x_1)$ *or* $\liminf_{x_1 \to -\infty} B(x_1) > \limsup_{x_1 \to \infty} B(x_1)$.

(B2) $B(x_1) = B_0$ *for some constant* B_0 *if* $|x_1|$ *is sufficiently large, and there is a point* $\overline{x_1}$ *such that* $B'(x_1) \leq 0$ *for* $x_1 \leq \overline{x_1}$ *and* $B'(x_1) \geq 0$ *for* $x_1 \geq \overline{x_1}$ $(B'(x_1) \geq 0$ *for* $x_1 \leq \overline{x_1}$ *and* $B'(x_1) \leq 0$ *for* $x_1 \geq \overline{x_1})$ *and* $B'(x_1)$ *is not identically 0.*

The absolute continuity of L signifies that the particle is not trapped in any bounded region and wanders off to infinity. In this case, the particle wanders off to infinity along straight lines parallel to the x_2 axis. Moreover, if $B(x_1)$ is nondecreasing, we can compute the spectrum of L:

$$\sigma(L) = \bigcup_{n=1}^{\infty} [(2n-1)B_-, (2n-1)B_+]$$

where $B_\pm \equiv \lim_{x_1 \to \pm\infty} B(x_1)$. Therefore each Landau level is broadened into a band.

Next let us consider the periodic fields. More precisely, we shall assume

(H4.1) There exist constants T_1, $T_2 > 0$ such that

$$\begin{aligned} V(x_1 + T_1, x_2) &= V(x_1, x_2 + T_2) = V(x_1, x_2) \\ B(x_1 + T_1, x_2) &= B(x_1, x_2 + T_2) = B(x_1, x_2) \end{aligned}$$

for all $x = (x_1, x_2) \in \boldsymbol{R}^2$.

As is well known, in the case where the system is free of magnetic field B, the spectral properties of $L = -\Delta + V$ are studied by using the Bloch wave, which is a periodic function multiplied by a plane wave, and the operator $-\Delta + V$ is absolutely continuous for *any* periodic electric potential (see [20] and [26]). However, in the presence of magnetic fields, the situation becomes more complicated.

It is known (see, e.g., [27]) that an analysis using magnetic Bloch waves similar to the case of $-\Delta + V$ is possible if the magnetic flux penetrating a fundamental domain $\Omega = [0, T_1] \times [0, T_2]$ is an integral multiple of 2π:

(H4.2) $\iint_\Omega B(x)dx = 2\pi N$, where N is an integer.

The choice of the magnetic vector potential $\boldsymbol{b} = (b_1, b_2)$ is arbitrary as far as it satisfies $B(x) = \partial_1 b_2(x) - \partial_2 b_1(x)$ by gauge invariance. Under the assumption (H4.1), it is known (see, e.g., [4]) that we can take

$$\begin{cases} b_1(x) = -\dfrac{B_0}{2} x_2 + a_1(x) \\ b_2(x) = \dfrac{B_0}{2} x_1 + a_2(x) \\ a_j(x_1 + T_1, x_2) = a_j(x_1, x_2 + T_2) = a_j(x_1, x_2) \quad (j = 1,\ 2) \end{cases}$$

where B_0 is the density of the magnetic flux,

$$B_0 T_1 T_2 = \iint_\Omega B(x)dx.$$

Note that, unless B_0 equals 0, the vector potential cannot be taken to be periodic, even though the magnetic field is periodic.

Next, consider the operators S_1, S_2:

$$\begin{cases} S_1u(x_1, x_2) \equiv e^{-\frac{i}{2}B_0T_1x_2}u(x_1+T_1, x_2), \\ S_2u(x_1, x_2) \equiv e^{\frac{i}{2}B_0T_2x_1}u(x_1, x_2+T_2), \end{cases} \tag{4.2}$$

which are known as magnetic translations ([27]). These operators commute with L:

$$LS_j = S_jL \quad (j = 1, 2),$$

while usual translations do not commute with L because of the terms $(-B_0x_2/2,\ B_0x_1/2)$ in the vector potential $\boldsymbol{b}$. Moreover we have $S_1S_2 = e^{iB_0T_1T_2}S_2S_1$ by direct calculation. The commutativity of S_1 and S_2 is essential to the construction of magnetic Bloch theory and we shall assume (H4.2) (which is equivalent to $S_1S_2 = S_2S_1$).

Define the space of the magnetic Bloch functions by

$$\mathcal{E}(p) \equiv \{\, u \in C^\infty(\boldsymbol{R}^2)\,;\ S_ju = e^{ip_jT_j}u\ (j = 1, 2)\,\}$$

with quasi-momenta $p = (p_1, p_2) \in \boldsymbol{R}^2$. $\mathcal{E}(p)$ satisfies the relation

$$\mathcal{E}(p_1, p_2) = \mathcal{E}(p_1 + \frac{2\pi}{T_1}, p_2) = \mathcal{E}(p_1, p_2 + \frac{2\pi}{T_2}).$$

Define further the space $\mathcal{E}$ by

$$\begin{aligned} \mathcal{E} \ \equiv\ \Big\{ & u \in C^\infty(\boldsymbol{R}^2_x \times \boldsymbol{R}^2_p)\,; u(x,p) \in \mathcal{E}(p) \text{ for all } p \in \boldsymbol{R}^2, \\ & u(x,p) = u(x, p_1 + \frac{2\pi}{T_1}, p_2) = u(x, p_1, p_2 + \frac{2\pi}{T_2}) \Big\} \end{aligned}$$

equipped with the norm

$$\|u\|^2_{\mathcal{E}} \equiv \int_\Omega dx \int_{\Omega^*} dp |u(x,p)|^2,$$

where $\Omega^* \equiv [0,\ 2\pi/T_1] \times [0,\ 2\pi/T_2]$. Then we have the following

Theorem 4.4 *Suppose that (H4.1) and (H4.2) hold. Then there exists a bijective correspondence* $U : \mathcal{S}(\boldsymbol{R}^2) \to \mathcal{E}$, *where* U *and* U^{-1} *are given by*

$$\begin{aligned} Uf(x,p) &= \frac{1}{\sqrt{T_1T_2}} \sum_{k,m\in\mathbf{Z}} e^{-i(kp_1T_1+mp_2T_2)} S_1^k S_2^m f(x) \\ U^{-1}u(x) &= \frac{1}{\sqrt{T_1T_2}} \int_{\Omega^*} u(x,p)dp, \end{aligned}$$

for $f \in \mathcal{S}(\boldsymbol{R}^2)$ *(= the space of rapidly decreasing functions in* $\boldsymbol{R}^2$*) and* $u \in \mathcal{E}$. *Moreover* U *is unitary, i.e.,* $\|Uf\|_{\mathcal{E}} = \|f\|_{L^2(\mathbf{R}^2)}$. *If we put* $\tilde{L} \equiv ULU^{-1}$, *we have*

$$(\tilde{L}u)(x,p) = \left\{\tilde{L}(p)u(\cdot, p)\right\}(x)$$

where $\tilde{L}(p)$ *is an operator on* $\mathcal{E}(p)$ *given by* $(\tilde{L}(p)u)(x) = Lu(x)$ *for* $u \in \mathcal{E}(p)$.

Because $\tilde{L}(p)$ is an elliptic operator acting essentially in a compact domain Ω, $\tilde{L}(p)$ has purely discrete spectrum. Hence the study of the spectral property of L is reduced to that of an eigenvalue problem of a family of the operators $\tilde{L}(p)$ with parameters $p = (p_1, p_2) \in \Omega^*$. While $\tilde{L}(p)$ have different domains of definition $\mathcal{E}(p)$, we can show the following

Proposition 4.5 *$e^{i(p_1x_1+p_2x_2)}u(x) \in \mathcal{E}(p)$ for $u \in \mathcal{E}(0)$. Let $A(p)$ be an operator in $\mathcal{E}(0)$ defined by $A(p)u(x) \equiv e^{-i(p_1x_1+p_2x_2)}\tilde{L}(p)\{e^{i(p_1x_1+p_2x_2)}u(x)\}$. Then we have*

$$\begin{aligned} A(p) &= -(D_1 + ip_1)^2 - (D_2 + ip_2)^2 + V \\ &= A(0) - 2i(p_1D_1 + p_2D_2) + p_1^2 + p_2^2 \end{aligned}$$

where $D_j = \partial_j - ib_j$ $(j = 1,\ 2)$. Moreover D_j are infinitesimally small with respect to $A(0)$, i.e., for all $\varepsilon > 0$, there is a constant $C_\varepsilon > 0$ such that

$$\|D_ju\| \leq \varepsilon\|A(0)u\| + C_\varepsilon\|u\|$$

for all $u \in \mathcal{E}(0)$ where $\|u\| = \{\int_\Omega |u(x)|^2dx\}^{1/2}$.

Hence the operator $A(p)$, which is unitarily equivalent to $\tilde{L}(p)$, has a common domain of definition $\mathcal{E}(0)$ and depends analytically on p (it forms an entire analytic family of type A in the sense of Kato).

Recently the author has shown that, generically for small perturbation by a periodic electric potential V, the operator $L = L_0(B_0) + V$ is absolutely continuous in the case where the magnetic field $B(x) \equiv B_0 =$ constant $\neq 0$, which relates to a fact known as Landau level broadening in the physics literature (see, e.g., [28]):

Theorem 4.6 ([15]) *Suppose that (H4.1) holds and $B_0T_1T_2 = 2\pi$ (i.e., (H4.2) hold with $N = 1$). Let all the Fourier coefficient of V do not vanish and $\|V\|_\infty \equiv \sup_x |V(x)| \leq B_0/4$. Then there exists a countable set Σ_V in the interval $[-1,\ 1]$ such that $L = L_\kappa = L_0(B_0) + \kappa V$ is absolutely continuous for $\kappa \in [-1,\ 1] \setminus \Sigma_V$.*

We give a remark as to the case where we assume (H4.1) but not (H4.2). Note that the Bloch wave analysis is applicable also to the case of rational $N = B_0T_1T_2/2\pi$ by considering a fundamental domain $[0,\ T_1q] \times [0,\ T_2]$ if $N = p/q$ (p, q are integers). In the case where N is irrational, it seems that not much has been studied so far. Recently, B. Helffer and J. Sjöstrand [8] showed that, in this case, the spectrum of L contains a part similar to the Cantor set under suitable additional conditions.

5 Purely discrete spectrum

In this section, we shall consider the property that L has purely discrete spectrum, i.e., that the resolvents $(L - z)^{-1}$ of L are compact operators in $L^2(\boldsymbol{R}^n)$ for $z \in \boldsymbol{C} \setminus \sigma(L)$.

First we give general criteria for this property under the assumption (H2.1), and we consider the operator L as the self-adjoint operator associated with the nonnegative form h defined by (2.1). Define the following quantity

$$e(\Omega) \equiv \inf\left\{\frac{(L\phi,\phi)}{(\phi,\phi)}\,; \phi \in C_0^\infty(\Omega),\ \phi \not\equiv 0\right\}$$

for open sets Ω in $\boldsymbol{R}^n$. Then we have

Theorem 5.1 ([13]) *Suppose that (H2.1) holds. Then the following four conditions are equivalent to each other:*

(a) L has compact resolvent.
(b) $e(\Omega_R) \to \infty$ *as* $R \to \infty$, *where* $\Omega_R = \{x \mid |x| > R\}$.
(c) $e(Q_x) \to \infty$ *as* $|x| \to \infty$, *where* $Q_x = \{y \mid |y - x| < 1\}$.
(d) There exists a real-valued continuous function $\lambda(x)$ *on* $\boldsymbol{R}^n$ *such that*

$$\begin{aligned} \lambda(x) &\to \infty \quad \text{as} \quad |x| \to \infty \\ (L\phi, \phi) &\geq \int \lambda(x)|\phi(x)|^2 dx \quad \text{for all } \phi \in C_0^\infty(\boldsymbol{R}^n). \end{aligned}$$

This theorem depends on the fact that an operator M of multiplication by a bounded function on $\boldsymbol{R}^n$ which tends to 0 at infinity is relatively compact with respect to L, i.e., $M(L+E)^{-1}$ is compact for $E > 0$, which in turn follows from the operator inequality (2.2) (see, e.g., [1]). The part (d) $\Rightarrow$ (a) is given in [1], while the inverse implication does not seem to have been well-known.

Proof of Theorem 5.1. (a) $\Rightarrow$ (b): First note that we have the compactness of the operator $(L+1)^{-1/2}$ in $L^2(\boldsymbol{R}^n)$ as follows: Consider the decomposition

$$(5.1) \qquad (L+1)^{-1/2} = \chi_R(L)(L+1)^{-1/2} + (1 - \chi_R(L))(L+1)^{-1/2},$$

where $\chi_R(t)$ is the characteristic function of $\{\, t\, ;\, t < R\,\}$ on $\boldsymbol{R}$. Then

$$\chi_R(L)(L+1)^{-1/2} = (L+1)^{1/2}\chi_R(L) \cdot (L+1)^{-1}$$

is compact by (a) since $(L+1)^{1/2}\chi_R(L)$ is a bounded operator in $L^2(\boldsymbol{R}^n)$ by the inequality $L \geq 0$. Therefore, since $\|(1-\chi_R(L))(L+1)^{-1/2}\| \to 0$ as $R \to \infty$, we have the compactness of $(L+1)^{-1/2}$ by (5.1). Suppose that (b) does not hold. Then there exist a sequence of real numbers $\{R_n\}$ and a sequence of functions $\{\phi_n\} \subset C_0^\infty(\boldsymbol{R}^n)$ such that $R_n \nearrow \infty$ as $n \to \infty$, $\operatorname{supp} \phi_n \subset \{\, x \mid R_n < |x| < R_{n+1}\,\}$, $\|\phi_n\| = 1$ and $\{(L\phi_n, \phi_n)\}$ is bounded. Put $u_n \equiv (L+1)^{1/2}\phi_n$. Then $\{u_n\}$ is bounded in $L^2(\boldsymbol{R}^n)$. Hence $\{\phi_n\} = \{(L+1)^{-1/2}u_n\}$ has a convergent subsequence, since $(L+1)^{-1/2}$ is compact as is noted above. This contradicts the fact that $\{\phi_n\}$ is orthonormal.

(b) $\Rightarrow$ (c): $e(Q_x) \geq e(\Omega_R)$ if $|x| > R+1$, since $e(\Omega) > e(\Omega')$ for $\Omega \subset \Omega'$. Therefore (b) implies (c).

(c) $\Rightarrow$ (d): Let $\Gamma = \{\, (l_1, \ldots, l_n) \mid l_j = k_j/\sqrt{n}\, ;\, k_j \in \boldsymbol{Z}\ (j = 1, \ldots, n)\}$ and let $\{\zeta_l\}_{l \in \Gamma}$ be a sequence of C^∞ real-valued functions such that

$$(5.2) \qquad \zeta_l(x) = \zeta_{(0,\ldots,0)}(x - l) \quad \text{for } l \in \Gamma,$$

$$(5.3) \qquad \sum_{l \in \Gamma} \zeta_l(x)^2 = 1,$$

$$(5.4) \qquad \operatorname{supp} \zeta_l \subset Q_l.$$

Since we have by direct computation

$$\begin{aligned} &\operatorname{Re}((\partial_j - ib_j)\phi\, \overline{(\partial_j - ib_j)(\zeta^2\phi)}) \\ &\qquad = \zeta^2|(\partial_j - ib_j)\phi|^2 + 2\zeta(\partial_j\zeta)\operatorname{Re}((\partial_j - ib_j)\phi\,\overline{\phi}) \\ &\qquad = |(\partial_j - ib_j)(\zeta\phi)|^2 - |(\partial_j\zeta)\phi|^2 \end{aligned}$$

for $\phi \in C_0^\infty(\boldsymbol{R}^n)$ and for a real-valued C^∞ function ζ, where Re means the real part, we obtain

$$\begin{aligned}(L\phi,\phi) &= \mathrm{Re}(\sum_l (L\phi,\zeta_l^2\phi)) \qquad \text{(by (5.3))}\\ &= \sum_l (L(\zeta_l\phi),\zeta_l\phi) - \sum_l \|(\nabla\zeta_l)\phi\|^2.\end{aligned}$$

Thus we have by (5.4) and the definition of $e(\Omega)$,

$$\begin{aligned}(L\phi,\phi) &\geq \sum_l e_l\|\zeta_l\phi\|^2 - \sum_l \|(\nabla\zeta_l)\phi\|^2\\ &= \int \lambda(x)|\phi(x)|^2 dx,\end{aligned}$$

where $e_l = e(Q_l)$ and $\lambda(x) = \sum_l e_l\zeta_l(x)^2 - \sum_l |\nabla\zeta_l(x)|^2$. Therefore we have (d) since $\lambda(x) \to \infty$ as $|x| \to \infty$ by (5.2), (5.3) (5.4) and (c).

(d) $\Rightarrow$ (a): Choose a sufficiently large number $E > 0$ so that $\inf_{x\in\mathbf{R}^n} \lambda(x) + E > 0$. First note that (d) implies that

(5.5) $\qquad (\lambda + E)^{1/2}(L+E)^{-1/2}$ is a bounded operator in $L^2(\boldsymbol{R}^n)$.

Consider the decomposition

(5.6) $$(L+E)^{-1} = \chi_R(L)(L+E)^{-1} + (1-\chi_R(L))(L+E)^{-1},$$

where $\chi_R(t)$ is as in (5.1). From

$$\begin{aligned}\chi_R(L)(L+E)^{-1} &= (L+E)^{-1}(\lambda+E)^{-1/2}\cdot\\ &\qquad (\lambda+E)^{1/2}(L+E)^{-1/2}\cdot(L+E)^{1/2}\chi_R(L),\end{aligned}$$

we have the compactness of $\chi_R(L)(L+E)^{-1}$ in view of (5.5), because $(L+E)^{-1}(\lambda+E)^{-1/2} = \{(\lambda+E)^{-1/2}(L+E)^{-1}\}^*$ is compact as is remarked after Theorem 5.1 and because $(L+E)^{1/2}\chi_R(L)$ is bounded by the inequality $L \geq 0$. Therefore we have the compactness of $(L+E)^{-1}$ by (5.6) since $\|(1-\chi_R(L))(L+E)^{-1}\| \to 0$ as $R \to \infty$. □

It is well known that, in the absence of the magnetic field, $L = -\Delta + V$ has compact resolvent if $V(x) \to \infty$ as $|x| \to \infty$, which is also known from (d) of Theorem 5.1. As for the magnetic field, the situation is not so simple. Namely, in the absence of the electric potential, the condition

(5.7) $$|B(x)| \to \infty \quad \text{as} \quad |x| \to \infty$$

does not necessarily imply the compactness of the resolvent of L (see [5]). In $\boldsymbol{R}^2$, (5.7) is sufficient for the compactness of the resolvents of L. However, if the space dimension $n \geq 3$, (5.7) is not sufficient for the compactness of the resolvents of L, no matter how fast $|B(x)|$ grows at infinity. In addition to (5.7), one needs some other conditions which assure some mildness of $B(x)$ near infinity: Consider the condition

(A_δ) $\nabla B_{jk}(x) = o(|B(x)|^\delta)$ for $j,k = 1,\ldots, n$.

Then (A_δ) and (5.7) imply the compactness of the resolvent of L if $0 < \delta \leq 2$, while (A_δ) and (5.7) do not if $\delta > 2$ and $n \geq 3$ (see [13]).

We would like to conclude this section by giving a remark concerning the asymptotic distribution of eigenvalues. Let $N(\lambda)$ be the number of eigenvalues less than λ of L. In the case where the magnetic field is absent, $N(\lambda)$ obeys the following asymptotic formula under suitable assumptions on $V(x)$:

$$N(\lambda) = (2\pi)^{-n}\mathrm{vol}\left[\left\{ (x,\xi) \in \boldsymbol{R}^n_x \times \boldsymbol{R}^n_\xi \; ; \; |\xi|^2 + V(x) < \lambda\right\}\right] \cdot (1 + o(1))$$

as $\lambda \to \infty$. This formula does not make sense for the operator L with $V \equiv 0$, even if L has compact resolvent, since

$$\mathrm{vol}\left[\left\{ (x,\xi) \in \boldsymbol{R}^n_x \times \boldsymbol{R}^n_\xi \; ; \; \sum_j |\xi_j - b_j(x)|^2 < \lambda\right\}\right] = \infty.$$

Tamura [25] has given an asymptotic formula by examining the heat kernel for L. Colin de Verdière [3] also gives an asymptotic formula by using the Weyl method, i.e., by cutting the space into cubes and estimating the number of eigenvalues in each cube. We do not give here a precise statement of their results, but only give a remark according to [3] that, in the case where $V \equiv 0$, there is the "density of states" $\nu_B(\lambda)$ dependent on the value of the magnetic field at each point such that

$$N(\lambda) \sim \int_{\mathbf{R}^n} \nu_{B(x)}(\lambda)dx.$$

$\nu_B(\lambda)$ is a quantity related to the Schrödinger operator $L(B)$ with uniform magnetic field $B(x) \equiv B$ as follows: If $N_{\Omega,B}(\lambda)$ denotes the number of eigenvalues $< \lambda$ of $L(B)|_{L^2(\Omega)}$ with the Dirichlet boundary condition for a bounded domain $\Omega \in \boldsymbol{R}^n$, then

$$\frac{N_{\Omega,B}(\lambda)}{\mathrm{vol}\,(\Omega)} \sim \nu_B(\lambda) \quad (\mathrm{vol}\,(\Omega)\text{: large}).$$

In $\boldsymbol{R}^3$ and $\boldsymbol{R}^2$, $\nu_B(\lambda)$ depends only on the magnitude $|B| = B_0$ of the magnetic field B and given by

$$\begin{aligned} \nu_{B_0}(\lambda) &= \frac{2}{(2\pi)^2} B_0 \sum_{j\geq 1} (\lambda - (2j-1)B_0)_+^{1/2} \quad \text{in } \boldsymbol{R}^3, \\ \nu_{B_0}(\lambda) &= \frac{1}{2\pi} B_0 \cdot \sharp\{\, j \mid (2j-1)B_0 < \lambda,\ j \geq 1\} \quad \text{in } \boldsymbol{R}^2, \end{aligned}$$

where $t_+ = \max(t, 0)$. To get some feel for the last formula, take as an example $\tilde{L}(0)$ in the preceding section. Namely consider the Schrödinger operator in uniform magnetic field $B(x) \equiv B_0$ with "periodic" boundary condition $S_1 u = S_2 u = u$ (S_j is as in (4.2)) in $\Omega = [0, T_1] \times [0, T_2]$, where the periods T_1, T_2 satisfy the condition (H4.2), i.e., $B_0 T_1 T_2 \in 2\pi\boldsymbol{Z}$. Then by the same reasoning as used to obtain the spectral property of the harmonic oscillator (or the Schrödinger operator with uniform magnetic field in the whole space $\boldsymbol{R}^2$ (see Theorem 4.1)), it is known that $\sigma(\tilde{L}(0)) = \{\, (2n-1)B_0 \mid n\text{: integer} \geq 1 \,\}$ and all the eigenspaces have the same multiplicity $\equiv \ell$. If $N(\lambda)$ is the number of eigenvalues $< \lambda$

of $\tilde{L}(0)$, $N(\lambda) \sim \dfrac{\lambda \ell}{2B_0}$ as $\lambda \to \infty$. On the other hand, as is known for general second order elliptic operators $N(\lambda) \sim \dfrac{\pi\lambda}{4\pi^2} T_1 T_2$ as $\lambda \to \infty$. By comparing these two asymptotic formulae, we have $\ell = \dfrac{B_0 T_1 T_2}{2\pi} = N$ and thus $N(\lambda)$ is exactly equal to $\nu_{B_0}(\lambda) \cdot \mathrm{vol}\,(\Omega)$.

References

[1] J. Avron, I. Herbst and B. Simon, Schrödinger operators with magnetic fields, I., General Interactions, *Duke Math. J.* **45** (1978), 847–883.

[2] H. L. Cycon, R. G. Froese, W. Kirsch and B. Simon, *Schrödinger Operators with Application to Quantum Mechanics and Global Geometry*, Texts and Monographs in Physics, Springer-Verlag, New York/Berlin, 1987.

[3] Y. Colin de Verdiere, L'asymptotique de Weyl pour les bouteilles magnétiques, *Comm. Math. Phys.* **105** (1986), 327–335.

[4] B. A. Dubrovin and S. P. Novikov, Ground states in a periodic field, Magnetic Bloch functions and vector bundles, *Soviet Math. Dokl.* **22** (1980), 240–244.

[5] A. Dufresnoy, Un exemple de champ magnétique dans $\boldsymbol{R}^\nu$, *Duke Math. J.* **50** (1983), 729–734.

[6] H. Hess, R. Schrader and D. A. Uhlenbrock, Domination of semigroups and generalization of Kato's inequality, *Duke Math. J.* **44** (1977), 893–904.

[7] B. Helffer, Effet d'Aharonov Bohm sur un état borné de l'équation de Schrödinger, *Comm. Math. Phys.* **119** (1988), 315–329.

[8] B. Helffer and J. Sjöstrand, Equation de Schrödinger avec champ magnétique fort et équation de Harper, preprint.

[9] W. Hunziker, Schrödinger operators with electric or magnetic fields, in *Mathematical Problems in Theoretical Physics*, Lecture Notes in Physics, Vol. 116, ed. by K. Osterwalder, Springer-Verlag, New York/Berlin, 1979, pp. 25–44.

[10] T. Ikebe and T. Kato, Uniqueness of the self-adjoint extension of singular elliptic differential operators, *Arch. Rational Mech. Anal.* **9** (1962), 77–92.

[11] A. Iwatsuka, The essential spectrum of two-dimensional Schrödinger operators with perturbed constant magnetic fields, *J. Math. Kyoto Univ.* **23** (1983), 475–480.

[12] A. Iwatsuka, Examples of absolutely continuous Schrödinger operators in magnetic fields, *Publ. RIMS, Kyoto Univ.* **21** (1985), 385–401.

[13] A. Iwatsuka, Magnetic Schrödinger operators with compact resolvent, *J. Math. Kyoto Univ.* **26** (1986), 357–374.

[14] A. Iwatsuka, Essential self-adjointness of the Schrödinger operators with magnetic fields diverging at infinity, in preparation.

[15] A. Iwatsuka, Landau level broadening by periodic perturbation in uniform magnetic fields, in preparation.

[16] T. Kato, Remarks on Schrödinger operators with vector potentials, *Integral Equations and Operator Theory* **1** (1978), 103–113.

[17] H. Leinfelder, Gauge invariance of Schrödinger operators and related spectral properties, *J. Op. Theory* **9** (1983), 163–179.

[18] R. Lavine and M. O'Carroll, Ground state properties and lower bounds for energy levels of a particle in a uniform magnetic field and external potential, *J. Math. Phys.* **18** (1977), 1908–1912.

[19] H. Leinfelder and C. G. Simader, Schrödinger operators with singular magnetic potentials, *Math. Z.* **176** (1981), 1–19.

[20] F. Odeh and J. B. Keller, Partial differential equation with periodic coefficients and Bloch waves in crystals, *J. Math. Phys.* **5** (1964), 1499–1504.

[21] M. Reed and B. Simon, *Methods of Modern Mathematical Physics, Vol. II*, Academic Press, New York, 1975.

[22] M. Reed and B. Simon, *Methods of Modern Mathematical Physics, Vol. IV*, Academic Press, New York, 1978.

[23] B. Simon, Maximal and minimal Schrödinger forms, *J. Op. Theory* **1** (1979), 37–47.

[24] B. Simon, Kato's inequality and the comparison of semigroups, *J. Funct. Anal.* **32** (1979), 97–101.

[25] H. Tamura, Asymptotic distribution of eigenvalues for Schrödinger operators with magnetic fields, *Nagoya Math. J.* **105** (1987), 49–69.

[26] L. E. Thomas, Time dependent approach to scattering from impurities in a crystal, *Comm. Math. Phys.* **33** (1973), 335–343.

[27] J. Zak, Magnetic translation groups I. II., *Phys. Rev.* **134–A** (1966), 1602–1611.

[28] J. Zak, Group theoretical consideration of Landau level broadening in crystals, *Phys. Rev.* **136–A** (1964), 776–780.

Existence of Bound States for Double Well Potentials and the Efimov Effect

Hideo Tamura
Department of Mathematics, Ibaraki University
Mito, Ibaraki, 310 Japan

1. Introduction

In the present work we study the Efimov effect of three-particle systems. The Efimov effect is one of the most interesting results in the spectral analysis for three-particle Schrödinger operators. Roughly speaking, this effect can be explained as follows : If all three two-particle subsystems do not have negative bound state energies but a zero resonance energy, then the three-particle system under consideration has an infinite number of negative bound state energies accumulating at zero. This remarkable spectral property was first discovered by Efimov [1]. Since then, the problem has been studied in detail in several physical journals. For related references, see, for example, the book [5]. The rigorous mathematical proof of the effect has been done by Yafaev [8] and Ovchinnikov - Sigal [6]. In [8], the method based on the Faddeev equation has been used to prove the Efimov effect. Roughly speaking, the class of pair potentials considered there requires the decaying property $V(x) = O(|x|^{-\rho})$, $\rho > 3$, as $|x| \to \infty$, $x \in R^3$, and the $C^{1+\theta}$ - class, $\theta > 1/2$, smoothness restriction. On the other hand, an interesting variational method has been introduced by [6] to prove the Efimov effect for 2 H (heavy) - 1 L (light) particle systems under the assumption that only H - L subsystems with spherically symmetric interactions having the decaying property $V(|x|) = O(|x|^{-\rho})$, $\rho > 2$, have a zero resonance energy. All three two-particle subsystems are not necessarily assumed to have a zero resonance energy.

The aim of this work is to put forward the variational method initiated by [6] to prove the Efimov effect for general three-particle systems without mass restriction under the assumptions that pair potentials (not necessarily spherically symmetric) have the decaying property $V(x) = O(|x|^{-\rho})$, $\rho > 2$, and that all three two-particle subsystems have a zero resonance energy.

The precise formulation of the obtained result requires several notations and assumptions. We consider a system of three particles with masses $m_j > 0$, $1 \leq j \leq 3$, moving in R^3 through an interaction given by the sum of pair potentials V_{jk}, $1 \leq j < k \leq 3$. After elimination of the kinetic energy of the center of mass, the energy Hamiltonian H (Schrödinger operator) for such a system takes the form

(1.1) $$H = H_0 + V, \quad V = \Sigma_{1 \leq j < k \leq 3} V_{jk},$$

where H_0 is the free Hamiltonian and the pair potential V_{jk} between the j-th and k-th particles is a real function of the relative coordinates $x_j - x_k \in R^3$; $V_{jk} = V_{jk}(x_j - x_k)$. The operators H_0 and H act on the space $L^2(R^6)$.

We first assume that

(A.1) $|V_{jk}(x)| \leq C(1 + |x|)^{-\rho}$, $x \in R^3$, for some $\rho > 2$.

For given pair $\delta = (j,k)$, $1 \leq j < k \leq 3$, we define the reduced mass μ_δ as $\mu_\delta = m_j m_k/(m_j + m_k)$ and the two-particle subsystem Hamiltonian h_δ acting on the space $L^2(R^3)$ as $h_\delta = -(2\mu_\delta)^{-1}\Delta + V_{jk}$. We further assume that : For all pairs $\delta = (j,k)$,

(A.2) h_δ has no negative bound state energies ;

(A.3) h_δ has a zero resonance energy.

Roughly speaking, assumption (A.3) means that there exists a solution ψ_δ not in $L^2(R^3)$ such that $h_\delta \psi_\delta = 0$. The precise definition of zero resonance energy is given in subsection 2.2.

If assumptions (A.1) and (A.2) are satisfied, then it follows from the HVZ (Hunziker - Van Winter - Zhislin) theorem that H has essential spectrum beginning at zero and discrete spectrum in $(-\infty, 0]$.

Theorem 1 (Efimov's effect). *Assume that (A.1) ∿ (A.3) are satisfied. Then the three-particle Hamiltonian H defined by (1.1) has an infinite number of negative bound state energies accumulating at zero.*

Remark. In general, the Efimov effect cannot be expected without assuming that all three two-particle subsystems have a zero resonance energy. In fact, consider the Hamiltonian H with masses (m,m,∞), $0 < m < \infty$. Assume that only the two-particle subsystem Hamiltonian h_δ with pairs $\delta = (1,3)$ and $(2,3)$ have a zero resonance energy. If V_{12}

≥ 0, then H is non-negative and hence H has no negative bound state energies.

2. Existence of bound states for double well potentials

As is pointed out in [6], the Efimov effect is closely related to the problem on the binding of particles through conspiracy of potential wells.

Consider the Schrödinger operator $H(R)$, $R = (R_1, R_2, R_3) \in R^3$, with double well potential ;

$$H(R) = -\Delta + V_1(x) + V_2(x-R) \quad \text{on} \quad L^2(R^3).$$

The real potential $V_j(x)$, $1 \leq j \leq 2$, is assumed to have the decaying property (A.1) and also the operator $H_j = -\Delta + V_j$ is assumed to have the following spectral properties :

(H.1) H_j has no negative bound state energies ;

(H.2) H_j has a zero resonance energy.

These assumptions mean that the single well potential $V_j(x)$ itself cannot produce a bound state and that the double wells of potential $V_1(x) + V_2(x-R)$, $|R| \gg 1$, do not interact with each other at negative energies in classical mechanics. Nevertheless we have the following theorem.

Theorem 2. *Let the notations be as above. Assume that (A.1), (H.1) and (H.2) are satisfied. Then the operator $H(R)$ has at least one negative bound state energy for $|R| \gg 1$ large enough.*

Remark. In a different way, a similar result has been already proved by [3] for a class of compactly supported non-positive potentials.

2.1. Before going into the rigorous proof of Theorem 2, we first present an idea how the negative bound state energy of $H(R)$ is approximately determined. The idea developed here plays a basic role in proving the Efimov effect by the variational method.

For brevity, we assume $V_j(x)$ to be of compact support. Consider the eigenvalue problem

$$H(R)\phi = -E\phi, \quad E = E(R) > 0. \tag{2.1}$$

We construct an approximate normalized eigenfunction $\phi = \phi(x;R)$ in the form

$$\phi = A_1(R)f_1(x;R) + A_2(R)f_2(x-R;R)$$

with normalization constant $A_j(R)$, $1 \leqq j \leqq 2$. The construction of such an eigenfunction is based on the adiabatic approximation method due to [6] and on the low energy analysis of resolvents developed by [2] and [4].

Let $\phi(x)$ be as above. Then, (2.1) can be put into

$$\begin{aligned}(-\Delta + E)&(A_1f_1(x) + A_2f_2(x-R)) \\ &+ V_1(x)(A_1f_1(x) + A_2f_2(x-R)) \\ &\quad + V_2(x-R)(A_1f_1(x) + A_2f_2(x-R)) = 0.\end{aligned} \tag{2.2}$$

The coupling terms $V_1(x)f_2(x-R)$ and $V_2(x-R)f_1(x)$ make it difficult to analyze the problem (2.2). We approximate these coupling terms by use of the adiabatic approximation method.

We fix $K \gg 1$ large enough and define $f_j(x) = f_j(x;R)$, $1 \leqq j \leqq 2$, as

$$f_j(x;R) = |x|^{-1}\exp(-\sqrt{E}|x|) \tag{2.3}$$

for $|x| > K$. The function f_j satisfies $-\Delta f_j + Ef_j = 0$ and changes slowly in $|x| \gg 1$. Thus the coupling terms above can be approximated as follows :

$$V_1(x)f_2(x-R) \sim f_2(R)V_1(x)$$

$$V_2(x-R)f_1(x) \sim f_1(R)V_2(x-R),$$

so that (2.2) is decoupled as

$$A_1(H_1 + E)f_1 = -A_2f_2(R)V_1,$$

$$A_2(H_2 + E)f_2 = -A_1f_1(R)V_2.$$

Let K be as above. We may assume that $E = E(R) \to 0$ as $|R| \to \infty$. Next, we use the low energy analysis of resolvent $(H_j + E)^{-1}$ to study the behavior as $|R| \to \infty$ of $f_j(x;R)$ in x, $|x| < 2K$. By the resonance assumption (H.2), there exists a real function (zero resonance function) ψ_j, $1 \leq j \leq 2$, not in L^2 such that $H_j\psi_j = 0$ and hence ψ_j satisfies the integral equation

$$\psi_j(x) = -(4\pi)^{-1} \int |x-y|^{-1} V_j(y)\psi_j(y)\,dy,$$

where the integration with no domain attached is taken over the whole space R^3. Denote by $\langle\ ,\ \rangle$ the L^2 scalar product. The several basic properties of zero resonance function are summarized in subsection 2.2 below. One of the most important properties is that ψ_j satisfies $\langle V_j, \psi_j\rangle \neq 0$. If $\langle V_j, \psi_j\rangle = 0$, then ψ_j is in L^2 and hence ψ_j becomes the eigenfunction of H_j associated with zero eigenvalue. As is easily seen, the converse is also true. Thus we can normalize ψ_j as $\langle V_j, \psi_j\rangle = (4\pi)^{1/2}$. Then this normalized zero resonance function behaves like

$$\psi_j(x) = -(4\pi)^{-1/2}|x|^{-1}(1+o(1)), \qquad |x| \to \infty.$$

We now use the result on the low energy behavior as $E \to 0$ of resolvent $(H_j + E)^{-1}$ ([2, 4]). Then it follows that

$$A_1 f_1(x) = -A_2 f_2(R) E^{-1/2} \langle V_1, \psi_1\rangle \psi_1(x) + O(1), \qquad E \to 0,$$

locally uniformly in x (but not globally). Hence we have by (2.3) and by normalization that

$$A_1 f_1(x;R) \sim A_2 |R|^{-1} \exp(-\sqrt{E}|R|) E^{-1/2} |x|^{-1} \tag{2.4}$$

as $|R| \to \infty$ for x, $K < |x| < 2K$. Similarly

$$A_2 f_2(x;R) \sim A_1 |R|^{-1} \exp(-\sqrt{E}|R|) E^{-1/2} |x|^{-1}. \tag{2.5}$$

On the other hand, it follows from (2.3) that

$$A_j f_j(x;R) = A_j |x|^{-1}(1+o(1)), \qquad |R| \to \infty, \tag{2.6}$$

for x as above.

We are now in a position to determine approximately the negative bound state energy $-E = -E(R)$, $|R| \gg 1$, in question. This is done by matching the asymptotic behaviors (2.4) $\sim$ (2.6). We obtain the following matching condition :

$$\sqrt{E}|R|A_1 = \exp(-\sqrt{E}|R|)A_2,$$
$$\sqrt{E}|R|A_2 = \exp(-\sqrt{E}|R|)A_1.$$

Since $|A_1| + |A_2| \neq 0$, $E = E(R)$ must satisfy the equation

$$\sqrt{E}|R| = \exp(-\sqrt{E}|R|) \tag{2.7}$$

and hence it takes the approximate form

$$E = E(R) = \kappa^2|R|^{-2}, \quad |R| \gg 1, \tag{2.8}$$

where $\kappa \sim 0.567143$ is the unique root of equation $s = \exp(-s)$.

The rigorous proof of Theorem 2 is done by constructing a normalized test function $\phi = \phi(x;R)$ such that

$$\langle H(R)\phi, \phi\rangle = -\kappa^2|R|^{-2}(1 + o(1)), \quad |R| \to \infty.$$

We end this subsection by making a comment that the argument developed here can be easily extended to the case of multiple well potentials. In particular, the argument applied to the case of triple well potentials is used in proving the Efimov effect by the variational method.

2.2. We here give the precise definition of zero resonance function and state the several basic properties which are used to prove Theorem 2. These properties can be easily verified.

Assume that $V(x)$ satisfies (A.1). Without loss of generality, we may assume that $2 < \rho < 3$. Consider the Schrödinger operator $H = -\Delta + V$ acting on $L^2(R^3)$. The operator H is assumed to have a zero resonance energy. Then the zero resonance function $\psi(x)$ of H, $H\psi = 0$, is defined as a non-trivial solution not in L^2 to the integral equation

$$\psi(x) = -(4\pi)^{-1} \int |x - y|^{-1} V(y)\psi(y)\,dy \tag{2.9}$$

when considered in the weighted L^2 space

$$L^2_{-s} = \{ f(x) : (1 + |x|)^{-s} f(x) \in L^2\}, \quad 1/2 < s \leqq 3/2.$$

As is easily seen, the solution $\psi \in L^2_{-s}$ to (2.9) behaves like

$$\psi(x) = -(4\pi)^{-1}\langle V, \psi\rangle |x|^{-1} + O(|x|^{-\rho+1})$$

as $|x| \to \infty$. If $\langle V, \psi\rangle = 0$, then $\psi(x) = O(|x|^{-\rho+1})$ and hence it follows from (2.9) that $\psi(x) = O(|x|^{-\rho+1-\varepsilon})$, $\varepsilon = \rho - 2 > 0$. Thus we can prove by repeated use of the argument above that ψ is in L^2 and hence $\psi(x)$ is the eigenfunction of H associated with zero eigenvalue. Therefore, the zero resonance function ψ not in L^2 must satisfy $\langle V,\psi\rangle \neq 0$ and also such a solution to (2.9) is nondegenerate. We normalize ψ as $\langle V,\psi\rangle = (4\pi)^{1/2}$.

Lemma 2.1. The normalized zero resonance function ψ has the following asymptotic properties as $|x| \to \infty$:

(i) $\psi(x) = -(4\pi)^{-1/2}|x|^{-1} + O(|x|^{-\rho+1})$;

(ii) $\nabla\psi(x) = (4\pi)^{-1/2}|x|^{-2}(x/|x|) + O(|x|^{-\rho})$.

2.3. We shall prove Theorem 2.

Proof of Theorem 2. The proof is done through a series of lemmas.

(1) Let $\chi(x) \in C_0^\infty(R^3)$, $0 \leqq \chi \leqq 1$, be a smooth cut-off function such that $\chi = 1$ for $|x| < 1$ and $\chi = 0$ for $|x| > 2$. We fix θ as

(2.10) $$2/\rho < \theta < 1$$

and define

$$\chi_0(x;R) = \chi(x/|R|^\theta), \quad \chi_\infty(x;R) = 1 - \chi_0(x;R).$$

Let $\psi_j(x)$, $1 \leqq j \leqq 2$, be the zero resonance function of $H_j = -\Delta + V_j$ with normalization $\langle V_j,\psi_j\rangle = (4\pi)^{1/2}$ and let $E = E(R) = \kappa^2|R|^{-2}$, $\kappa \sim 0.567143$, be defined by (2.8), so that E satisfies (2.7). We further define f_j, $1 \leqq j \leqq 2$, as

$$f_j(x;R) = \chi_0(x;R)g_j(x;R) + \chi_\infty(x;R)F(x;R),$$

where

$$F(x;R) = |x|^{-1}\exp(-\sqrt{E}|x|)$$

and

$$g_j(x;R) = -(4\pi)^{1/2}\psi_j(x) - E^{1/2}.$$

(2) By definition, we have

Lemma 2.2. (i) For $|x| < |R|^\theta$,

$$(H_j + E)f_j = -(4\pi)^{1/2}E\psi_j - E^{1/2}V_j - E^{3/2}.$$

(ii) For $|x| > 2|R|^\theta$, $(H_j + E)f_j = Vf_j$.

By Lemma 2.1, it follows that

$$g_j = F(x;R) + O(|R|^{-2+\theta}),$$
$$\nabla g_j = \nabla F(x;R) + O(|R|^{-2}),$$

as $|R| \to \infty$ for x, $|R|^\theta \leqq |x| \leqq 2|R|^\theta$. Hence we have

Lemma 2.3. For x, $|R|^\theta \leqq |x| \leqq 2|R|^\theta$,

$$(H_j + E)f_j = O(|R|^{-2-\theta}), \quad |R| \to \infty .$$

(3) We now define the normalized test function $\phi = \phi(x;R)$, $|R| \gg 1$, as

$$\phi = A_1(R)f_1(x;R) + A_2(R)f_2(x-R;R), \tag{2.11}$$

where the normalization constant $A_j(R)$ satisfies the relation $A_1(R) = A_2(R)$ by the matching condition. We can easily prove that $A_j(R)$ behaves like

$$A_1(R) = A_2(R) = \omega|R|^{-1/2}(1 + o(1)), \quad |R| \to \infty,$$

for some $\omega > 0$. The lemma below completes the proof of Theorem 2.

Lemma 2.4. Let $\phi(x;R)$ be as above. Then

$$\langle H(R)\phi,\phi\rangle = -\kappa^2|R|^{-2}(1 + o(1)), \quad |R| \to \infty .$$

Proof. Set $A(R) = A_1(R) = A_2(R)$. We write

$$\langle (H(R) + E)\phi,\phi\rangle = T_1(R) + T_2(R) + \text{remainder term},$$

where

$$T_1 = A^2(\langle (H_1 + E)f_1(x),f_1(x)\rangle + \langle V_1(x)f_2(x-R),f_1(x)\rangle),$$
$$T_2 = A^2(\langle (H_2 + E)f_2(x),f_2(x)\rangle + \langle V_2(x)f_1(x+R),f_2(x)\rangle).$$

It follows from Lemmas 2.2 and 2.3 that the remainder term is of order $O(|R|^{-2})$. By (2.7), $E^{1/2} = E(R)^{1/2} = f_j(R;R)$ and hence

$$f_2(x-R) - E^{1/2} = O(|R|^{-2+\theta})$$

for x, $|x| < |R|^\theta$. This, together with Lemmas 2.2 and 2.3, yields that $T_1(R) = o(|R|^{-2})$. Similarly, we have $T_2(R) = o(|R|^{-2})$. Thus the lemma and hence Theorem 2 are proved. □

3. The Efimov effect

The idea of the proof of Theorem 1 is simple but the proof requires a rather tedious calculation. We here give only a sketch of the proof. For details, see Tamura [7].

3.1. We first consider the case of three-particle systems with three identical masses $m_j = m$, $1 \leq j \leq 3$. The general cases without mass restriction are reduced to the case with identical masses.

(1) We begin by representing the Hamiltonian H with three identical masses $m_j = m$, $1 \leq j \leq 3$, in terms of the Jacobi coordinates $(r,R) \in R^{3\times 2}$ defined by

$$(3.1) \qquad r = (r_1,r_2,r_3) = x_1 - x_2, \qquad R = (R_1,R_2,R_3) = \frac{1}{2}(x_1 + x_2) - x_3.$$

Let Δ_r and Δ_R denote the Laplace operators with respect to the variables r and R, respectively. Then the Hamiltonian H under consideration is represented as

$$H = -\frac{3}{4m}\Delta_R - \frac{1}{m}\Delta_r + V_{23}(R-r/2) + V_{13}(R+r/2) + V_{12}(r).$$

We denote by $\langle\ ,\ \rangle_r$ the L^2 scalar product in $L^2(R_r^3)$ and introduce a real function $\phi = \phi(r,R)$ with the following properties : (i) $\langle\phi(\cdot,R),\phi(\cdot,R)\rangle_r = 1$ for all $R \in R^3$; (ii) $(\Delta_r\phi)(\cdot,R) \in L^2(R_r^3)$; (iii) $\phi(r,R)$ is of C^2-class as a function of R with values in $L^2(R_r^3)$. The variables R are regarded as parameters. For such a function ϕ, we define the projection $P_\phi : L^2(R^6) \to L^2(R^6)$ by

$$(P_\phi u)(r,R) = \langle u(\cdot,R),\phi(\cdot,R)\rangle_r\phi(r,R).$$

Then the operator $P_\phi H P_\phi$ can be regarded as an operator acting on $L^2(R_R^3)$ and has the following representation

$$P_\phi H P_\phi \simeq -\frac{3}{4m}\Delta_R + U_\phi(R),$$

where the effective potential $U_\phi(R)$ takes the form

$$U_\phi(R) = \langle H_0(R)\phi,\phi\rangle_r + \langle -\frac{3}{4m}\Delta_R\phi,\phi\rangle_r$$

with

$$H_0(R) = -\frac{1}{m}\Delta_r + V_{23}(R-r/2) + V_{13}(R+r/2) + V_{12}(r).$$

If we can find a test function ϕ with the above properties to satisfy

$$\frac{4m}{3} U_\phi(R)|R|^2 < -\frac{1}{4}, \quad |R| \gg 1, \tag{3.2}$$

then the Hamiltonian H under consideration has an infinte number of negative bound state energies. In fact, this follows immediately from the well-known fact that the Schrödinger operator $-\Delta + V$ with potential $V(x)$, $x \in R^3$, behaving like $V \sim -c|x|^{-2}$, $c > 1/4$, at infinity has an infinite number of negative bound state energies.

(2) We consider only R with $|R| \gg 1$ large enough. As such a test function, we choose

$$\phi = A_1(R)f_1(R-r/2;R) + A_2(R)f_2(R+r/2;R) + A_3(R)f_3(r;R),$$

so that ϕ satisfies approximately the eigenvalue problem

$$H_0(R)\phi - \frac{3}{4m}\{A_1(\Delta_r f_1)(R-r/2)+A_2(\Delta_r f_2)(R+r/2)\} = -\frac{1}{m}E\,\phi.$$

More precisely, the problem above can be written as

$$\begin{aligned} -\frac{1}{m}\{ A_1(\Delta_r f_1)(R-r/2) + A_2(\Delta_r f_2)(R+r/2) + A_3(\Delta_r f_3)(r)\} \\ + \{V_{23}(R-r/2) + V_{13}(R+r/2) + V_{12}(r)\}\phi + \frac{1}{m}E\,\phi = 0. \end{aligned} \tag{3.3}$$

By the resonance condition, the two-particle subsystem Hamiltonian $h_\delta = -(1/m)\Delta + V_\delta$ has a zero resonance energy for all pairs $\delta = (j,k)$, $1 \le j < k \le 3$. Thus we can apply the same argument as in section 2 to problem (3.3) with triple well potential. In fact, the eigenvalue $E = E(R)$ is determined by the following matching condition ;

$$\begin{aligned} \sqrt{E}\,A_1 &= (2|R|)^{-1}\exp(-\sqrt{E}|2R|)(A_2 + A_3), \\ \sqrt{E}\,A_2 &= (2|R|)^{-1}\exp(-\sqrt{E}|2R|)(A_1 + A_3), \\ \sqrt{E}\,A_3 &= |R|^{-1}\exp(-\sqrt{E}|R|)(A_1 + A_2), \end{aligned}$$

so that $E = E(R)$ takes the form

$$E = E(R) = \kappa^2|R|^{-2}, \quad |R| \gg 1, \tag{3.4}$$

where $\kappa \sim 0.538808$ is the unique root of equation

$$1 = (2s)^{-1}\exp(-2s) + s^{-2}\exp(-3s).$$

We can also see that the normalization constant $A_j(R)$, $1 \leqq j \leqq 3$, must satisfy the relation

$$(3.5) \qquad A_2(R) = A_1(R), \quad A_3(R) = 2\kappa^{-1}\exp(-\kappa)A_1(R).$$

(3) The test function $\phi = \phi(r,R)$ constructed above takes a form similar to (2.11). Let $\chi_0(r;R)$ and $\chi_\infty(r;R)$ be as in section 2. Denote by ψ_1, ψ_2 and ψ_3 the normalized zero resonance function of h_δ with $\delta = (2,3)$, $(1,3)$ and $(1,2)$, respectively, where ψ_1 is normalized as $m\langle V_{23},\psi_1\rangle = (4\pi)^{1/2}$; similarly for ψ_2 and ψ_3. We further define $f_j(r;R)$, $1 \leqq j \leqq 3$, as

$$f_j(r;R) = \chi_0(r;R)g_j(r;R) + \chi_\infty(r;R)F(r;R),$$

where

$$F(r;R) = |r|^{-1}\exp(-\sqrt{E}|r|)$$

with $E = \kappa^2|R|^{-2}$ defined by (3.4), and

$$g_j(r;R) = -(4\pi)^{1/2}\psi_j(r) - \kappa|R|^{-1}.$$

Then the test function ϕ is defined by

$$\phi(r,R) = A_1(R)f_1(R-r/2;R) + A_2(R)f_2(R+r/2;R) + A_3(R)f_3(r;R),$$

where the normalization constant $A_j(R)$, $1 \leqq j \leqq 3$, satisfies relation (3.5). This test function ϕ yields

$$U_\phi(R) = -\frac{1}{m}\kappa^2|R|^{-2} + \text{remainder term} + o(|R|^{-2})$$

as $|R| \to \infty$. The remainder term is not of order $o(|R|^{-2})$ but of order $O(|R|^{-2})$. Hence, to prove that ϕ satisfies (3.2), we have to evaluate the remainder term precisely. This requires a tedious calculation. We here omit carrying out such a calculation.

<u>3.2</u>. Next we consider the case of 2 L - 1 H particle systems with two light identical and one heavy different masses ; $m = m_1 = m_2 < M = m_3 \leqq \infty$.

Let $(r,R) \in R^{3\times 2}$ be the Jacobi coordinates introduced by (3.1). Then the Hamiltonian H under consideration is represented as

$$H = -\frac{1}{2\nu}\Delta_R - \frac{1}{m}\Delta_r + V_{23}(R-r/2) + V_{13}(R+r/2) + V_{12}(r)$$

with $\nu = 2mM/(2m+M)$. We can rewrite H as $H = \beta_1 H_1 + \gamma_1 T_1$ with

$$\beta_1 = \frac{2m}{3\nu}, \quad \gamma_1 = \frac{2(M-m)}{3M} > 0,$$

where

$$H_1 = -\frac{3}{4m}\Delta_R - \frac{1}{m}\Delta_r + W_{23}(R-r/2) + W_{13}(R+r/2) + W_{12}(r)$$

with

$$W_{23} = \frac{2M}{m+M}V_{23}, \quad W_{13} = \frac{2M}{m+M}V_{13}, \quad W_{12} = V_{12},$$

and

$$T_1 = -\frac{1}{m}\Delta_r + \frac{1}{4}W_{23}(R-r/2) + \frac{1}{4}W_{13}(R+r/2) + W_{12}(r).$$

The Hamiltonian H_1 can be regarded as the energy operator for the system with three identical masses $m_j = m$, $1 \leqq j \leqq 3$, and with pair interactions W_{jk}, $1 \leqq j < k \leqq 3$. By the resonance assumption (A.3), the two-particle subsystem Hamiltonian $-(1/m)\Delta + W_{jk}$ acting on $L^2(R^3)$ has a zero resonance energy. Hence we can find a test function $\phi = \phi(r,R)$ such that the operator $P_\phi H_1 P_\phi$ has an infinite number of negative eigenvalues. For such a test function ϕ, we can prove that

$$\langle T_1\phi, \phi\rangle_r < 0, \quad |R| \gg 1.$$

This proves the Efimov effect for the case of $2L-1H$ particle systems.

3.3. Finally we consider the case of general three-particle systems without mass restriction. Consider the three-particle system with masses ; $m_1 = m$, $m_2 = n$, $m_3 = M$. We may assume that $0 < n < m < M \leqq \infty$. Let $(r,R) \in R^{3\times 2}$ be again the coordinates defined by (3.1). For the above system, the coordinates (r,R) do not necessarily become the Jacobi coordinates but the center of mass can be separated by these coordinates. In fact, the Hamiltonian H is represented as

$$H = -\frac{1}{2}c_1\Delta_R - \frac{1}{2}c_2\Delta_r - \frac{1}{2}c_3\nabla_R\cdot\nabla_r + V_{23}(R-r/2) + V_{13}(R+r/2) + V_{12}(r)$$

with

$$c_1 = \frac{1}{4m} + \frac{1}{4n} + \frac{1}{M}, \quad c_2 = \frac{1}{m} + \frac{1}{n}, \quad c_3 = \frac{1}{m} - \frac{1}{n}.$$

We decompose H as the sum of the Hamiltonian H_2 for the $2L-1H$

particle system with masses (m,m,M), $m < M$, and the remainder operators ;

$$H = \beta_2 H_2 + \gamma_2 \{T_2 + \frac{1}{2m} \nabla_R \cdot \nabla_r\}$$

with

$$\beta_2 = \frac{4mn + M(m+n)}{2n(2m+M)}, \quad \gamma_2 = \frac{m-n}{n} > 0,$$

where

$$H_2 = -\frac{1}{2}\frac{2m+M}{2mM}\Delta_R - \frac{1}{m}\Delta_r + W_{23}(R-r/2) + W_{13}(R+r/2) + W_{12}(r)$$

with

$$W_{23} = \frac{n(m+M)}{m(n+M)} V_{23}, \quad W_{13} = V_{13}, \quad W_{12} = \frac{2n}{m+n} V_{12},$$

and

$$T_2 = -\frac{1}{2m+M}\Delta_r + U_{23}(R-r/2) - U_{13}(R+r/2) + U_{12}(r)$$

with

$$U_{23} = \frac{M}{2(m+M)}\frac{3m+M}{2m+M} W_{23}, \quad U_{13} = \frac{M}{2(2m+M)} W_{13}, \quad U_{12} = \frac{m}{2m+M} W_{12}.$$

By assumption, the Hamiltonian H_2 satisfies the assumptions (A.1) $\sim$ (A.3) when considered as the energy operator for the 2 L - 1 H particle system with masses (m,m,M). Hence we can find a test function $\phi = \phi(r,R)$ such that $P_\phi H_2 P_\phi$ has an infinite number of negative eigenvalues. For such a test function ϕ, we can prove that

$$\langle T_2\phi + \frac{1}{2m} \nabla_R \cdot \nabla_r \phi, \phi \rangle_r < 0, \quad |R| \gg 1,$$

which proves the Efimov effect for general three-particle systems without mass restriction.

References

[1] Efimov, V. : Energy levels arising from resonant two-body forces in a three-body system, Phys. Lett., B 33 (1970), 563-564.

[2] Jensen, A. and Kato, T. : Spectral properties of Schrödinger operators and time-decay of the wave functions, Duke J. Math., 46 (1979), 583-611.

[3] Klaus, M. and Simon, B. : Binding of Schrödinger particles through conspiracy of potential wells, Ann. Inst. H. Poincaré, Sect. A 30 (1979), 83-87.

[4] Murata, M. : Asymptotic expansion in time for solutions of Schrödinger-type equations, J. Func. Anal., 49 (1982), 10-56.

[5] Newton, R. G. : Scattering Theory of Waves and Particles, 2-nd edition, Springer-Verlag, 1982.

[6] Ovchinnikov, Yu. N. and Sigal, I. M. : Number of bound states of three-body systems and Efimov's effect, Ann. of Phys., 123 (1979), 274-295.

[7] Tamura, H. : The Efimov effect of three-body Schrödinger operators, Preprint, 1989, Ibaraki University (to be published in Func. Anal.).

[8] Yafaev, D. R. : On the theory of the discrete spectrum of the three-particle Schrödinger operator, Math. USSR Sb., 23 (1974), 535-559.

High Energy Asymptotics for the Total Scattering Phase in Potential Scattering Theory

Arne Jensen
Department of Mathematics and Computer Science
Institute for Electronic Systems, Aalborg University
Strandvejen 19, DK-9000 Aalborg, Denmark

1 Introduction. Statement of Results.

An asymptotic expansion for the total scattering phase in potential scattering is derived under minimal decay and regularity assumptions on the potential. The result is obtained from the representation in stationary scattering theory of the derivative of the total scattering phase with respect to energy proved in [9], combined with a perturbation argument.

Let $H_0 = -\Delta$ denote the free Schrödinger operator in $L^2(\mathbf{R}^3)$. The following assumption on the potential is introduced:

Assumption $V(\beta, m)$: Let $m \geq 0$ be an integer and $\beta > 0$ a real number. Let $V \in C^m(\mathbf{R}^3)$ be a real-valued function such that for some $\delta > 0$ and all multi-indices α with $|\alpha| \leq m$

$$|\partial_x^\alpha V(x)| \leq c_\alpha (1 + |x|)^{-\beta-\delta-|\alpha|}, \qquad x \in \mathbf{R}^3.$$

Under Assumption $V(1,0)$ (short range potential) the operator $H = H_0 + V$ is self-adjoint with domain $\mathcal{D}(H) = \mathcal{D}(H_0)$. The wave operators $W_\pm = s - \lim_{t\to\pm\infty} e^{itH}e^{-itH_0}$ exist and are complete, i.e. their ranges $\operatorname{Ran}(W_\pm)$ equal the orthogonal complement of the subspace spanned by the L^2-eigenvectors of H. The scattering operator $S = W_+^* W_-$ is unitary and commutes with H_0, hence it has a direct integral decomposition in the spectral representation of H_0. Write $S = \{S(\lambda)\}_{\lambda>0}$ for this decomposition. Here $S(\lambda)$, $\lambda > 0$, is a unitary operator on $\mathcal{P} = L^2(\mathbf{S}^2)$ ($\mathbf{S}^2$ denotes the unit sphere in $\mathbf{R}^3$), called the scattering matrix. See for example [15, 16] for these results.

Under Assumption $V(3,0)$ the operator $S(\lambda) - I$ is a trace class operator (see [15]). The total scattering phase $\theta(\lambda)$ is given by

$$\det S(\lambda) = e^{2i\theta(\lambda)}, \qquad \lambda > 0. \tag{1.1}$$

This relation defines $\theta(\lambda)$ modulo an integer multiple of π. To define it uniquely, we use Krein's spectral shift function $\xi(\lambda)$, see [13, 14, 2]. There exists a real-valued function $\xi \in L^1_{loc}(\mathbf{R})$, such that $\xi(\lambda) = 0$ for $\lambda < \inf \sigma(H)$ ($\sigma(H)$ denotes the spectrum of H) and such that the trace formula

$$\operatorname{tr}(\phi(H) - \phi(H_0)) = \int_{-\infty}^{\infty} \phi'(\lambda)\xi(\lambda)\, d\lambda \tag{1.2}$$

holds for all $\phi \in C_0^\infty(\mathbf{R}^3)$. The connection between $\theta(\lambda)$ and $\xi(\lambda)$ is given by

$$\theta(\lambda) = -\pi\xi(\lambda), \qquad \lambda > 0, \tag{1.3}$$

which then defines $\theta(\lambda)$ uniquely, see [10]. We shall prove the following two theorems.

Theorem 1.1 *Let V satisfy Assumption $V(3,2)$. Then*

$$\theta(\lambda) = \theta_0\lambda^{1/2} + \theta_1 + \theta_2\lambda^{-1/2} + o(\lambda^{-1/2}) \quad \text{as } \lambda \to \infty. \tag{1.4}$$

Here

$$\theta_0 = \frac{-1}{4\pi}\int_{\mathbf{R}^3} V(x)\,dx, \tag{1.5}$$

θ_1 is equal to π times the number of nonpositive eigenvalues of H plus $1/2$, if zero is a resonance for H, and

$$\theta_2 = \frac{-1}{16\pi}\int_{\mathbf{R}^3} V(x)^2\,dx. \tag{1.6}$$

Remark 1.2 Concerning the coefficient θ_1 we note that zero is said to be a resonance for H, if the differential equation $(-\Delta + V)u = 0$ has a solution $u(x) \neq 0$, such that $u \in L^{2,s}(\mathbf{R}^3)$ for some s, $-3/2 \leq s < -1/2$, and $u \notin L^2(\mathbf{R}^3)$. See [11] for a complete discussion of zero resonances. Here $L^{2,s}(\mathbf{R}^3) = \{u \in L^2_{loc}(\mathbf{R}^3) \,|\, (1+x^2)^{s/2}u \in L^2(\mathbf{R}^3)\}$.

Theorem 1.3 *Let V satisfy Assumption $V(3,m)$ for $m \geq 2$. Then*

$$\theta(\lambda) = \sum_{j=0}^{m} \theta_j\lambda^{(1-j)/2} + o(\lambda^{(1-m)/2}) \tag{1.7}$$

as $\lambda \to \infty$.

These results are well known for $V \in C_0^\infty(\mathbf{R}^3)$. It is the proofs under the minimal regularity and decay assumptions that constitute our contribution.

Let us briefly describe previous results. The first results on the high energy asymptotics were obtained by Buslaev. In [3] the leading term θ_0 (see (1.5)) is found, and in [4] a complete asymptotic expansion is announced. It seems that complete proofs of these results have not been published. The first two terms in the expansion (1.4) have been obtained by Newton in [18]. His proof is based on the modified Fredholm determinant. For $V \in C_0^\infty(\mathbf{R}^3)$ a complete asymptotic expansion has been obtained by Colin de Verdière [5]. He also computes the first four terms explicitly and gives results on the structure of the general term. These results are obtained by relating the coefficients in the expansion of $\theta(\lambda)$ as $\lambda \to \infty$ to the coefficients in the expansion of $\mathrm{tr}(\exp(-tH) - \exp(-tH_0))$ as $t \downarrow 0$ via Krein's formula (1.2). It follows from this method that the coefficients in (1.7) satisfy $\theta_j = 0$ for $j \geq 3$, j odd.

For $V \in C_0^\infty(\mathbf{R}^n)$, n odd, existence of a complete asymptotic expansion for $\theta(\lambda)$ has been obtained by Guillopé in [6, 7]. The case of even n has been treated by Popov in [20] using wave equation methods. See also [8] for an exposition of these results.

Recently, the full asymptotic expansion of $\theta'(\lambda)$ has been obtained for general $V \in C^\infty(\mathbf{R}^n)$ satisfying

$$|(\partial_x^\alpha V)(x)| \leq c_\alpha(1+|x|)^{-\rho-|\alpha|}, \qquad \rho > n, \tag{1.8}$$

for all multi-indices α by Robert in [21, 22]. In [21] a general Yang-Mills type operator is treated. His approach is based on the formula for $\xi'(\lambda)$ described in Section 2 (see also Remark 2.3) and a calculus of pseudodifferential operators with a precise control of parameter dependence.

For obstacle scattering results on the high energy asymptotics of the total scattering phase have been obtained by Majda-Ralston [17], Jensen-Kato [10], and Petkov-Popov [19].

2 A Stationary Formula for $\xi'(\lambda)$.

In this section we briefly recall the formula for $\xi'(\lambda)$ obtained in [9] using the stationary Kato-Kuroda scattering theory.

The bounded operators from a Hilbert space $\mathcal{H}$ to another Hilbert space $\mathcal{K}$ are denoted $\mathcal{B}(\mathcal{H},\mathcal{K})$. We write $\mathcal{B}(\mathcal{H},\mathcal{H}) = \mathcal{B}(\mathcal{H})$. The trace ideal or von Neumann-Schatten class is denoted $\mathcal{B}_p(\mathcal{H},\mathcal{K})$, $0 < p \leq \infty$. See [24] for the results on trace ideals needed here. To simplify the notation, we write $L^{2,s} = L^{2,s}(\mathbf{R}^3)$ for the weighted L^2-space.

The trace operator $\Gamma(\lambda)$, $\lambda > 0$, is defined for $f \in L^{2,s}$, $s > 1/2$, by

$$(\Gamma(\lambda)f)(\omega) = 2^{-1/2}\lambda^{1/4}\hat{f}(\lambda^{1/2}\omega), \qquad \lambda > 0, \quad \omega \in \mathbf{S}^2. \tag{2.1}$$

Here $\hat{f}$ denotes the Foureir transform. We have for any $\lambda > 0$

$$\Gamma(\lambda) \in \mathcal{B}_p(L^{2,s},\mathcal{P}), \qquad s > 1/2, \quad p > 4/(2s-1). \tag{2.2}$$

Thus for $s > 3/2$ we have $\Gamma(\lambda) \in \mathcal{B}_2(L^{2,s},\mathcal{P})$, and then (2.1) shows

$$\|\Gamma(\lambda)\|^2_{\mathcal{B}_2(L^{2,s},\mathcal{P})} = 2\pi\sqrt{\lambda}\int_{\mathbf{R}^3}\langle x\rangle^{-2s}\,dx, \tag{2.3}$$

where $\langle x\rangle = (1+x^2)^{1/2}$.

Let $R_0(z) = (H_0 - z)^{-1}$, $\operatorname{Im} z \neq 0$. The limiting absorption principle states that the boundary values $\lim_{\varepsilon\to 0} R_0(\lambda \pm i\varepsilon) = R_0(\lambda \pm i0)$ exist in operator norm on $\mathcal{B}(L^{2,s},L^{2,-s})$, $s > 1/2$, $\lambda > 0$. Furthermore, we have the estimate

$$\|R_0(\lambda \pm i0)\|_{\mathcal{B}(L^{2,s},L^{2,-s})} \leq c\lambda^{-1/2}, \qquad \lambda > 1, \quad s > 1/2. \tag{2.4}$$

Let $E_0(\lambda)$ denote the spectral family for H_0. It is differentiable as a function of $\lambda > 0$ with values in $\mathcal{B}(L^{2,s},L^{2,-s})$, $s > 1/2$. The trace operator $\Gamma(\lambda)$ and the derivative $E_0'(\lambda)$ are related by

$$E_0'(\lambda) = \Gamma(\lambda)^*\Gamma(\lambda), \tag{2.5}$$

see [15, 16]. It follows from (2.2) and (2.3) that $E_0'(\lambda) \in \mathcal{B}_1(L^{2,s},L^{2,-s})$, $s > 3/2$, and

$$\|E_0'(\lambda)\|_{\mathcal{B}_1(L^{2,s},L^{2,-s})} = 2\pi\sqrt{\lambda}\int_{\mathbf{R}^3}\langle x\rangle^{-2s}\,dx. \tag{2.6}$$

We also note the relation

$$E_0'(\lambda) = \frac{1}{2i\pi}(R_0(\lambda + i0) - R_0(\lambda - i0)). \tag{2.7}$$

Let V satisfy Assumption $V(1,0)$. Then the limiting absorption principle and the estimate (2.4) also hold for $R(z)=(H-z)^{-1}$, $H=H_0+V$. Let us note that under our assumption on V the operator H has no positive eigenvalues, see [12, 1]. Using these results the modified trace operators are defined by

$$\Gamma_\pm(\lambda)=\Gamma(\lambda)(1-VR(\lambda\pm i0)). \tag{2.8}$$

We have $\Gamma_\pm(\lambda)\in\mathcal{B}(L^{2,s},\mathcal{P})$, $s>1/2$, $\lambda>0$.

As mentioned in the introduction, under Assumption $V(3,0)$ the total scattering phase $\theta(\lambda)$ is defined.

Lemma 2.1 *Let V satisfy Assumption $V(3,1)$. Then the total scattering phase $\theta(\lambda)$ is continuously differentiable with*

$$\theta'(\lambda)=\frac{\pi}{\lambda}\mathrm{tr}(\Gamma_-(\lambda)\tilde{V}\Gamma_-(\lambda)^*), \tag{2.9}$$

where $\tilde{V}=V+\frac{1}{2}x\cdot\nabla V$. The trace is in $\mathcal{B}_1(\mathcal{P})$.

PROOF: See [9, Lemma 3.2 and Remark 3.3]. Note that $\Gamma_-(\lambda)\tilde{V}\Gamma_-(\lambda)^*$ is a trace class operator on $\mathcal{P}$ due to the assumption on V, (2.2), and (2.8).

Lemma 2.2 *Let V satisfy Assumption $V(3,1)$. Then we have*

$$\theta'(\lambda)=\frac{\pi}{\lambda}\mathrm{tr}(\langle x\rangle^\beta\tilde{V}(1+R_0(\lambda+i0)V)^{-1}E_0'(\lambda)(1+VR_0(\lambda-i0))^{-1}\langle x\rangle^{-\beta}), \tag{2.10}$$

where $\beta=(3+\delta)/2$, δ from Assumption $V(3,1)$. The trace is in $\mathcal{B}_1(L^2)$ and can be computed as the integral of the kernel on the diagonal.

PROOF: Write

$$\Gamma_-(\lambda)\tilde{V}\Gamma_-(\lambda)^*=\left(\Gamma_-(\lambda)\langle x\rangle^{-\beta}\right)\cdot\left(\langle x\rangle^\beta\tilde{V}\Gamma_-(\lambda)^*\right).$$

By (2.2) and (2.8) the two factors above are both Hilbert-Schmidt operators. We can then exchange the order of the two factors and compute the trace in $\mathcal{B}(L^2)$ as the integral of the kernel on the diagonal, see [24]. We write with an obvious notation

$$\mathrm{tr}_{\mathcal{P}}(\Gamma_-(\lambda)\tilde{V}\Gamma_-(\lambda)^*)=\mathrm{tr}_{L^2}(\langle x\rangle^\beta\tilde{V}\Gamma_-(\lambda)^*\Gamma_-(\lambda)\langle x\rangle^{-\beta}). \tag{2.11}$$

As operators in $\mathcal{B}(L^{2,s})$, $1/2<s<2\beta-(1/2)$,

$$(1-VR(\lambda-i0))=(1+VR_0(\lambda-i0))^{-1},\qquad \lambda>0, \tag{2.12}$$

and in $\mathcal{B}(L^{2,-s})$

$$(1-R(\lambda+i0)V)=(1+R_0(\lambda+i0)V)^{-1},\qquad \lambda>0. \tag{2.13}$$

The result (2.10) now follows from the equations (2.11), (2.8), (2.12), and (2.13).

Let $E(\lambda)$ denote the spectral family of H. We note the following results from stationary scattering theory for later use. The family $E(\lambda)$ is differentiable with values in $\mathcal{B}(L^{2,s},L^{2,-s})$, $s>1/2$, $\lambda>0$, and

$$\begin{aligned}E'(\lambda)&=\Gamma_-(\lambda)^*\Gamma_-(\lambda) &(2.14)\\ &=(1+R_0(\lambda+i0)V)^{-1}E_0'(\lambda)(1+VR_0(\lambda-i0))^{-1} &(2.15)\end{aligned}$$

Remark 2.3 Let V satisfy Assumption $V(3,1)$. Using (2.14) and the computations from the proof of Lemma 2.2 together with (1.3) we get

$$\xi'(\lambda) = -\frac{1}{\lambda}\mathrm{tr}_{L^{2,s}}(\tilde{V}E'(\lambda)), \tag{2.16}$$

where the trace is computed in $L^{2,s}$ for $1/2 < s < 2\beta - (1/2)$. Let $\phi \in C_0^\infty((0,\infty))$ and $\psi(\lambda) = -\phi(\lambda)/\lambda$. It follows from (1.2), Lemma 2.1, and (2.16) that we have

$$\mathrm{tr}_{L^2}(\phi(H) - \phi(H_0)) = \mathrm{tr}_{L^{2,s}}(\tilde{V}\psi(H)). \tag{2.17}$$

Related results have been obtained independently in [23].

3 Asymptotic Expansion for $\theta(\lambda)$

We first derive an asymptotic expansion for $\theta'(\lambda)$. This expansion is then integrated. The constant term is obtained using a result due to Guillopé [6], which in turn used the low energy expansion due to Jensen-Kato [11]. See also Newton [18].

Proposition 3.1 *Let V satisfy Assumption $V(3,0)$. Then for $\ell \geq 0$ we have the asymptotic expansion*

$$E'(\lambda) = \sum_{\substack{j+k\leq\ell \\ j\geq 0,\, k\geq 0}} (-1)^{j+k}(R_0(\lambda+i0)V)^j E_0'(\lambda)(VR_0(\lambda-i0))^k + o(\lambda^{-(\ell-1)/2}) \tag{3.1}$$

as $\lambda \to \infty$ in trace norm in $\mathcal{B}_1(L^{2,s}, L^{2,-s})$ for $s > 3/2$.

PROOF: This result is an immediate consequence of the finite Neumann series, the estimate (2.4), and the equations (2.6) and (2.15).

The uniqueness of the coefficients in an asymptotic expansion, together with Proposition 3.1, shows that in order to get an asymptotic expansion for $\theta'(\lambda)$ it suffices to consider a *finite* number of terms. In particular, we need no estimate on the magnitude of the coefficients in each term.

Lemma 3.2 *Let V satisfy Assumption $V(3,m)$, $m \geq 1$. Let $\beta = (3+\delta)/2$, δ from the assumption. For ℓ, $0 \leq \ell \leq m-1$, we have*

$$\mathrm{tr}_{L^{2,\beta}}\Big(\sum_{\substack{j+k\leq\ell \\ j\geq 0,\, k\geq 0}} \tilde{V}(R_0(\lambda+i0)V)^j E_0'(\lambda)(VR_0(\lambda-i0))^k\Big) = \sum_{\nu=-1}^{\ell} c_\nu^\ell \lambda^{-\nu/2} + o(\lambda^{-\ell/2}) \tag{3.2}$$

as $\lambda \to \infty$.

PROOF: Introduce the shorthand notation $R_0^\pm = R_0(\lambda \pm i0)$. Equation (2.7) and straightforward computations yield

$$\mathrm{tr}_{L^{2,\beta}}\Big(\sum_{\substack{j+k=\ell \\ j\geq 0,\, k\geq 0}} \tilde{V}(R_0^+V)^j E_0'(\lambda)(VR_0^-)^k\Big) = \frac{1}{2i\pi}\mathrm{tr}_{L^{2,\beta}}(\tilde{V}(R_0^+V)^\ell R_0^+ - \tilde{V}(R_0^-V)^\ell R_0^-).$$

We now use the integral kernel of $R_0(z)$, which is given by

$$\frac{e^{i\sqrt{z}|x-y|}}{4\pi|x-y|}, \qquad \mathrm{Im}\sqrt{z} \geq 0, \quad x, y \in \mathbf{R}^3$$

to write this expression as

$$\frac{1}{4(4\pi)^\ell}\int \cdots \int \frac{\tilde{V}(x_1)V(x_2)\ldots V(x_\ell)}{|x_1-x_2|\ldots|x_\ell - x_1|} \cdot \\ \cdot \sin(\sqrt{\lambda}(|x_1-x_2| + |x_2-x_3| + \ldots + |x_\ell - x_1|))dx_1 \cdots dx_\ell.$$

This integral has an asymptotic expansion in powers of $\sqrt{\lambda}$ as $\lambda \to \infty$ with remainder $o(\lambda^{-\ell/2})$, $0 \leq \ell \leq m-1$. This can be seen using the same change of variables as in [5, page 33] and then integration by parts. See also the proof of the next lemma.

Lemma 3.3 *Let V satisfy Assumption $V(3,m)$, $m \geq 1$. We have for ℓ, $0 \leq \ell \leq m-1$, the asymptotic expansion*

$$\mathrm{tr}_{\mathcal{P}}(\Gamma_-(\lambda)\tilde{V}\Gamma_-(\lambda)^*) = \sum_{j=-1}^{\ell} a_j \lambda^{-j/2} + o(\lambda^{-\ell/2}) \tag{3.3}$$

as $\lambda \to \infty$. For the coefficients we have

$$a_{-1} = -\frac{1}{8\pi^2}\int V(x)\,dx, \tag{3.4}$$

$$a_0 = 0, \tag{3.5}$$

$$a_1 = \frac{1}{32\pi^2}\int V(x)^2\,dx. \tag{3.6}$$

PROOF: The existence of the asymptotic expansion is a consequence of the computations in section 2, Proposition 3.1, and Lemma 3.2. Carrying out the computations in detail, we find the following leading term:

$$\mathrm{tr}_{L^{2,\beta}}(\tilde{V}E_0'(\lambda)) = \int \tilde{V}(x)\,dx \cdot \frac{\sqrt{\lambda}}{4\pi^2}. \tag{3.7}$$

Now $x \cdot \nabla V = \nabla \cdot (xV) - 3V$, so this term equals

$$-\frac{1}{8\pi^2}\int V(x)\,dx \cdot \sqrt{\lambda}. \tag{3.8}$$

The next term is given by

$$\mathrm{tr}_{L^{2,\beta}}(\tilde{V}R_0(\lambda+i0)VE_0'(\lambda) + \tilde{V}E_0'(\lambda)VR_0(\lambda - i0)) \tag{3.9}$$

$$= \frac{1}{16\pi^3}\int\int \tilde{V}(x_1)V(x_2)\frac{\sin(2\sqrt{\lambda}|x_1-x_2|)}{|x_1-x_2|^2}dx_1dx_2. \tag{3.10}$$

Changing to the variables $u = x$, $v = y - x$, and integrating by parts we find that this term equals

$$\frac{1}{\sqrt{\lambda}}\frac{1}{8\pi^2}\int \tilde{V}(x)V(x)dx + o(\lambda^{-1/2}) \tag{3.11}$$

as $\lambda \to \infty$. Using $(x \cdot \nabla V)V = \frac{3}{2}V^2 - \frac{1}{2}\nabla \cdot (xV^2)$ we get the above expression for the coefficent a_1. It follows from these computations that we also have $a_0 = 0$, since the next term ($\ell = 2$) is of order $O(\lambda^{-1})$.

PROOF OF THEOREM 1.1:

Lemma 2.1 and Lemma 3.3 show that

$$\theta'(\lambda) = \pi a_{-1}\lambda^{-1/2} + \pi a_1 \lambda^{-3/2} + o(\lambda^{-3/2}). \tag{3.12}$$

Integration yields

$$\theta(\lambda) = 2\pi a_{-1}\lambda^{1/2} + c - 2\pi a_1 \lambda^{-1/2} + o(\lambda^{-1/2}). \tag{3.13}$$

The constant term c above is identified with π times the number of bound states plus $1/2$, if zero is a resonance for H, using the results in [5, 6, 18].

PROOF OF THEOREM 1.3:

From Lemma 2.1 and Lemma 3.3 we get

$$\theta'(\lambda) = \sum_{j=-1}^{\ell} a_j \lambda^{-j/2-1} + o(\lambda^{-\ell/2-1}). \tag{3.14}$$

Integration and the result $a_0 = 0$ give

$$\theta(\lambda) = \sum_{j=-1}^{\ell} (-\frac{2}{j}) a_j \lambda^{-j/2} + o(\lambda^{-\ell/2}), \tag{3.15}$$

which proves the result (1.7).

Remark 3.4 Our proof of the existence of the asymptotic expansion of the term (3.2) is quite close to the one given in [6], although our assumptions on the potential are different. The proof can be extended to any odd dimension n by using an integral representation formula for the Hankel functions instead of the simple explicit integral kernel available for $n = 3$, see [6]. In this general case the Assumption $V(\beta, m)$ must be replaced by the assumption $V \in C^m(\mathbf{R}^n)$ satisfying (1.8) for all multi-indices α with $|\alpha| \leq m$.

Acknowledgement

This paper was written while the author visited Département de Mathématiques et d'Informatiques, Université de Nantes and Department of Mathematics, Gakushuin University. The author wishes to thank the two departments for their hospitality.

References

[1] P. ALSHOLM, G. SCHMIDT, *Spectral and scattering theory for Schrödinger operators*, Arch. Rat. Mech. Anal. **40** (1971), 281–311.

[2] M. S. BIRMAN, M. G. KREIN, *On the theory of wave operators and scattering operators*, Dokl. Akad. Nauk. SSSR **144** (1962), 475–478; Soviet Math. Dokl. **3** (1962), 740–744.

[3] V. S. BUSLAEV, *Trace formulas for Schrödinger's operator in three-space*, Dokl. Akad. Nauk SSSR **143** (1962), 1067–1070; Soviet Phys. Dokl. **7** (1962), 295–297.

[4] V. S. BUSLAEV, *Scattered plane waves, spectral asymptotics and trace formulas in exterior problems*, Dokl. Akad. Nauk SSSR **197** (1971), 999–1002; Soviet Math. Dokl. **12** (1971), 591–595.

[5] Y. COLIN DE VERDIÈRE, *Une formule de traces pour l'opérateur de Schrödinger dans* $\mathbf{R}^3$, Ann. Scient. Éc. Norm. Sup. **14** (1981), 27–39.

[6] L. GUILLOPÉ, *Une formule de trace pour l'opérateur de Schrödinger dans* $\mathbf{R}^n$, Thèse de 3ème cycle, Grenoble,1981.

[7] L. GUILLOPÉ, *Asymptotique de la phase de diffusion pour l'opérateur de Schrödinger avec potentiel*, C. R. Acad. Sc. Paris Série I **293** (1981), 601–603.

[8] L. GUILLOPÉ, *Asymptotique de la phase de diffusion pour l'opérateur de Schrödinger dans* $\mathbf{R}^n$, Sem. Bony-Sjöstrand-Meyer 1984–1985, Exposé no. V.

[9] A. JENSEN, *A stationary proof of Lavine's formula for time-delay*, Lett. Math. Phys. **7** (1983), 137–143.

[10] A. JENSEN, T. KATO, *Asymptotic behavior of the scattering phase for exterior domains*, Comm. in. Partial Differential Equations, **3** (1978), 1165–1195.

[11] A. JENSEN, T. KATO, *Spectral properties of Schrödinger operators and time-decay of the wave functions*, Duke Math. J. **46** (1979), 583–611.

[12] T. KATO, *Growth properties of solutions of the reduced wave equation with a variable coefficient*, Comm. Pure Appl. Math. **12** (1959), 403–425.

[13] M. G. KREIN, *On the trace formula in the theory of perturbation*, Math. Sb. **33**(75) (1953), 597–626.

[14] M. G. KREIN, *On perturbation determinants and a trace formula for unitary and selfadjoint operators*, Dokl. Akad. Nauk SSSR **144** (1962), 268–271; Soviet Math. Dokl. **3** (1962), 707–710.

[15] S. T. KURODA, *Scattering theory for differential operators, I, II*, J. Math. Soc. Japan **25** (1973), 75–104, 222–234.

[16] S. T. KURODA, *An introduction to scattering theory*, Aarhus Universitet, Matematisk Institut, lecture notes series No. 51, 1978.

[17] A. MAJDA, J. RALSTON, *An analogue of Weyl's theorem for unbounded domains. I, II, and III*, Duke Math. J. **45** (1978), 183–196, 513–536, and **46** (1979), 725–731.

[18] R. G. NEWTON, *Noncentral potentials: The generalized Levinson theorem and the structure of the spectrum*, J. Math. Phys. **18** (1977), 1348–1357.

[19] V. PETKOV, G. ST. POPOV, *Asymptotic behavior of the scattering phase for non-trapping obstacles*, Ann. Inst. Fourier Grenoble **32** (1982), 111–149.

[20] G. ST. POPOV, *Asymptotic behavior of the scattering phase for the Schrödinger operator*, preprint.

[21] D. ROBERT, *Asymptotique de la phase de diffusion à haute énergie pour des perturbations du Laplacien*, Sem. E.D.P. Ecole Polytechnique 1988–1989, Exposé no. XVII.

[22] D. ROBERT, *Asymptotique à grande énergie de la phase de diffusion pour un potentiel*, preprint 1989.

[23] D. ROBERT, H. TAMURA, *Semi-classical asymptotics for local spectral densities and time delay problems in scattering processes*, J. Funct. Anal. **80** (1988), 124–147.

[24] B. SIMON, *Trace ideals and their applications*, London Mathematical Society Lecture Notes Series 35, Cambridge University Press, Cambridge 1979.

FEYNMAN PATH INTEGRAL TO RELATIVISTIC QUANTUM MECHANICS

Takashi Ichinose*
Department of Mathematics, Kanazawa University
Kanazawa, 920, Japan

Abstract. *This lecture makes a survey of some recent developments of path integral approach to relativistic quantum mechanics focussed on the* $1+1$ *-dimensional Dirac equation.*

1. Introduction.

The Feynman path integral (Feynman [7], Feynman-Hibbs [10]) gives an alternative formulation of nonrelativistic quantum mechanics. It gives a direct, though nonrigorous, expression of the fundamental solution $K(t,x;0,y)$ for the Schrödinger equation in d-dimensional space $\mathbf{R}^d$, which physically is the total probability amplitude for the event that a particle at position y at time 0 will be at position x at time t. Assigned to each path X is the probability amplitude $Ce^{iS(X)}$ with a constant C independent of X, where $S(X)$ is the *action* for the path X, and then $K(t,x;0,y)$ is the sum of $Ce^{iS(X)}$ or the formal "integral"

$$K(t,x;0,y) \sim \int_{X(0)=y,X(t)=x} e^{iS(X)}\mathcal{D}X \tag{1.1}$$

over all the paths $X : [0,t] \to \mathbf{R}^d$ satisfying $X(0) = y$ and $X(t) = x$. $\mathcal{D}X$ is a formal uniform "measure" proportional to something like $\prod_{0<\tau<t} dX(\tau)$, the product of the uncountably many copies of the Lebesgue measure $dX(\tau)$ on $\mathbf{R}^d$. Here and throughout, physical units are so chosen that the reduced Planck's constant $\hbar$ and the light velocity c are equal to 1. The mass of a particle will be denoted by m.

In relativistic quantum mechanics, too, Feynman himself seems to have tried such an approach (see Feynman's autographic notes photocopied in Schweber [32, pp.470-471, pp.482-483]). In Feynman-Hibbs [10, pp.34-36] there is a cryptic description for the $1+1$ -dimensional *free* Dirac equation. He assigned to each path X a probability amplitude rather different from the case for the Schrödinger equation. Supposing that a relativistic particle moving in one dimension can go only forward and backward at the light velocity, and that reversals can occur, with time t divided into small equal steps of length ε, only at the boundaries of these steps, he wrote the *free* fundamental solution $K_0(t,x;0,y) = (K_0(t,x,\sigma;0,y,\rho))_{\sigma,\rho=\pm1}$ as

$$K_0(t,x,\sigma;0,y,\rho) \sim \sum_{R=0}^{\infty} N_{\sigma\rho}(R)(i\varepsilon m)^R. \tag{1.2}$$

*Partially supported by Grant-in-Aid for Scientific Research (C) No.01540112, the Ministry of Education, Science and Culture.

Here $N_{\sigma\rho}(R)$ is the number of those possible paths X with R reversals which start from $X(0) = y$ in the direction ρ and end at $X(t) = x$ in the direction σ. On the right-hand side of (1.2) the limit $\varepsilon \to 0$ should be taken. Some further physical treatments in this context can be found in Riazanov [29] and Rosen [30, Appendix, pp.118-122].

In Ichinose [14-16] and Ichinose-Tamura [19-22], for the $1+1$ -dimensional Dirac equation with the vector and scalar potentials $A(t,x)$ and $\Phi(t,x)$ of an electromagnetic field, we have constructed a 2×2 -matrix-valued countably additive path space measure $\nu_{t;0}$ on the space of the Lipschitz continuous paths X to realize Feynman's original idea of path integral. If somewhat formally expressed, the fundamental solution $K(t,x;0,y)$ has the following representation

$$(1.3)\quad K(t,x;0,y) = \int_{X(0)=y,X(t)=x} \exp\left(i\int_0^t [A(s,X(s))dX(s) - \Phi(s,X(s))ds]\right) d\nu_{t;0}(X).$$

The path space measure $\nu_{t;0}$ obtained is concentrated on the set of the n-vertex zigzag paths X with slopes of plus or minus the light velocity 1 for $n = 0,1,2,\ldots$, as it turned out. In the meanwhile there has appeared a probabilistic approach through the Poisson process by Gaveau *et al.* [12] in the absence of electromagnetic potentials, and by Blanchard *et al.* [2] (*cf.* [3]) in their presence. The latter have given a representation of $K(t,x;0,y) = (K(t,x,\sigma;0,y,\rho))_{\sigma,\rho=\pm 1}$ such as, if expressed in a somewhat formal form ready to compare with (1.3),

$$(1.4)\quad K(t,x,\sigma;0,y,\rho) = e^{mt}\int_{\substack{X(0)=y,X(t)=x\\ J(0)=\rho,J(t)=\sigma}} (-i)^{N(t)} e^{i\int_0^t [J(s)A(s,X(s))-\Phi(s,X(s))]ds} d\mu^m(N),$$

where μ^m is the probability measure associated with the Poisson process $N(t)$ starting from $N(0) = 0$ with intensity m, and $J(s) = \sigma(-1)^{N(t-s)}$, $X(s) = x - \int_s^t J(u)du$. There is also a slightly different treatment based on the Poisson process by T.Zastawniak [35]. This approach may be considered to be an appropriate development of the idea in Feynman-Hibbs [10].

In this note we shall explain some of the basic ideas used to obtain the above results about the path integral formulas (1.3) and (1.4) for the 1+1 -dimensional Dirac equation.

For some results on path integral for the imaginary-time relativistic Schrödinger equation with the quantum spinless Hamiltonians $H_A + \Phi$ corresponding to the classical relativistic Hamiltonian $\sqrt{(p - A(x))^2 + m^2} + \Phi(x)$, we refer to [23], [17], [18] and [6].

2. Path Integral for the Dirac Equation.

We seek a path integral formula for the fundamental solution $K(t,x;0,y)$ of the Cauchy problem for the Dirac equation in $d+1$ -dimensional space-time

$$(2.1)\quad \partial_t\varphi(t,x) = -[\alpha\cdot(\partial_x - iA(t,x)) + im\beta + i\Phi(t,x)]\varphi(t,x), \qquad t>0, \quad x\in\mathbf{R}^d,$$

where $d+1$ is an even integer, and $\alpha = (\alpha_1, \ldots, \alpha_d)$ with $\alpha_j, 1 \le j \le d$, and β Dirac matrices.

Let us begin with a heuristic argument (cf. Feynman [8, pp.447-448]). It seems more appropriate to use the method of phase space path integral or Hamiltonian path integral (Feynman [9, p.125]. See also Garrod [11], Mizrahi [27]). We presume that the action for a Dirac particle of mass m interacting with the vector and scalar potentials $A(t,x)$ and $\Phi(t,x)$ be given by

$$(2.2) \qquad S(P,X) = \int_0^t [P(s)\dot{X}(s) - \alpha \cdot (P(s) - A(s,X(s))) - m\beta - \Phi(s,X(s))]ds,$$

with $\dot{X}(s) = dX(s)/ds$. Then this method assumes that the fundamental solution $K(t,x;0,y)$ for the Dirac equation (2.1) should be given as a formal double "integral"

$$(2.3) \qquad K(t,x;0,y) \sim \iint_{\substack{X(0)=y, X(t)=x \\ P(\cdot):unrestricted}} \mathsf{T}\exp[iS(P,X)]\,\mathcal{D}P\mathcal{D}X$$

over all the phase space paths $(P,X) : [0,t] \to \mathbf{R}^d \times \mathbf{R}^d$ satisfying $X(0) = y$ and $X(t) = x$ with $P(\cdot)$ unrestricted. Here $\mathsf{T}\exp$ is the time-ordered exponential; the time-ordering symbol T is introduced, because $S(P,X)$ is matrix-valued. $\mathcal{D}P\mathcal{D}X$ is a formal uniform "measure"

$$(2.4) \qquad \prod_{0<\tau<t} (2\pi)^{-d} dP(\tau)dX(\tau),$$

the product of the uncountably many copies of $(2\pi)^{-d}$ times the Lebesgue measure $dP(\tau)dX(\tau)$ on the phase space $\mathbf{R}^{2d}$.

By a formal change of variables $P'(s) = P(s) - A(s,X(s))$ and $X'(s) = X(s)$, the "integral" on the right-hand side of (2.3) is rewritten (with P and X instead of P' and X') as

$$(2.5) \qquad \iint_{\substack{X(0)=y, X(t)=x \\ P(\cdot):unrestricted}} \mathsf{T}\exp\Big(i\int_0^t [(P(s) + A(s,X(s)))\dot{X}(s)$$
$$- (\alpha \cdot P(s) + m\beta) - \Phi(s,X(s))]ds\Big)\mathcal{D}P\mathcal{D}X.$$

However, (2.4) solely is an infinite quantity, which mathematically is not a measure. But never give up ! There might be a possibility of getting a mathematically significant measure if we combine it with some factor in the integrand of (2.5). That is, we expect the part

$$(2.6) \qquad \mathsf{T}\exp\Big(i\int_0^t [P(s)\dot{X}(s) - (\alpha \cdot P(s) + m\beta)]ds\Big)\mathcal{D}P\mathcal{D}X.$$

to create something finite. In fact, it is from this part that mathematical rigor has constructed a countably additive path space measure $\nu_{t;0}$ in two space-time dimensions, so as to justify the path integral representation (2.3) and/or (2.5) for the fundamental solution $K(t,x;0,y)$.

The path space measure obtained from (2.6) differs from the Wiener measure. Daletskii [4, 5] treated related problems, but did not construct a countably additive path space measure. By the way, it is just in the same way that the Wiener measure was constructed for the heat equation or imaginary-time Schrödinger equation (Nelson [28]. See also Simon [33]).

Further note the time-ordered exponential in (2.6) is expanded as

$$\sum_{n=0}^{\infty}(-im)^n\int_0^t dt_n\int_0^{t_n}dt_{n-1}\cdots\int_0^{t_2}dt_1 \\ e(t,t_n;P,X)\beta e(t_n,t_{n-1};P,X)\beta\cdots\beta e(t_1,0;P,X),$$

where $e(t,s;P,X)=\mathsf{T}\exp(i\int_s^t[P(u)\dot{X}(u)-\alpha\cdot P(u)]du)$. Hence we get, since the $\alpha_j, 1\le j\le d$, anticommute with β,

$$\begin{aligned} &\mathsf{T}\exp\left(i\int_0^t[P(s)\dot{X}(s)-(\alpha\cdot P(s)+m\beta)]ds\right)\mathcal{D}P\mathcal{D}X \\ &=\sum_{n=0}^{\infty}\Big[(-im\beta)^n\int_0^t dt_n\int_0^{t_n}dt_{n-1}\cdots\int_0^{t_2}dt_1 \\ &\qquad \mathsf{T}\exp\left(i\int_0^t[P(s)\dot{X}(s)-\varepsilon(s;t_0,t_1,\dots,t_{n+1})\alpha\cdot P(s)]ds\right)\Big]\mathcal{D}P\mathcal{D}X \\ &\equiv\sum_{n=0}^{\infty}E^n(P,X)\mathcal{D}P\mathcal{D}X, \end{aligned}\tag{2.7}$$

where $\varepsilon(s;t_0,t_1,\dots,t_{n+1})=(-1)^{k-1}, t_{k-1}<s<t_k, 1\le k\le n+1$, with $t_0=0$ and $t_{n+1}=t$. The expansion (2.7) will give an intuitive knowledge of the support property of our path space measure. We can similarly show that each term $E^n(P,X)\mathcal{D}P\mathcal{D}X$ in (2.7) also produces a countably additive path space measure.

Now we come to our mathematical result. The two-space-time-dimensional Dirac equation we consider is of the form

$$\partial_t\varphi(t,x)=-[\alpha(\partial_x-iA(t,x))+im\beta+i\Phi(t,x)]\varphi(t,x),\qquad t>0,\quad x\in\mathbf{R},\tag{2.8}$$

where α and β are 2×2 Hermitian matrices satisfying $\alpha^2=\beta^2=I$ and $\alpha\beta+\beta\alpha=0$. Without loss of generality we may take

$$\alpha=\begin{pmatrix}1&0\\0&-1\end{pmatrix},\qquad \beta=\begin{pmatrix}0&-1\\-1&0\end{pmatrix}.\tag{2.9}$$

We assume that both $A:\mathbf{R}^2\to\mathbf{R}$ and $\Phi:\mathbf{R}^2\to\mathbf{R}$ are continuous.

Then the fundamental solution $K(t,x;0,y)$ of the Cauchy problem for the Dirac equation (2.8) admits the following path integral representation. $M_2(\mathbf{C})$ denotes the linear space of complex 2×2 matrices.

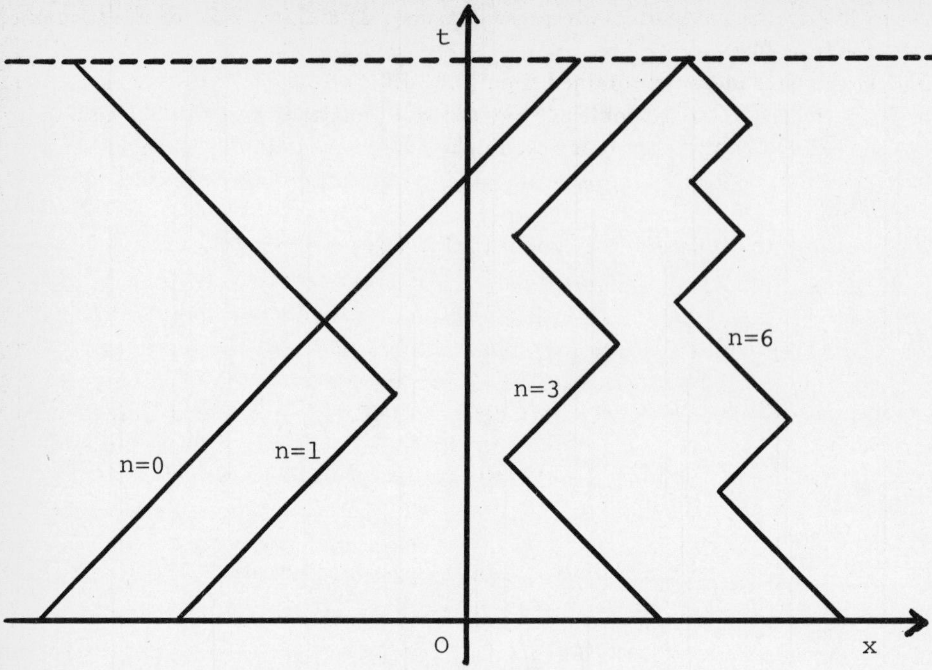

Fig. *The zigzag paths*

THEOREM. *There exists a unique $\mathcal{S}'(\mathbf{R}\times\mathbf{R}; M_2(\mathbf{C}))$-valued countably additive path space measure $\nu_{t;0}$ on the space $C([0,t]\to\mathbf{R})$ of the continuous paths $X(\cdot)$ such that, for $f,g\in\mathcal{S}(\mathbf{R};\mathbf{C}^2)$,*

$$(2.10)\quad \int\int_{\mathbf{R}\times\mathbf{R}} {}^t\overline{f(x)}K(t,x;0,y)g(y)dxdy = \int \exp\left(i\int_0^t [A(s,X(s))dX(s)-\Phi(s,X(s))ds]\right)(f,d\nu_{t;0}(X)g).$$

The path space measure $\nu_{t;0}$ is concentrated on the set of those Lipschitz continuous paths $X:[0,t]\to\mathbf{R}$ which are n-vertex zigzag paths with slopes ±1 and with $n=0,1,2,\ldots$ (when $m=0$, on the set of the straight lines $X(s)-X(0)=\pm s$, $0\le s\le t$). In particular, for every $f=\begin{pmatrix} f_1\\ f_2\end{pmatrix}$ and $g=\begin{pmatrix} g_1\\ g_2\end{pmatrix}$ in $\mathcal{S}(\mathbf{R};\mathbf{C}^2)$,

$$(2.11)\qquad (f,\nu_{t;0}(\cdot)g)=\sum_{j,k=1}^{2} <\nu_{t;0}^{jk}(\cdot),\overline{f_j}\otimes g_k>,$$

with $\nu_{t;0}(\cdot)=\left(\nu_{t;0}^{jk}(\cdot)\right)_{1\le j,k\le 2}$, is a complex-valued countably additive path space measure on $C([0,t]\to\mathbf{R})$ concentrated on the same set of the paths $X:[0,t]\to\mathbf{R}$ as mentioned above with the additional condition $X(0)\in\operatorname{supp}g$ and $X(t)\in\operatorname{supp}f$.

We make here comments on the support property of the path space measure obtained in this theorem.

The measure-theoretic support of this path space measure $\nu_{t;0}$ is on the set of the Lipschitz continuous paths $X : [0,t] \to \mathbf{R}$ which admit a finite partition $0 = t_0 < t_1 < \cdots < t_n = t$ of the interval $[0,t]$ with

$$X(s) - X(t_{j-1}) = \pm(s - t_{j-1}), \qquad t_{j-1} \le s \le t_j, \quad 1 \le j \le n. \tag{2.12}$$

In other words, a.e. path $X(\cdot)$ has differential coefficients of magnitude equal to the light velocity 1 in each finite time interval except at most finitely many instants of time. Therefore the trajectories of the particle shuttle back and forth in one-dimensional space with slopes of ± 1. This reminds us of the *Zitterbewegung* of a Dirac particle such as an electron (see Schrödinger [31], Bjorken-Drell [1]). This fact can also be seen from the result (1.4) of Blanchard *et al.* [2] with the basic property of the Poisson process.

The result on the support property of $\nu_{t;0}$ that we first showed in [14-16] and [19] is that $\nu_{t;0}$ is concentrated on the set of the Lipschitz continuous paths X satisfying $|\dot{X}(s)| \le 1$ a.e. Next, it was improved in [20] to show that $\nu_{t;0}$ is concentrated on the set of the Lipschitz continuous paths X satisfying $|\dot{X}(s)| = 1$ a.e. Here "a.e." is with respect to the Lebesgue measure on $[0,t]$. Then we see that these results are unsatisfactory, because we believe that a Dirac particle or an electron is certainly *not* aware of the Lebesgue measure, so that the statement with Lebesgue measure is not adequate. The final result as in Theorem has been obtained in [21, 22]. In the course of getting this result, we are stimulated by the work of Blanchard *et al.* [2] with Poisson process in probability theory. Our proof is a direct derivation from the path space measure $\nu_{t;0}$, as a development of our theory, but not passing through the Poisson process.

Remarks. 1) With a *formal* path space measure

$$\nu_{t,x;0,y}(\cdot) = \left(\nu^{jk}_{t,x;0,y}(\cdot)\right)_{1 \le j,k \le 2} = \left(< \nu^{jk}_{t;0}(\cdot), \delta_x \otimes \delta_y >\right)_{1 \le j,k \le 2}$$

obtained by formally taking in (2.11) $f = \begin{pmatrix} \delta_x \\ \delta_x \end{pmatrix}$ and $g = \begin{pmatrix} \delta_y \\ \delta_y \end{pmatrix}$, where $\delta_x(\cdot) = \delta(\cdot - x)$ and $\delta_y(\cdot) = \delta(\cdot - y)$, are the delta distributions at x and y, the formula (2.10) looks like (1.3). Compare these with (2.5).

2) What about 4 space-time dimensions ? The present theory does not in general apply to the $3+1$ dimensional Dirac equation. The key fact used in the proof of Theorem is the L^∞ well-posedness of the Cauchy problem for the hyperbolic system of the first order in two independent variables. For the 4-space-time-dimensional Dirac equation this does not hold. However, there are three special cases to which the theory applies: (i) the free Dirac equation, (ii) the Dirac equation for a central electric field, and (iii) the Dirac equation for parallel electric and uniform magnetic fields, because these three cases are reduced with suitable transforms of the variables to equations with two independent variables, in fact, by use of (i) the Radon transform, (ii) the spherical coordinates and (iii) the Fourier transform in one variable together with an Hermite function expansion in another variable (For the details see [22]).

3. Proof of Theorem.

We give an outline of proof. We shall first construct the path space measure $\nu_{t;0}$ and then establish the path integral formula (2.10).

To construct $\nu_{t;0}$, let $\dot{\mathbf{R}} = \mathbf{R} \cup \{\infty\}$ be the one-point compactification of $\mathbf{R}$, and, for $t > 0, \mathbf{X}_t = \prod_{[0,t]} \dot{\mathbf{R}} = \dot{\mathbf{R}}^{[0,t]}$ the infinite product of the uncountably many copies of $\dot{\mathbf{R}}$. By the Tychonoff theorem $\mathbf{X}_t$ is a compact Hausdorff space in the product topology. It may be regarded as the space of all paths $X : [0,t] \to \dot{\mathbf{R}}$, possibly discontinuous and possibly passing through ∞. Let $C(\mathbf{X}_t)$ be the Banach space of the complex-valued continuous functions Ψ on $\mathbf{X}_t$, and $C_{fin}(\mathbf{X}_t)$ its subspace of all $\Psi \in C(\mathbf{X}_t)$ for which there exist a finite partition $0 = s_0 < s_1 < \cdots < s_k = t$ of the interval $[0,t]$ and a continuous function $F(x_0, \ldots, x_k)$ on $\dot{\mathbf{R}}^{k+1}$ such that $\Psi(X) = F(X(s_0), \ldots, X(s_k))$. By the Stone-Weierstrass theorem $C_{fin}(\mathbf{X}_t)$ is dense in $C(\mathbf{X}_t)$.

Define a functional $L(\Psi; f, g)$ which is linear in $\Psi \in C_{fin}(\mathbf{X}_t)$ and sesquilinear in $(f, g) \in \mathcal{S}(\mathbf{R}; \mathbf{C}^2) \times \mathcal{S}(\mathbf{R}; \mathbf{C})$ by

$$(3.1) \quad L(\Psi; f, g) = \overbrace{\int_{\mathbf{R}} \cdots \int_{\mathbf{R}}}^{k} {}^t\overline{f(x_k)} K_0(s_k - s_{k-1}, x_k - x_{k-1}) \cdots K_0(s_1 - s_0, x_1 - x_0) \times F(x_0, \ldots, x_k) g(x_0)\, dx_0 \cdots dx_k.$$

Here $K_0(t, x)$ is the fundamental solution of the free Dirac equation. Intuitively, $L(\cdot; \cdot, \cdot)$ is a functional associated with the part (2.6) with the $\mathcal{D}P$ integration performed. Put

$$(3.2) \qquad C = A + B = -\alpha\partial_x - im\beta.$$

Then (3.1) is rewritten as

$$(3.1)' \qquad L(\Psi; f, g) = (f(x_k), e^{\Delta s_k C} \cdots e^{\Delta s_1 C} F(x_0, \ldots, x_k) g(x_0)),$$

with $\Delta s_j = s_j - s_{j-1}$, $j = 1, 2, \ldots, k$. The kernel of $e^{\Delta s_j C}$ is $K_0(s_j - s_{j-1}, x_j - x_{j-1})$, and so $e^{\Delta s_j C}$ is an operator transforming the functions of x_{j-1} to the functions of x_j. Since e^{tC} satisfies the integral equation

$$e^{tC} = e^{tA} + \int_0^t ds\, e^{(t-s)A} B e^{sC},$$

we have the Dyson series expansion

$$\begin{aligned} e^{tC} &= e^{tA} + \sum_{n=1}^{\infty} \int_0^t dt_n \int_0^{t_n} dt_{n-1} \cdots \int_0^{t_2} dt_1\, e^{(t-t_n)A} B e^{(t_n - t_{n-1})A} \cdots B e^{t_1 A} \\ (3.3) \qquad &= e^{tA} + \sum_{n=1}^{\infty} B^n \int_0^t dt_n \int_0^{t_n} dt_{n-1} \cdots \int_0^{t_2} dt_1\, e^{[\sum_{j=0}^{n} (-1)^j (t_{j+1} - t_j)]A} \\ &\equiv \sum_{n=0}^{\infty} C_n(t) \end{aligned}$$

in the L^∞ operator norm, where for n fixed we understand $t_0 = 0$ and $t_{n+1} = t$. Note that A and B anticommute. Substituting (3.3) with Δs_j in place of t into $e^{\Delta s_j C}$ of (3.1)′ yields

$$
\begin{aligned}
L(\Psi; f, g) &= \sum_{n=0}^{\infty} \sum_{\substack{n_1,\dots,n_k \geq 0 \\ n_1+\dots+n_k=n}} (f(x_k), C_{n_k}(\Delta s_k) \cdots C_{n_1}(\Delta s_1) F(x_0, \cdots, x_k) g(x_0)) \\
&\equiv \sum_{n=0}^{\infty} L^n(\Psi; f, g).
\end{aligned} \tag{3.4}
$$

Obviously, this expansion (3.4) corresponds to the expansion (2.7), the n-th functional $L^n(\cdot;\cdot,\cdot)$ to the n-th term $E^n(P,X)\mathcal{D}P\mathcal{D}X$ in (2.7).

Then of crucial importance is the following lemma.

LEMMA. *i) $L(\Psi; f, g)$ and $L^n(\Psi; f, g), n = 0, 1, 2, \dots,$ are well-defined, i.e. independent of the choice of $F(x_0, \dots, x_k)$ corresponding to $\Psi \in C_{fin}(\mathbf{X}_t)$.*

ii) The following inequalities hold: For $\Psi \in C_{fin}(\mathbf{X}_t)$,

$$|L(\Psi; f, g)| \leq e^{mt} \| f \|_{L^1} \| g \|_{L^\infty} \| \Psi \|, \tag{3.5}$$

$$|L^n(\Psi; f, g)| \leq \sum_{\substack{n_1,\dots,n_k \geq 0 \\ n_1+\dots+n_k=n}} \frac{(mt)^n}{n_1! \cdots n_k!} \| f \|_{L^1} \| g \|_{L^\infty} \| \Psi \|. \tag{3.6}$$

The first statement of this lemma is due to the semigroup property. The second statement can be shown if we note that the hyperbolic system of the first order in two independent variables is L^∞ well-posed: $\| e^{tC} g \|_\infty \leq e^{mt} \| g \|_\infty$, and $\| e^{tA} g \|_\infty \leq \| g \|_\infty$, and that the L^∞ operator norm of B is m.

This lemma implies, since both (3.5) and (3.6) hold for all $\Psi \in C(\mathbf{X}_t)$ by denseness of $C_{fin}(\mathbf{X}_t)$ in $C(\mathbf{X}_t)$, that $L(\Psi; f, g)$ and $L^n(\Psi; f, g)$ are continuous linear forms on $C(\mathbf{X}_t)$ and continuous sesquilinear forms on $\mathcal{S}(\mathbf{R}; \mathbf{C}^2) \times \mathcal{S}(\mathbf{R}; \mathbf{C}^2)$. By the kernel theorem they are elements of the space $\mathcal{S}'(\mathbf{R} \times \mathbf{R}; M_2(\mathbf{C})) = (\mathcal{S}(\mathbf{R}; C^2) \otimes \mathcal{S}(\mathbf{R}; \mathbf{C}^2))'$ of the $M_2(\mathbf{C})$-valued tempered distributions on $\mathbf{R}^2$. It follows that L and L^n are weakly compact by the Grothendieck theorem (e.g. [26, Vol.II,42.2.(2), p.204]) and even compact (noting [26, Vol.I, 21.5(4), p.263 and 27.2, p.369]), as linear maps of $C(\mathbf{X}_t)$ into $\mathcal{S}'(\mathbf{R} \times \mathbf{R}; M_2(\mathbf{C}))$, which is a reflexive Montel space. Then by the Riesz-type representation theorem [34] there exist $\mathcal{S}'(\mathbf{R} \times \mathbf{R}; M_2(\mathbf{C}))$-valued regular Borel measures $\nu_{t;0}$ and $\nu^n_{t;0}$ on $\mathbf{X}_t$ such that for $\Psi \in C(\mathbf{X}_t)$,

$$L(\Psi; f, g) = \int_{\mathbf{X}_t} \Psi(X)(f, d\nu_{t;0}(X) g), \tag{3.7}$$

$$L^n(\Psi; f, g) = \int_{\mathbf{X}_t} \Psi(X)(f, d\nu^n_{t;0}(X)g). \tag{3.8}$$

It follows with (3.4) that

$$\nu_{t;0}(\cdot) = \sum_{n=0}^{\infty} \nu^n_{t;0}(\cdot). \tag{3.9}$$

Since the integral kernel of e^{tA} is

$$2^{-1}(1 + \alpha \mathrm{sgn}(x-y))\delta(t - \mid x - y \mid),$$

we can see, with the rather formal expansion (2.7) in mind, that each measure $\nu^n_{t;0}$ on the right of (3.9) is concentrated on the set of those Lipschitz continuous paths X which are n-vertex zigzag paths with slopes ± 1 and therefore $\nu_{t;0}$ on the left is concentrated on the union of these sets, to conclude the support property of our path space measure $\nu_{t;0}$ mentioned in Theorem.

Next we show the path integral formula (2.10), however, for simplicity, when $A(t,x) = A(x)$ and $\Phi(t,x) = \Phi(x)$ are t-independent continuous functions. The proof of the general case is referred to [15] and [22]. Define a bounded linear operator $T(t)$ on $L^\infty(\mathbf{R}; \mathbf{C}^2)$ by

$$(T(t)g)(x) = \int K_0(t, x-y) e^{i[A(x)(x-y) - \Phi(x)t]} g(y)\, dy.$$

Then for $f, g \in \mathcal{S}(\mathbf{R}; \mathbf{C}^2)$, we have with $s_j = jt/k$,

$$\begin{aligned}(f, T(t/k)^k g) &= \overbrace{\int_{\mathbf{R}} \cdots \int_{\mathbf{R}}}^{k} {}^t\overline{f(x_k)} \prod_{j=1}^{k} K_0(s_j - s_{j-1}, x_j - x_{j-1}) \\ &\quad \times \exp\left(i \sum_{j=1}^{k} [A(x_j)(x_j - x_{j-1}) - \Phi(x_j)t/k]\right) g(x_0)\, dx_0 \cdots dx_k \\ &= \int \exp\left(i \sum_{j=1}^{k} [A(X(s_j))(X(s_j) - X(s_{j-1})) - \Phi(X(s_j))t/k]\right)(f, d\nu_{t;0}(X)g).\end{aligned} \tag{3.10}$$

First suppose that both $A(x)$ and $\Phi(x)$ are in $C_0^\infty(\mathbf{R})$. The integrand in the last member of (3.10) is uniformly bounded and convergent to

$$\exp\left(i \int_0^t [A(X(s))dX(s) - \Phi(X(s))ds]\right)$$

as $k \to \infty$ for every Lipschitz continuous path X, i.e. for $\nu_{t;0}$ -a.e. X, because of the support property of $\nu_{t;0}$. Then by the Lebesgue bounded convergence theorem, the last member of (3.10) converges to the right-hand side of (2.10). As for the first member of (3.10), we can show that $T(t/k)^k g$ converges to $e^{-itH}g$ in L^∞, where

$$H = \alpha(-i\partial_x - A(x)) + m\beta + \Phi(x). \tag{3.11}$$

Now we consider the case where both $A(x)$ and $\Phi(x)$ are continuous. Choose sequences $\{A^{(k)}(x)\}$ and $\{\Phi^{(k)}(x)\}$ in $C_0^\infty(\mathbf{R})$ such that $A^{(k)}(x)$ and $\Phi^{(k)}(x)$ are, on each compact set in $\mathbf{R}$, uniformly bounded and convergent to $A(x)$ and $\Phi(x)$, respectively. Define the selfadjoint operator $H^{(k)}$ by (3.11) with $A^{(k)}(x), \Phi^{(k)}(x)$ in place of $A(x)$, $\Phi(x)$. Then

$$(3.12)\quad (f, e^{-itH^{(k)}}g)$$
$$= \int \exp\left(i\int_0^t [A^{(k)}(X(s))dX(s) - \Phi^{(k)}(X(s))ds]\right)(f, d\nu_{t;0}(X)g).$$

The right-hand side of (3.12) converges to that of (2.10), as $k \to \infty$, by the Lebesgue bounded convergence theorem, similarly to above. As for the left-hand side, since $H^{(k)}$ and H are essentially selfadjoint on $C_0^\infty(\mathbf{R};\mathbf{C}^2)$, $H^{(k)}g$ is, for $g \in C_0^\infty(\mathbf{R};\mathbf{C}^2)$, convergent to Hg in $L^2(\mathbf{R};\mathbf{C}^2)$. It follows by the Trotter-Kato theorem [25] that for g in $L^2(\mathbf{R};\mathbf{C}^2)$, $e^{-itH^{(k)}}g$ converges to $e^{-itH}g$ in L^2 as $k \to \infty$. This proves (2.10) when both $A(t,x) = A(x)$ and $\Phi(t,x) = \Phi(x)$ are t-independent continuous functions.

4. Probabilistic Approach through Poisson Process.

We briefly sketch the probabilistic approach based on the Poisson process to path integral for the two-space-time-dimensional Dirac equation (2.8), made by Blanchard *et al.* [2] (*cf.* [3]) and by T.Zastawniak [35]. In [24] M.Kac already gave such a treatment for the telegrapher's equation.

Introduce the dichotomic variable $\sigma = \pm 1$ and put

$$\varphi(t,x) = (\varphi(t,x,\sigma))_{\sigma=\pm 1}. \tag{4.1}$$

Then the Dirac equation (2.8) with α and β as in (2.9) is rewritten as

$$\partial_t\varphi(t,x,\sigma) = im\varphi(t,x-\sigma) - \sigma\partial_x\varphi(t,x,\sigma) + i(\sigma A(t,x) - \Phi(t,x))\varphi(t,x,\sigma), \quad t>0, x\in\mathbf{R}. \tag{4.2}$$

We wish the Dirac equation (4.2) in this form were a Kolmogoroff equation for a Markov process. But unfortunately, it is not. The strategy of Blanchard *et al.* [2] is to transform (4.2) to a backward Kolmogoroff equation for a Markov jump process. We perform a time reversal with $T > 0$ and introduce a new variable y to put

$$\psi(t,x,y,\sigma) = e^{-m(T-t)}e^{-iy}\varphi(T-t,x,\sigma). \tag{4.3}$$

If $\varphi(t,x,\sigma)$ satisfies (4.2) with initial condition $\varphi(0,x,\sigma) = g(x,\sigma)$, then $\psi(t,x,y,\sigma)$ satisfies

$$\begin{aligned}\partial_t\psi(t,x,y,\sigma) &= -m[\psi(t,x,y+\frac{\pi}{2},-\sigma)-\psi(t,x,y,\sigma)]+\sigma\partial_x\psi(t,x,y,\sigma)\\ &\quad+(\sigma A(T-t,x)-\Phi(T-t,x))\partial_y\psi(t,x,y,\sigma),\\ &\qquad\qquad t<T,\quad (x,y)\in\mathbf{R}\times\mathbf{R},\end{aligned}\tag{4.4}$$

with final condition $\psi(T,x,y,\sigma)=e^{-iy}g(x,\sigma)$. It turns out that (4.4) is a backward Kolomogoroff equation for a Markov jump process with values in $\mathbf{R}\times\mathbf{R}\times\{-1,+1\}$. Then by the general theory (e.g. [13, pp.173-178]), the solution $\psi(t,x,y,\sigma)$ of (4.4) admits the following probabilistic representation

$$\psi(t,x,y,\sigma)=\int e^{-iY_t(T)}g(X_t(T),J_t(T))\,d\mu^m(N),\tag{4.5}$$

with

$$\begin{aligned}X_t(s)&=x-\sigma\int_t^s(-1)^{N(u)-N(t)}du,\\ Y_t(s)&=y+\frac{\pi}{2}(N(s)-N(t))\\ &\quad-\int_t^s[\sigma(-1)^{N(u)-N(t)}A(T-u,X_t(u))-\Phi(T-u,X_t(u))]du,\\ J_t(s)&=\sigma(-1)^{N(s)-N(t)},\qquad\qquad t\le s\le T.\end{aligned}\tag{4.6}$$

Here μ^m is the probability measure associated with the Poisson process $N(t)$ starting from $N(0)=0$ with intensity m. Namely, it is a probability measure on the space of the nondecreasing, right-continuous, step paths $N:[0,\infty)\to\mathbf{Z}$ with $N(0)=0$ such that $\int_{N(t)=n}d\mu^m(N)=e^{-mt}\frac{(mt)^n}{n!}$, $n=0,1,2,\ldots$. Hence we have for the solution $\varphi(t,x,\sigma)$ of the Cauchy problem for the Dirac equation (4.2) with initial condition $\varphi(0,x,\sigma)=g(x,\sigma)$

$$\begin{aligned}\varphi(t,x,\sigma)&=e^{mt}\int e^{-i\frac{\pi}{2}N(t)+i\int_0^t[J_0(s)A(t-s,X_0(s))-\Phi(t-s,X_0(s))]ds}\\ &\qquad\qquad\times g(X_0(t),J_0(t))\,d\mu^m(N)\\ &=e^{mt}\int(-i)^{N(t)}e^{i\int_0^t[J_0(t-s)A(s,X_0(t-s))-\Phi(s,X_0(t-s))]ds}\\ &\qquad\qquad\times g(X_0(t),J_0(t))\,d\mu^m(N).\end{aligned}\tag{4.7}$$

This can also be written as

$$\varphi(t,x,\sigma)=e^{mt}\int(-i)^{N(t)}e^{i\int_0^t[J(s)A(s,X(s))-\Phi(s,X(s))]ds}g(X(0),J(0))\,d\mu^m(N),\tag{4.8}$$

with

$$(4.9)\quad X(s) := X_0(t-s) = x - \sigma \int_s^t (-1)^{N(t-u)} du, \quad J(s) := J_0(t-s) = \sigma(-1)^{N(t-s)},$$

which is what is meant by (1.4).

The probabilistic representation of the solution $\varphi(t,x,\sigma)$ of (4.2) given by T. Zastawniak [35] is slightly different from (4.7) and/or (4.8):

$$(4.10)\qquad \varphi(t,x,\sigma) = e^t \int (im)^{N(t)} e^{i\int_0^t [J(s)A(s,X(s)) - \Phi(s,X(s))]ds} g(X(0),J(0))\, d\mu(N).$$

Here we understand $(im)^{N(t)} = 1$ for $m = 0$. μ is the probability measure associated with the Poisson process $N(t)$ starting from $N(0) = 0$ with intensity 1 and

$$(4.11)\qquad X(s) = x - \int_s^t J(u)du, \quad J(s) = \sigma(-1)^{N(t)-N(s)}.$$

To prove (4.10) expand $\varphi(t,x,\sigma)$ as

$$(4.12)\qquad \begin{aligned} \varphi(t,x,\sigma) = g(x,\sigma) + \sum_{n=1}^{\infty} (im)^n \int_0^t dt_n \int_0^{t_n} dt_{n-1} \cdots \int_0^{t_2} dt_1 \\ e^{i\int_0^t [J(s;t_0,t_1,\dots,t_{n+1})A(s,X(s;t_0,t_1,\dots,t_{n+1})) - \Phi(s,X(s;t_0,t_1,\dots,t_{n+1}))]ds} \\ \times g(X(0;t_0,t_1,\dots,t_{n+1}), J(0;t_0,t_1,\dots,t_{n+1})), \end{aligned}$$

where

$$(4.13)\qquad \begin{aligned} X(s;t_0,t_1,\dots,t_{n+1}) &= x - \int_s^t J(u;t_0,t_1,\dots,t_{n+1})du, \\ J(s;t_0,t_1,\dots,t_{n+1}) &= \sigma(-1)^{n-k}, \qquad t_{k-1} \le s < t_k, \quad 1 \le k \le n, \end{aligned}$$

with $t_0 = 0$ and $t_{n+1} = t$. This is due to an analogous argument used to get the expansion (3.3). For each $N(t)$ put

$$(4.14)\qquad t_j(N) = \inf\{s \ge 0;\ N(s) = j\}, \qquad j = 0,1,2,\dots.$$

For every continuous function $F(x_1,\dots,x_n)$ on $\mathbf{R}^n$ we have

$$(4.15)\qquad \int_{N(t)=n} F(s_1(N),\dots,s_n(N))\, d\mu(N) = e^{-t} \int_0^t dt_n \int_0^{t_n} dt_{n-1} \cdots \int_0^{t_2} dt_1\, F(t_1,\dots,t_n).$$

Then the expression (4.10) follows from (4.12) together with (4.15).

It is seen by the basic property of the Poisson process that a.e. $N(\cdot)$ jumps only a finite number of times in each finite time interval, so that the typical path $X(\cdot)$ in (4.9) and (4.11) satisfies (2.12).

References

[1] J.D.Bjorken and S.D.Drell, "Relativistic Quantum Mechanics," MacGraw-Hill, New York, 1964.

[2] Ph.Blanchard, Ph.Combe, M.Sirugue and M.Sirugue-Collin, *Probabilistic solution of the Dirac equation, Path integral representation for the solution of the Dirac equation in presence of an electromagnetic field*, Bielefeld, BiBoS Preprint Nos. 44, 66(1985).

[3] Ph.Combe, M.Sirugue and M.Sirugue-Collin, *Point processes and quantum physics: Some recent developments and results*, in "Proc. the 8th Internat. Congress on Math. Phys. ($M \cap \Phi$) Marseille 1986," World Scientific, Singapore, 1987, pp. 421-430.

[4] Yu.L.Daletskii, *Continual integrals and the characteristics connected with a group of operators*, Dokl. Akad. Nauk SSSR **141(6)**, 1290-1293(1961); *English transl.*, Soviet Math. Dokl. **2**, 1634-1637(1961).

[5] Yu.L.Daletskii, *Functional integrals connected with operator evolution equations*, Uspehi Mat. Nauk. **17(5)**, 3-115(1962); *English transl.*, Russian Math. Surveys **17**, 1-107(1962).

[6] G.F.De Angelis and M.Serva, *Jump processes and diffusions in relativistic stochastic mechanics, On the relativistic Feynman-Kac formula*, Camerino-l'Aquila-Roma"La Sapienza"-Roma"Tor Vergata", Preprints Nos.10, 23(1989).

[7] R.P.Feynman, *Space-time approach to non-relativistic quantum mechanics*, Rev. Mod. Phys. **20**, 367-387(1948).

[8] R.P.Feynman, *Mathematical formulation of the quantum theory of electromagnetic interaction*, Phys. Rev. **80**, 440-457(1950).

[9] R.P.Feynman, *An operator calculus having applications in quantum electrodynamics*, Phys.Rev. **84**, 108-128(1951).

[10] R.P.Feynman and A.P.Hibbs, "Quantum Mechanics and Path Integrals," McGraw-Hill, New York, 1965.

[11] C.Garrod, *Hamiltonian path-integral methods*, Rev. Mod. Phys. **38**, 483-493(1966).

[12] B.Gaveau, T.Jacobson, M.Kac and L.S.Schulman, *Relativistic extension of the analogy between quantum mechanics and Brownian motion*, Phys. Rev. Lett. **53**, **No.5**, 419-422(1984).

[13] I.I.Gihman and A.V.Skorohod, "The Theory of Stochastic Processes," Vol.III, Springer, Berlin-Heidelberg-New York, 1979.

[14] T.Ichinose, *Path integral for the Dirac equation in two space-time dimensions*, Proc. Japan Acad. **58A**, 290-293(1982).

[15] T.Ichinose, *Path integral for a hyperbolic system of the first order*, Duke Math. J. **51**, 1-36(1984).

[16] T.Ichinose, *Path integral formulation of the propagator for a two-dimensional Dirac particle*, Physica **124A**, 419-426 (1984).

[17] T.Ichinose, *The nonrelativistic limit problem for a relativistic spinless particle in an electromagnetic field*, J. Functional Analysis **73**, 233-257(1987).

[18] T.Ichinose, *Essential selfadjointness of the Weyl quantized relativistic Hamiltonian*, Ann. Inst. H. Poincaré, Phys. Théor.1989(to appear).

[19] T.Ichinose and H.Tamura, *Propagation of a Dirac particle. A path integral approach*, J. Math. Phys. **25**, 1810-1819(1984).

[20] T.Ichinose and H.Tamura, *A remark on path integral for the Dirac equation*, Suppl. Rend. Circ. Mat. Palermo (2),**No. 17**, 237-248(1987).

[21] T.Ichinose and H.Tamura, *The Zitterbewegung of a Dirac particle in two-dimensional space-time*, J. Math. Phys. **29**, 103-109(1988).

[22] T.Ichinose and H.Tamura, *Path integral approach to relativistic quantum mechanics - Two-dimensional Dirac equation*, Suppl. Prog. Theor. Phys., **No. 92**, 144-175(1987).

[23] T.Ichinose and H.Tamura, *Imaginary-time path integral for a relativistic spinless particle in an electromagnetic field*, Commun. Math. Phys. **105**, 239-257(1986).

[24] M.Kac, *A stochastic model to the telegrapher's equation*, Rocky Mountain J. Math. **4**, 497-509 (1974); Reprinted from Magnolia Petroleum Company Colloquium Lectures in the Pure and Applied Sciences, **No. 2**, 1956, "*Some stochastic problems in physics and mathematics*".

[25] T.Kato, "Perturbation Theory for Linear Operators, 2nd ed.," Springer, Berlin-Heidelberg-New York, 1976.

[26] G.Köthe, "Topological Vector Spaces," Vol. I, II, Springer, Berlin-Heidelberg-New York, 1969, 1979.

[27] M.M.Mizrahi, *Phase space path integrals, without limiting procedure*, J. Math. Phys. **19**, 298-308 (1978); *Erratum*, ibid. **21**, 1965 (1980).

[28] E.Nelson, *Feynman integrals and the Schrödinger equation*, J. Math. Phys. **5**, 332-343(1964).

[29] G.V.Riazanov, *The Feynman path integral for the Dirac equation*, Soviet Phys. JETP **6(33)**, 1107-1113(1958).

[30] G.Rosen, "Formulations of Classical and Quantum Dynamical Theory," Academic, New York, 1969.

[31] E.Schrödinger, *Über die kräftefreie Bewegung in der relativistischen Quantenmechanik*, Sitzungsber. Preuss. Akad. Wiss. Phys.-Math. Kl. **24**, 418-428(1930).

[32] S.S.Schweber, *Feynman's visualization of space-time processes*, Rev.Mod.Phys. **58**, 449-508 (1986).

[33] B.Simon, "Functional Integration and Quantum Physics," Academic, New York, 1979.

[34] K.Swong, *A representation theory of continuous linear maps*, Math. Ann. **155**, 270-291(1964).

[35] T.Zastawniak, *Path integrals for the Dirac equation - Some recent developments in mathematical theory*, "Stochastic Analysis, Path Integration and Dynamics," Pitman Research Notes in Math. **No. 206**, Longman Scientific & Technical, 1989, pp. 243-263.

ON THE DISTRIBUTION OF POLES OF THE SCATTERING MATRIX FOR SEVERAL CONVEX BODIES

MITSURU IKAWA

Department of Mathematics
Osaka University
Toyonaka, Osaka 560, Japan

1. Introduction. Concerning relationships between the geometry of scatterers and the distribution of poles of scattering matrices, Lax and Phillips gave in [10, page 158] the following conjecture:

In the case of a trapping obstacle there is an infinite sequence of poles of the scattering matrix z_n *such that* $\operatorname{Im} z_n \to 0$ *and* $|\operatorname{Re} z_n| \to \infty$.

In this note we shall consider problems related to this conjecture. Let $\mathcal{O}$ be an open bounded set in $\mathbf{R}^3$ with smooth boundary Γ. Assume that

$$\Omega = \mathbf{R}^3 - \overline{\mathcal{O}} \qquad \text{is connected,} \tag{1.1}$$

and consider the following acoustic problem

$$\begin{cases} \Box u = \dfrac{\partial^2 u}{\partial t^2} - \Delta u = 0 & \text{in } \Omega \times \mathbf{R} \\ u = 0 & \text{on } \Gamma \times \mathbf{R} \\ u(x,0) = f_1(x), \quad \dfrac{\partial u}{\partial t}(x,0) = f_2(x). \end{cases} \tag{1.2}$$

Regarding (1.2) as a system perturbed by an obstacle $\mathcal{O}$ of the wave equation in the free space $\mathbf{R}^3$, we define the scattering matrix $\mathcal{S}(z)$, which is an $\mathcal{L}(L^2(S^2)), L^2(S^2))$–valued function defined for $z \in \mathbf{C}$. For the definition, see [10]. Concerning the fundamental properties of the scattering matrix we refer to two theorems in Chapter V of [10]:

THEOREM 5.1. *The scattering matrix $\mathcal{S}(z)$ is holomorphic on the real axis and meromorphic in the whole plane, having a pole at exactly those points z for which there is a nontrivial eventually outgoing local solution of*

$$Af = izf.$$

THEOREM 5.6. *The scattering matrix determines uniquely the scatterer.*

The second part of Theorem 5.1 can be read as follows: Consider the boundary value problem with parameter $\mu \in \mathbf{C}$

$$\begin{cases} (-\Delta - \mu^2)w(x) = 0 & \text{in } \Omega \\ w(x) = g(x) & \text{on } \Gamma \end{cases} \tag{1.3}$$

for $g(x) \in C^\infty(\Gamma)$. It is well known that for $\operatorname{Re}\mu > 0$ (1.3) has a unique solution in $H^2(\Omega)$. Denote by $U(\mu)$ the operator which maps $g(x)$ to $w(x)$, that is,

$$w(x) = (U(\mu)g(\cdot))(x).$$

Then $U(\mu)$ is an $\mathcal{L}(L^2(\Gamma), H^2(\Omega))-$valued holomorphic function in $\operatorname{Re}\mu > 0$. By the regularity theorem $U(\mu)$ can be regarded as $\mathcal{L}(C^\infty(\Gamma), C^\infty(\overline{\Omega}))-$valued function. If we regard $U(\mu)$ as $\mathcal{L}(C^\infty(\Gamma), C^\infty(\overline{\Omega}))-$valued function it can be prolonged analytically into the whole complex plane as a meromorphic function. Concerning the poles of $\mathcal{S}(z)$ and $U(\mu)$ we have that

$$z \text{ is a pole of } \mathcal{S}(z) \text{ if and only if } \mu = iz \text{ is a pole of } U(\mu).$$

Theorem 5.6 shows us that all the geometric properties of the obstacle $\mathcal{O}$ is contained in the scattering matrix $\mathcal{S}(z)$. It is an important and interesting problem to extract the geometic properties of $\mathcal{O}$ from the analytic properties of $\mathcal{S}(z)$. In regard to this problem it is natural to pose the following problem:

PROBLEM. *How the geometry of $\mathcal{O}$ reflects in the distribution of the poles of $\mathcal{S}(z)$?*

The Lax and Phillips conjecture quoted in the beginning of this note is on this subject. Even though it is more than 20 years since the conjecture was given, there are only a few examples for which its validity was proved. To my best knowledge, only the obstacles $\mathcal{O}$ consisting of two disjoint convex bodies are considered. We present here briefly these results.

Let

$$\mathcal{O} = \mathcal{O}_1 \cup \mathcal{O}_2, \qquad \overline{\mathcal{O}_1} \cap \overline{\mathcal{O}_2} = \phi.$$

First consider the case that $\mathcal{O}_j$, $j = 1, 2$ are strictly convex. Then, there exists a constant $c > 0$, which is determined by the distance of $\mathcal{O}_1$ and $\mathcal{O}_2$ and their curvatures, such that for any $\varepsilon > 0$

(i) $\{z; \operatorname{Im} z \le c - \varepsilon\}$ contains a finite number of poles,

(ii) $\{z; c - \varepsilon < \operatorname{Im} z < c + \varepsilon\}$ contains an infinite number of poles.

In Ikawa[5] and Gérard[3] more precise informations on the distribution of poles are given.

Next, consider the case that $\mathcal{O}_1$ and $\mathcal{O}_2$ are not strictly convex. Let

$$|a_1 - a_2| = \text{distance}(\mathcal{O}_1, \mathcal{O}_2), \quad a_j \in \Gamma_j (= \partial\mathcal{O}_j), \ j = 1, 2.$$

Suppose that the principale curvatures of Γ_j vanish at a_j and do not vanish elsewhere. Then, for any $\varepsilon > 0$ there exist an infinite number of poles in $\{z; 0 < \operatorname{Im} z < \varepsilon\}$(Ikawa[6], Soga[18]).

Thus the result for two strictly convex bodies indicates us that the original Lax and Phillips conjecture is not valid in general and must be modified. We would like to propose the following modification:

Modified Lax and Phillips Conjecture. *If $\mathcal{O}$ is trapping, there exists $\alpha > 0$ such that $\mathcal{S}(z)$ has an infinite number of poles in $\{z;\ \mathrm{Im}\, z \leq \alpha\}$.*

Hereafter, we say that MLPC(abbreviation of the Modified Lax and Phillips Conjecture) is valid for obstacle $\mathcal{O}$, when there is $\alpha > 0$ such that the scattering matrix corresponding to $\mathcal{O}$ has an infinite number of poles in $\{z;\ \mathrm{Im}\, z \leq \alpha\}$.

We should mention here on a result on distribution of poles of scattering matrices for nontrapping obstacles: *if $\mathcal{O}$ is non-trapping, there exist positive constants a and b such that the domain $\{z;\ \mathrm{Im}\, z \leq a \log(|z|+1) + b\}$ is free from poles*(Lax and Phillips[11], Morawets, Ralston and Strauss[15], Melrose and Sjöstand[13,14]).

This result implies that for any $\alpha > 0$ the slab domain $\{z;\ 0 \leq \mathrm{Im}\, z \leq \alpha\}$ contains only a finite number of poles of $\mathcal{S}(z)$. Therefore, if the MLPC is correct, the existence of such α becomes a characterization of trapping obstacles by means of the distribution of poles.

The main purpose of this note is to extend the results in [5] and [3] to obstacles consisting of more than two strictly convex bodies. The difficulties of the problem comes from the existence of an infinite number of primitive periodic rays in Ω. In order to get informations on poles it is necessary to controle the complexity comming from the infiniteness of the number of primitive periodic rays, but we cannot do it for general obstacles consisting of more than two bodies. Here, we shall show that MLPC is valid for $\mathcal{O}$ consisting of small balls(Theorem 4.1 in Section 4).

For these obstacles consisting of several small balls, we can use the ergodic theory in order to controle the complexity of the geormetry of the periodic rays in Ω, more precisely we can apply the singular perturbation theory of symbolic flows developed in [9].

2. A general theorem for several strictly convex bodies

In this section we present a theorem in [8], which reduces the validity of MLPC for several strictly convex bodies to the verification of the existence of poles of a function determined by the geometry of the periodic rays in Ω. First we give notations and the statement of the theorem.

Let $\mathcal{O}_j$, $j = 1, 2, \cdots, L$, be bounded open sets with smooth boundary Γ_j satisfying

(H.1) every $\mathcal{O}_j$ is strictly convex,

(H.2) for every $\{j_1, j_2, j_3\} \in \{1, 2, \cdots, L\}^3$ such that $j_l \neq j_{l'}$ if $l \neq l'$,

$$(\text{convex hull of } \overline{\mathcal{O}_{j_1}} \text{ and } \overline{\mathcal{O}_{j_2}}) \cap \overline{\mathcal{O}_{j_3}} = \phi.$$

We set

$$\mathcal{O} = \cup_{j=1}^{L} \mathcal{O}_j, \quad \Omega = \mathbf{R}^3 - \overline{\mathcal{O}} \text{ and } \Gamma = \partial\Omega. \tag{2.1}$$

Denote by γ a periodic ray in Ω, and we shall use the following notations:

d_γ : the length of γ,

T_γ : the primitive period of γ,

i_γ : the number of the reflecting points of γ,

P_γ : the Poincaré map of γ.

We define a function $F_D(\mu)$ ($\mu \in \mathbf{C}$) by

$$F_D(\mu) = \sum_{\gamma} (-1)^{i_\gamma} T_\gamma |I - P_\gamma|^{-1/2} e^{-\mu d_\gamma} \tag{2.2}$$

where the summation is taken over all the oriented periodic rays in Ω and $|I - P_\gamma|$ denotes the determinant of $I - P_\gamma$.

Concerning the periodic rays in Ω we have easily

$$\#\{\gamma; \text{periodic ray in } \Omega \text{ such that } d_\gamma < r\} < e^{a_0 r}, \tag{2.3}$$

$$|I - P_\gamma| \geq e^{2a_1 d_\gamma}, \tag{2.4}$$

where a_0 and a_1 are positive constants depending on $\mathcal{O}$. By using the estimates (2.3) and (2.4) we see that the right hand side of (2.2) converges absolutely in $\{\mu \in \mathbf{C}; \operatorname{Re}\mu > a_0 - a_1\}$. Thus $F_D(\mu)$ is well defined in $\{\mu \in \mathbf{C}; \operatorname{Re}\mu > a_0 - a_1\}$, and holomorphic in this domain. Now consider the analytic continuation of $F_D(\mu)$ beyond the line $\operatorname{Re}\mu = a_0 - a_1$. The existence of singularities of $F_D(\mu)$ is closely related to MLPC. Namely, we have

THEOREM 2.1. *Let $\mathcal{O}$ be an obstacle given by (2.1) satisfying (H.1) and (H.2). If $F_D(\mu)$ cannot be prolonged analytically to an entire function, then MLPC is valid for $\mathcal{O}$.*

Since we did not give even a sketch of the proof in [8], we shall explain the outline of the proof.

Our proof rests on the basis of the trace formula due to Bardos, Guillot and Ralston[1]:

$$\begin{aligned} &\operatorname{Trace}_{L^2(\mathbf{R}^3)} \int \rho(t) \left(\cos t\sqrt{-A} \oplus 0 - \cos t\sqrt{-A_0}\right) dt \\ &= \frac{1}{2} \sum_{j=1}^{\infty} \hat{\rho}(z_j), \qquad \text{for all } \rho \in C_0^\infty(0,\infty) \end{aligned} \tag{2.5}$$

where

$$\hat{\rho}(z) = \int e^{izt} \rho(t) dt,$$

$\{z_j\}_{j=1}^\infty$ is a numbering of all the poles of $\mathcal{S}(z)$, A is the selfajoint realization in $L^2(\Omega)$ of the Laplacian with the Dirichlet boundary condition and A_0 the one in $L^2(\mathbf{R}^3)$, and $\oplus 0$ indicates the extension into $\mathcal{O}$ by 0. Remark that the above trace formula is valid for all smooth bounded obstacles satisfying (1.1). Let $\rho \in C_0^\infty(-1,1)$ satisfy

$$\begin{aligned} &\rho(t) \geq 0, \quad \rho(-t) = \rho(t) && \text{for all } t \in \mathbf{R}, \\ &\rho(t) > 1 && \text{on } [-1/2, 1/2], \\ &\hat{\rho}(k) \geq 0 && \text{for all } k \in \mathbf{R}. \end{aligned}$$

Let $\{l_q\}_{q=1}^\infty$ and $\{m_q\}_{q=1}^\infty$ be infinite sequences such that

$$l_q \to \infty, \quad m_q \to \infty \qquad \text{as } q \to \infty.$$

We denote by $\rho_q(t)$ $(q = 1, 2, \cdots)$ the functions defined by

$$\rho_q(t) = \rho(m_q(t - l_q)).$$

LEMMA 2.2. *Let $\alpha > 0$ and assume that*

$$\#\{j; \operatorname{Im} z_j \le \alpha\} = P(\alpha) < \infty.$$

Then we have an estimate

$$\sum_{j=1}^{\infty} |\hat{\rho}_q(z_j)| \le C_\alpha {m_q}^4 e^{-\alpha l_q} + P(\alpha) {m_q}^{-1} \qquad \text{for all } q, \tag{2.6}$$

where the constant C_α is independent of sequences $\{l_q\}_{q=1}^{\infty}$ and $\{m_q\}_{q=1}^{\infty}$.

PROOF: Suppose that $\mathcal{O} \subset \{x; |x| \le R\}$. Recall the estimate due to Bardos, Guillot and Ralston: for each $T > 2R$, it holds that

$$\sum_{\text{poles}} |\hat{\psi}(z_j)| \le C(T) \|\psi\|_{H^4(0,\infty)} \quad \text{for all } \psi \in C_0^\infty(2R, T), \tag{2.7}$$

where $C(T)$ is a constant depending on T. We set for $m \in [1, \infty)$ and $l > 0$

$$\rho_{m,l}(t) = \rho(m(t - l)).$$

Take l_0 in such a way $l_0 - 1 \ge 2R$. Then for all $m \in [1, \infty)$ we have $\rho_{m,l_0} \in C_0^\infty(2R, T)$ $(T = 2R + 2)$. Thus from (2.7) we have for all $m \in [1, \infty)$

$$\sum_{\text{poles}} |\hat{\rho}_{m,l_0}(z_j)| \le C(T) m^4.$$

Now we classify the poles $\mathcal{S}(z)$ into two groups:

$$G_1 = \{z_j; \operatorname{Im} z_j \ge \alpha\}, \quad G_2 = \{z_j; \operatorname{Im} z_j < \alpha\}.$$

By using the relation

$$\hat{\rho}(m(\cdot - l))(z) = e^{i(l - l_0)z} \hat{\rho}(m(\cdot - l_0))(z)$$

we have

$$\begin{aligned}
\sum_{z_j \in G_1} |\hat{\rho}_{m,l}(z_j)| &= \sum_{z_j \in G_1} |e^{i(l - l_0)z_j}| \, |\hat{\rho}_{m,l_0}(z_j)| \\
&\le e^{-(l - l_0)\alpha} \sum_{z_j \in G_1} |\hat{\rho}_{m,l_0}(z_j)| \\
&\le e^{-(l - l_0)\alpha} C(T) m^4.
\end{aligned}$$

On the other hand, since $|\hat{\rho}_{m,l}(z)| \leq m^{-1}$ for all $\operatorname{Im} z \geq 0$, we have

$$\sum_{z_j \in G_2} |\hat{\rho}_{m,l}(z_j)| \leq P(\alpha) m^{-1}.$$

Combining the above two estimates we get

$$\sum_{\text{poles}} |\hat{\rho}_{m,l}(z_j)| \leq C_\alpha e^{-\alpha l} m^4 + P(\alpha) m^{-1},$$

where $C_\alpha = C(t) e^{\alpha l_0}$. Substituting l_q and m_q for the places of l and m respectively, we have the desired assertion. Q.E.D.

Denote by $\hat{F}_D(t)$ the element of $\mathcal{D}'(0, \infty)$ defined by

$$\hat{F}_D(t) = \sum_\gamma \frac{(-1)^{i_\gamma} T_\gamma}{|I - P_\gamma|^{1/2}} \delta(t - d_\gamma). \tag{2.8}$$

Remark that $F_D(\mu)$ is the Laplace transformation of $\hat{F}_D(t)$.

PROPOSITION 2.3. *Suppose that $F_D(\mu)$ cannot be prolonged to an entire function. Then there exists $\alpha_0 > 0$ such that, for any $\beta > \alpha_0$, we can find sequences $\{l_q\}_{q=1}^\infty$ and $\{m_q\}_{q=1}^\infty$ with the following properties:*

(1) $$l_q \to \infty \quad \text{as } q \to \infty.$$

(2) $$e^{\beta l_q} \leq m_q \leq e^{2\beta l_q} \quad \text{for all } q.$$

(3) $$|\langle \rho_q, \hat{F}_D \rangle_{\mathcal{D}(0,\infty) \times \mathcal{D}'(0,\infty)}| \geq e^{-\alpha_0 l_q} \quad \text{for all } q.$$

PROOF: Let $\beta > a_0$. Then for all large p there is s_p satisfying

$$p - \frac{2}{3} < s_p < p - \frac{1}{3} \quad \text{and} \quad \text{distance}(s_p, \Xi) \geq e^{-\beta p}, \tag{2.9}$$

where $\Xi = \{d_\gamma; \gamma \text{ is a periodic ray in } \Omega\}$. Indeed, consider the intervals of the form $[(j-1)e^{-\beta p}, (j+1)e^{-\beta p}]$ ($j \in \mathbf{Z}$) which is contained in $[p - \frac{2}{3}, p - \frac{1}{3}]$. Then the number of such intervals is greater than $\frac{1}{6} e^{\beta p}$. Thus by taking account of (2.3) and $\beta > a_0$, we see that for large p the number of intervals is greater than that of d_γ's. Thus we have an interval of the form $[(j-1)e^{-\beta p}, (j+1)e^{-\beta p}]$ contained in $[p - \frac{2}{3}, p - \frac{1}{3}]$ where there is no point of Ξ. Then, $s_p = j e^{-\beta p}$ satisfies (2.9).

Similarly we have t_p satisfying

$$p + \frac{1}{3} < t_p < p + \frac{2}{3} \quad \text{and} \quad \text{distance}(t_p, \Xi) \geq e^{-\beta p}. \tag{2.10}$$

We set

$$T_p(t) = \left(\int_{-\infty}^\infty \rho(t)\, dt \right)^{-1} \int_{s_p e^{\beta p}}^{t_p e^{\beta p}} \rho(e^{\beta p} t - \tau)\, d\tau.$$

It is easy to see that

$$\begin{cases} \operatorname{supp} T_p \subset [s_p - e^{-\beta p}, t_p + e^{-\beta p}], \\ T_p(t) = 1 \quad \text{for all } t \in [s_p + e^{-\beta p}, t_p - e^{-\beta p}]. \end{cases} \tag{2.11}$$

We define $F_p(\mu)$ by

$$\begin{aligned} F_p(\mu) =& \langle T_p(t), e^{-\mu t} \hat{F}_D(t) \rangle_{\mathcal{D}(0,\infty)\times\mathcal{D}'(0,\infty)} \\ =& \sum_{\gamma:\text{periodic}} T_p(d_\gamma) \frac{(-1)^{i_\gamma} T_\gamma}{|I - P_\gamma|^{1/2}} e^{-\mu d_\gamma}. \end{aligned}$$

Then by using (2.9), (2.10) and (2.11) in the second equality we have

$$F_p(\mu) = \sum_{s_p < d_\gamma \le t_p} \frac{(-1)^{i_\gamma} T_\gamma}{|I - P_\gamma|^{1/2}} e^{-\mu d_\gamma}. \tag{2.12}$$

On the other hand, an interchange in the order of integration with respect to t and τ in the first equality gives

$$F_p(\mu) = \int_{s_p\, e^{\beta p}}^{t_p\, e^{\beta p}} \langle \rho(e^{\beta p} t - \tau),\, \hat{F}_D(t) e^{-\mu t} \rangle \, d\tau.$$

Since $\tau \in [s_p\, e^{\beta p}, t_p\, e^{\beta p}]$ and $\langle \rho(e^{\beta p} t - \tau), \hat{F}_D(t) e^{-\mu t} \rangle \neq 0$ imply that

$$\tau \in I_p = \cup_{s_p < d_\gamma \le t_p} [e^{\beta p} d_\gamma - 1, e^{\beta p} d_\gamma + 1],$$

we have

$$F_p(\mu) = \int_{I_p} \langle \rho(e^{\beta p} t - \tau),\, \hat{F}_D(t) e^{-\mu t} \rangle \, d\tau,$$

from which it follows that

$$|F_p(\mu)| \le \sum_{s_p < d_\gamma \le t_p} \int_{d_\gamma e^{\beta p} - 1}^{d_\gamma e^{\beta p} + 1} |\langle \rho(e^{\beta p} t - \tau), \hat{F}_D(t) e^{-\mu t} \rangle| \, d\tau.$$

Suppose that

$$|\langle \rho(e^{\beta p}(t - \sigma)),\, \hat{F}_D(t) \rangle| \le C e^{-\alpha p} \quad \text{for all } \sigma \in [s_p, t_p]. \tag{2.13}$$

Let $\tau \in I_p$ be fixed, and choose a periodic ray γ_0 such that

$$\rho(e^{\beta p} d_{\gamma_0} - \tau) \neq 0.$$

Then we have

$$
\begin{aligned}
&\langle \rho(e^{\beta p}t-\tau),\, e^{-\mu t}\hat{F}_D(t)\rangle \\
=&\langle \rho(e^{\beta p}t-\tau),\, e^{-\mu d_{\gamma 0}}\hat{F}_D(t)\rangle \\
+&\sum_{d_{\gamma 0}-2e^{-\beta p}\le d_\gamma \le d_{\gamma 0}+2e^{-\beta p}} \rho(e^{\beta p}d_\gamma-\tau)(e^{-\mu d_\gamma}-e^{-\mu d_{\gamma 0}})\frac{(-1)^{i_\gamma}T_\gamma}{|I-P_\gamma|^{1/2}} \\
=&I+II.
\end{aligned}
$$

Since $I = e^{-\mu d_{\gamma 0}}\,\langle \rho(e^{\beta p}t-\tau),\, \hat{F}_D(t)\rangle$, we have from (2.13)

$$|I| \le e^{|\mu|(p+1)}\, Ce^{-\alpha p}.$$

Next consider II. Since γ in the summation of II satisfies $|d_\gamma - d_{\gamma 0}| \le 2e^{-\beta p}$, we have from (2.3) and (2.4)

$$|II| \le |\mu|(p+1)2e^{-\beta p}e^{|\mu|(p+1)}e^{(a_0-a_1)(p+1)}.$$

Thus under the assumption (2.13) it holds that

$$|F_p(\mu)| \le C|\mu|(p+1)\, e^{a_0-a_1+|\mu|}\, e^{(-\mathrm{mim}(\alpha,\beta)+a_0-a_1+|\mu|)p}.$$

If

(2.13) holds for all sufficiently large p with the same constant C,

we have

$$\sum_{p=1}^{\infty} F_p(\mu) \ \text{ converges absolutely in } \ \{\mu; |\mu| < \min(\alpha,\beta)-a_0+a_1\}.$$

As remarked earlier the right hand side of (2.2) converges absolutely for $\mathrm{Re}\,\mu > a_1 - a_0$. Then, by using (2.12) and by changing the order of summation of the right hand side of (2.2) we have $F_D(\mu) = \sum_{p=1}^{\infty} F_p(\mu)$ for $\mathrm{Re}\,\mu > a_1 - a_0$. Since $F_p(\mu)$ is entire, the above assertion implies that $F_D(\mu)$ is holomorphic in $\{\mu; |\mu| < \min(\alpha,\beta) - a_0 + a_1\}$.

Now assume that $F_D(\mu)$ is not entire. Then there is a constant $R > 0$ such that $F_D(\mu)$ is not holomorphic in $\{\mu; |\mu| \le R\}$. Take $\alpha_0 = R + a_1 - a_0 + 1$, and let $\beta > \alpha_0$. Assume that (2.13) holds for all large p with the same constant C. Then the above result implies that $F_D(\mu)$ is holomorphic in $\{\mu; |\mu| < R+1\}$. This contradicts the assumption on $F_D(\mu)$. That is, for any $\beta > \alpha_0$, (2.13) does not hold for an infinite number of p. Namely, there exist sequences $\{p_n\}_{n=1}^{\infty}$ and $\{\sigma_n\}_{n=1}^{\infty}$ such that $\sigma_n \in [p_n - 1, p_n + 1]$ and $p_n \to \infty$ as $n \to \infty$ and

$$|\langle \rho(e^{\beta p_n}(t-\sigma_n)),\, \hat{F}_D(t)\rangle| \ge Ce^{-\alpha_0 p_n} \quad \text{for all } \ n.$$

Thus, if we take

$$l_n = \sigma_n, \quad m_n = e^{\beta p_n}$$

the assertions of Proposition are satisfied. Q.E.D.

The next theorem is on the estimate from the below of the left hand side of (2.5) by using the configuration of the bodies.

THEOREM 2.4. *Suppose that the obstacle $\mathcal{O}$ defined by (2.1) satisfies the assumptions (H.1) and (H.2). Then for any sequences $\{l_q\}_{q=1}^{\infty}$ and $\{m_q\}_{q=1}^{\infty}$, it holds for all q that*

$$(2.14) \qquad |\mathrm{Trace}_{L^2(\mathbf{R}^3)} \int \rho_q(t) \left(\cos t\sqrt{-A} \oplus 0 - \cos t\sqrt{-A_0}\right) dt| \geq |\langle \rho_q, \hat{F}_D \rangle_{\mathcal{D}(0,\infty)\times\mathcal{D}'(0,\infty)}| - C e^{a_0 l_q} m_q^{-\varepsilon_0},$$

where the constants C and ε_0 are independent of sequences $\{l_q\}_{q=1}^{\infty}$ and $\{m_q\}_{q=1}^{\infty}$.

The proof of Theorem 2.4 can be done by applying the procedures of [7] and [6]. Since this proof is fairly long we omit it.

Now, Theorem 2.1 follows from the combination of Lemma 2.2, Proposition 2.3 and Theorem 2.4. Suppose that $F_D(\mu)$ is not entire. Let α_0 be the constant in Proposition 2.3. Take α as

$$\alpha = 5(\alpha_0 + a_0 + 1)/\varepsilon_0,$$

and choose $\{l_q\}$ and $\{m_q\}$ as the assertions (1), (2) and (3) of Proposition 2.3 hold for $\beta = \alpha/5 > \alpha_0$. Suppose that $P(\alpha) < \infty$. Then, by substituting in the right hand side of (2.14) the estimates (2) and (3) of Proposition 2.3, and in the left hand side the estimate of Lemma 2.2 we get the inequality

$$(C_\alpha + CP(\alpha))\, e^{-\beta l_q} \geq (1 - Ce^{-l_q})\, e^{-\alpha_0 l_q} \quad \text{for all } q.$$

This inequality shows a contradiction because $\alpha_0 < \beta$ and $l_q \to \infty$. Thus $P(\alpha)$ cannot be finite.

3. Relationships between $F_D(\mu)$ and the zeta functions of symbolic flows

Let $A = (A(i,j))_{i,j=1,2,\cdots,L}$ be a zero-one $L \times L$ matrix. Following Parry and Pollicott[17] we set

$$\Sigma_A = \{\xi = (\cdots, \xi_{-1}, \xi_0, \xi_1, \cdots) \in \prod_{i=-\infty}^{\infty} \{1, 2, \cdots, L\}; A(\xi_j, \xi_{j+1}) = 1 \text{ for all } j\}.$$

Denote by σ the shift transformation defined by

$$(\sigma\xi)_j = \xi_{j+1}.$$

We denote by σ_A the restriction of σ on Σ_A. For $r \in C(\Sigma_A)$ we define $\mathrm{var}_n\, r$ and $\|r\|_\infty$ by

$$\mathrm{var}_n\, r = \sup\{|r(\xi) - r(\psi)|;\ \xi, \psi \in \Sigma_A \text{ and } \xi_j = \psi_j \text{ for } -n \leq j \leq n\},$$
$$\|u\|_\infty = \sup\{|r(\xi)|; \xi \in \Sigma_A\}.$$

We set for $0 < \theta < 1$

$$\|u\|_\theta = \sup_{n\geq 1} \frac{\mathrm{var}_n r}{\theta^n}, \quad |||\, r\,|||_\theta = \max\{\|r\|_\infty, \|u\|_\theta\},$$
$$\mathcal{F}_\theta(\Sigma_A) = \{r \in C(\Sigma_A); |||\, r\,|||_\theta < \infty\}.$$

Let $r(\xi,\mu)$ be a $\mathcal{F}_\theta(\Sigma_A)$–valued holomorphic function of μ defined in a domain of $\mathbf{C}$, and define $Z(\mu)$ by

$$Z(\mu) = \exp\left(\sum_{n=1}^{\infty}\frac{1}{n}\sum_{\sigma_A^n\xi=\xi}\exp S_n r(\xi,\mu)\right)$$

where

$$S_n r(\xi,\mu) = r(\xi,\mu) + r(\sigma_A\xi,\mu) + \cdots + r(\sigma_A^{n-1}\xi,\mu).$$

Note that $Z(\mu)$ is nothing but the zeta function $\zeta(r(\cdot,\mu))$ in the sense of Parry[16, Section 3], and we call $Z(\mu)$ the zeta function of a symbolic flow (Σ_A,σ_A) associated to $r(\cdot,\mu)$.

Now consider relationships between Σ_A and oriented bounded broken rays in the outside of $\mathcal{O}_j$'s satisfying (H.1) and (H.2). Let $X(s)$ $(s\in\mathbf{R})$ be a representation of an orientated broken ray by the arc length such that $X(0)\in\Gamma$ and $X(s)$ moves in the orientation as s increases. When $\{|X(s)|; s\in\mathbf{R}\}$ is bounded, $X(s)$ repeats reflections on the boundary Γ infinitely many times as s tends to $\pm\infty$. Let the j-th reflection point X_j be on Γ_{ξ_j}. Then an oriented bounded broken ray gives an infinite sequence $\xi=\{\cdots,\xi_{-1},\xi_0,\xi_1,\cdots\}$, which is called the reflection order of $X(s)$. Evidently there is a one to one correspondance between the set of all such representations and Σ_A, where the $L\times L$ matrix $A=(A(i,j))_{i,j=1,\cdots,L}$ is given by

$$A(i,j) = \begin{cases} 1, & \text{if } i\neq j \\ 0, & \text{if } i=j. \end{cases}$$

Remark that, for a bounded oriented broken ray, there is freedom of such representation, that is, the freedom of the choice of $X(0)$. Therefore the correspondance between bounded broken rays and Σ_A is not one to one. For example, let γ be an orientated periodic ray with n reflection points, and suppose that γ has a representation ξ. Then there are n choices of $X(0)$, which imply that $\sigma^j\xi$, $j=1,2,\cdots,n$, are also representaions of γ in Σ_A, and if γ is a primitive periodic ray, $\sigma^j\xi$'s are all different. Note that a periodic ray in Ω corresponds to a periodic element $\xi\in\Sigma_A$, that is, $\sigma^n\xi=\xi$ for some n. We set

$$f(\xi) = |X_0X_1|$$

where X_j denote the j-th reflection point of the oriented broken ray corresponding to ξ.

Denote by $\lambda_1(\xi)$ and $\lambda_2(\xi)$ the eigenvalues of P_γ greater than 1, and by $\kappa_l(\xi), l=1,2$, the principal curvatures at X_0 of the wave front of the phase function $\varphi_{\mathbf{i},0}^{\infty}$ defined in [7, Section 5], where $\mathbf{i}=(\xi_0,\cdots,\xi_{n-1})$. Then we have

$$\text{(3.1)}\qquad \lambda_1(\xi)\lambda_2(\xi) = \prod_{j=1}^{n}(1+f(\sigma^j\xi)\kappa_1(\sigma^j\xi))(1+f(\sigma^j\xi)\kappa_2(\sigma^j\xi)).$$

It is easy to check that

$$\text{(3.2)}\qquad \lambda_1(\xi)\lambda_2(\xi) \geq e^{cn} \quad (c>0).$$

Since the other eigenvalues of P_γ are $\lambda_1^{-1}$and λ_2^{-1}, it holds that

$$\text{(3.4)} \qquad \big|\,\lambda_1\lambda_2 - |I - P_\gamma|\,\big| \leq C(\lambda_1 + \lambda_2) \quad \text{for all} \quad \gamma.$$

Define $g(\xi)$ for an periodic element ξ by

$$g(\xi) = -\frac{1}{2}\log(1 + f(\xi)\kappa_1(\xi))(1 + f(\xi)\kappa_2(\xi)).$$

Then $g(\xi)$ can be extended to a function in $\mathcal{F}_\theta(\Sigma_A)$. Define $\zeta(\mu)$ by

$$\zeta(\mu) = \exp\Big(\sum_{n=1}^{\infty}\frac{1}{n}\sum_{\sigma_A^n\xi=\xi}\exp S_n(-\mu f(\xi) + g(\xi) + \pi i)\Big).$$

The estimates (3.2)~(3.4) imply that both $F_D(\mu)$ and $\zeta(\mu)$ converge absolutely for $\operatorname{Re}\mu$ large. Denote by ν_0 the abscissa of convergence of $\zeta(\mu)$, that is,

$$\nu_0 = \inf\{\nu; \zeta(\mu) \text{ converges absolutely for } \operatorname{Re}\mu > \nu\}.$$

Then it holds that for $\operatorname{Re}\mu > \nu_0$

$$\begin{aligned}
-\frac{d}{d\mu}\log\zeta(\mu) &= \sum_{n=1}^{\infty}\frac{1}{n}\sum_{\sigma_A^n\xi=\xi} S_n f(\xi)\exp(S_n(-\mu f(\xi) + g(\xi) + \pi i)) \\
&= \sum_{n=1}^{\infty}\sum_{\sigma_A^n\xi=\xi}\frac{S_n f(\xi)}{n}(-1)^n\exp(S_n g(\xi))\exp(-\mu S_n f(\xi)).
\end{aligned}$$

Obviousely we have

$$S_n(\xi) = d_\gamma, \quad n = i_\gamma, \quad (\lambda_1(\xi)\lambda_2(\xi))^{-1/2} = \exp S_n g(\xi).$$

Taking account of the number of elements $\xi \in \Sigma_A$ corresponding to γ, we have

$$\sum_{\xi\in(\gamma)}\frac{S_n f(\xi)}{n} = T_\gamma$$

where the summation is taken over all ξ corresponding to γ. By using the relations

$$S_n f(\xi) = d_\gamma, \quad n = i_\gamma, \quad (\lambda_1(\xi)\lambda_2(\xi))^{-1/2} = \exp S_n g(\xi).$$

we have

$$\begin{aligned}
\text{(3.5)} \qquad & F_D(\mu) - (-\frac{d}{d\mu}\log\zeta(\mu)) \\
&= \sum_\gamma T_\gamma(-1)^n\{|I - P_\gamma|^{-1/2} - (\lambda_1\lambda_2)^{-1/2}\}\exp(-\mu d_\gamma).
\end{aligned}$$

Since $\left| |I-P_\gamma|^{-1/2}-(\lambda_1\lambda_2)^{-1/2}\right| \le C(\lambda_1\lambda_2)^{-1/2}(\lambda_1^{-1}+\lambda_2^{-1})$ the left hand side of (3.5) is absolutely convergent in $\operatorname{Re}\mu \ge \nu_0 - a_3$ ($a_3 > 0$). Therefore the singularities of $F_D(\mu)$ and $-\dfrac{d}{d\mu}\log\zeta(\mu)$ coincide in $\{\mu; \operatorname{Re}\mu \ge \nu_0 - a_3\}$. Namely, if we can show the existence of poles of $-\dfrac{d}{d\mu}\log\zeta(\mu)$ in $\{\mu; \operatorname{Re}\mu \ge \nu_0 - a_3\}$, we get the existence of poles of $F_D(\mu)$.

4. Main theorem

In order to state our main theorem, we shall give the conditions required on the configuration of centers of the balls.

Let $P_j, j = 1, 2, \cdots, L$, be points in $\mathbf{R}^3$. The first condition we assume on the configuration of P_j's is

(A.1) any triad of P_j's does not lie on a straight line.

Set

$$d_{\max} = \max_{i \ne j} |P_iP_j|$$

and

$$B(i,j) = \begin{cases} 1 & \text{if } |P_iP_j| = d_{\max}, \\ 0 & \text{if } |P_iP_j| < d_{\max}. \end{cases} \tag{4.1}$$

By changing the numbering of the points if necessary, we may suppose that

$$\begin{aligned} B(i,j) &= 0 \qquad \text{for all } j \quad \text{if } \ i \ge K+1, \\ B(i,j) &= 1 \qquad \text{for some } j \quad \text{if } \ i \le K, \end{aligned}$$

hold for some $2 \le K \le L$. Denote by C the $K \times K$ matrix defined by

$$C = [B(i,j)]_{i,j=1,2,\cdots,K}.$$

The second condition on the configuration of P_j's is

(A.2) $\qquad C^N > 0 \quad$ for some positive integer N,

which means that all the entries of the matrix C^N are positive. We assume one more condition:

$$\min_{\substack{1\le i\le K, 1\le j\le L \\ i\ne j}} |P_iP_j| \ge \max_{i,j \ge K+1} |P_iP_j|. \tag{A.3}$$

We denote by $\mathcal{O}_{j,\varepsilon}$ the open ball of radius ε with center P_j, and set

$$\mathcal{O}_\varepsilon = \cup_{j=1}^L \mathcal{O}_{j,\varepsilon}. \tag{4.2}$$

The following is our main theorem.

THEOREM 4.1. *Suppose that* $\{P_j; j = 1, 2, \cdots, J\}$ *satisfies (A.1), (A.2) and (A.3). Then there exists* $\varepsilon_0 > 0$ *such that, for all* $0 < \varepsilon \leq \varepsilon_0$, *MLPC holds for* $\mathcal{O}_\varepsilon$.

Concerning the proof of the main theorem, as was shown in Sections 2 and 3, it suffices to prove the existence of singularities of the zeta function $\zeta(\mu)$ attached to the obstacle. To this end, we prepare a theorem on singular perturbations of symbolic flows.

As far as we consider singular perturbations, matrices B and C are not necessarily the ones appeared in the statement of Theorem 4.1. Let L be an integer ≥ 2, and let $A = [A(i,j)]_{i,j=1,2,\cdots,L}$ and $B = [B(i,j)]_{i,j=1,2,\cdots,L}$ be zero-one $L \times L$ matrices satisfying

$$B(i,j) = 1 \ \text{ implies } \ A(i,j) = 1. \tag{4.3}$$

Suppose that there is $2 \leq K \leq L$ such that $B(i,j) = 0$ for all j if $i \geq K+1$ and the $K \times K$ matrix C defined by $C = [B(i,j)]_{i,j=1,2,\cdots,K}$ satisfies

$$C^N > 0 \ \text{ for some integer } \ N. \tag{4.4}$$

Let $\varepsilon_1 > 0$, and let $f_\varepsilon, h_\varepsilon \in C(\Sigma_A)$ satisfy

$$f_\varepsilon, h_\varepsilon \in \mathcal{F}_\theta(\Sigma_A) \ \text{ for all } 0 \leq \varepsilon \leq \varepsilon_1 \quad (0 < \theta < 1)$$

and

$$|||f_\varepsilon - f_0|||_\theta, \ |||h_\varepsilon - h_0|||_\theta \to 0 \qquad \text{as } \varepsilon \to 0. \tag{4.5}$$

Let $k \in \mathcal{F}_\theta(\Sigma_A)$ be a real valued function satisfying

$$\begin{cases} k(\xi) = 0 & \text{for all } \xi \text{ such that } B(\xi_0, \xi_1) = 1, \\ k(\xi) > 0 & \text{for all } \xi \text{ such that } B(\xi_0, \xi_1) = 0, \\ 0 < c_0 = \sup\limits_{\xi \in \Sigma(1)} k(\xi) \leq \inf\limits_{\xi \in \Sigma(2)} k(\xi), \end{cases} \tag{4.6}$$

where

$$\Sigma(1) = \{\xi \in \Sigma_A; B(l, \xi_0) = 1 \ \text{ for some } 1 \leq l \leq K\},$$
$$\Sigma(2) = \{\xi \in \Sigma_A; B(l, \xi_0) = 0 \ \text{ for all } 1 \leq l \leq K\}.$$

For $0 < \varepsilon \leq \varepsilon_1$ we define $Z_\varepsilon(\mu)$ by

$$Z_\varepsilon(\mu) = \exp\left(\sum_{n=1}^{\infty} \frac{1}{n} \sum_{\sigma_A^n \xi = \xi} \exp(S_n r_\varepsilon(\xi, \mu))\right), \tag{4.7}$$

where

$$r_\varepsilon(\xi, \mu) = -\mu f_\varepsilon(\xi) + h_\varepsilon(\xi) + k(\xi) \log \varepsilon.$$

Now we have a theorem on the existence of poles of the zeta function $Z_\varepsilon(\mu)$.

THEOREM 4.2. *Suppose that* $(4,3) \sim (4.6)$, *and that*

$$f_0(\xi) > 0 \quad \text{for all } \xi \in \Sigma_A,$$
$$h_0(\xi) \text{ is real for all } \xi \in \Sigma_A \text{ satisfying } B(\xi_0, \xi_1) = 1$$

and

$$0 < \mathrm{Im}\, h_0(\xi) < \pi \quad \text{for all } \xi \in \Sigma_A \text{ satisfying } B(\xi_0, \xi_1) = 0.$$

Then there exist $s_0 \in \mathbf{R}$, D *a neighborhood of* s_0 *in* $\mathbf{C}$ *and* $0 < \varepsilon_0 \le \varepsilon_1$ *such that, for all* $0 < \varepsilon \le \varepsilon_0$, $Z_\varepsilon(\mu)$ *is meromorphic in* D *and it has a pole* μ_ε *in* D *with*

$$\mu_\varepsilon \to s_0 \qquad \text{as} \quad \varepsilon \to 0.$$

We omit the proof because the detailed one is given in [9].[1]

Now we turn to considerations on the singularities of $\zeta(\mu)$ corresponding to $\mathcal{O}_\varepsilon$ of (4.2). Remark that (A.2) implies (H.2) for $\mathcal{O}_\varepsilon$ when ε is small.

We denote $f(\xi), g(\xi)$ and $\zeta(\mu)$ attached to $\mathcal{O}_\varepsilon$ by $f_\varepsilon(\xi), g_\varepsilon(\xi)$ and $\zeta_\varepsilon(\mu)$ respectively. It is easy to see that, by setting $f_0(\xi) = |P_{\xi_0} P_{\xi_1}|$,

$$|\log \varepsilon| \; |||f_\varepsilon - f_0|||_\theta \to 0 \qquad \text{as} \quad \varepsilon \to 0. \tag{4.8}$$

From the relationship between the curvatures of the wave fronts of incident and reflected waves we have

$$\kappa_1(\xi) = \frac{2}{\varepsilon}(\cos \frac{\Theta(\xi)}{2})^{-1} + O(1), \quad \kappa_2(\xi) = \frac{2}{\varepsilon} + O(1)$$

where $\Theta(\xi) = \angle P_{\xi_{-1}} P_{\xi_0} P_{\xi_1}$. Thus we have immediately

$$|||\, g_\varepsilon(\xi) - \left(\log \varepsilon + \frac{1}{2} \log \frac{1}{4}(\cos \frac{\Theta(\xi)}{2})\right) |||_\theta \to 0 \qquad \text{as} \quad \varepsilon \to 0.$$

Then, by setting $\tilde{g}_\varepsilon(\xi) = g_\varepsilon(\xi) - \log \varepsilon$ and $\tilde{g}_0(\xi) = \frac{1}{2} \log \frac{1}{4}(\cos \frac{\Theta(\xi)}{2})$ we have

$$|||\tilde{g}_\varepsilon - \tilde{g}_0|||_\theta \to 0 \qquad \text{as} \quad \varepsilon \to 0. \tag{4.9}$$

Define $k(\xi)$ by

$$k(\xi) = 1 - f_0(\xi)/d_{\max}.$$

Then, by taking account of (A.3), we see easily that $k(\xi)$ satisfies (4.6). By putting $s' = s - (\log \varepsilon + \sqrt{-1}\,\pi)/d_{\max}$ we have

$$-sf_\varepsilon + g_\varepsilon + \sqrt{-1}\,\pi = -s' f_\varepsilon + h_\varepsilon + k \log \varepsilon,$$

[1] Even though only the zeta functions of symbolic flows in Σ_A^+ are considered in [9], the zeta functions of Σ_A can be reduced immediately to those of Σ_A^+(See, Bowen[2]).

where

$$h_\varepsilon = \tilde{g}_\varepsilon + \sqrt{-1}\,\pi\,k + (\log\varepsilon + \sqrt{-1}\,\pi)\frac{(f_0 - f_\varepsilon)}{d_{\max}}.$$

Evidently it follows from (4.6) that

$$h_0 = \tilde{g}_0 + \sqrt{-1}\,\pi\,k,$$

hence we have

$$h_0(\xi) = \tilde{g}_0(\xi) \quad \text{for } \xi \text{ satisfying } B(\xi_0, \xi_1) = 1.$$

Since $\operatorname{Im} h_0(\xi) = \pi k(\xi)$,

$$a \leq \operatorname{Im} h_0(\xi) \leq \pi - a \quad (a > 0) \quad \text{for all } \xi \text{ satisfying } B(\xi_0, \xi_1) = 0$$

follows from (4.6). Thus $h_\varepsilon, h_\varepsilon, k$ satisfy the conditions required in Theorem 4.6. Let $Z_\varepsilon(\mu)$ be the zeta function defined by (4.7) with these $h_\varepsilon, h_\varepsilon, k$. Note that we have the relation

$$\zeta_\varepsilon(\mu) = Z_\varepsilon(\mu - (\log\varepsilon + \sqrt{-1}\,\pi)/d_{\max}).$$

On the other hand, Theorem 4.2 says that there exist $\varepsilon_0 > 0, s_0 \in \mathbf{R}$ and D_0 such that $Z_\varepsilon(\mu)$ has a pole in D_0, which implies that $\zeta_\varepsilon(\mu)$ is meromorphic in $D_\varepsilon = \{\mu = z + (\log\varepsilon + \sqrt{-1}\,\pi)/d_{\max}; z \in D_0\}$ and has a pole near $s_0 + (\log\varepsilon + \sqrt{-1}\,\pi)/d_{\max}$. It is evident that this pole of $\zeta_\varepsilon(\mu)$ stays in the domain where the singularities of $\zeta_\varepsilon(\mu)$ and of $F_{D,\varepsilon}(\mu)$ coincide. Thus the existence of singularities of $F_{D,\varepsilon}(\mu)$ is proved.

References

1. C.Bardos, J.C.Guillot and J.Ralston, *La relation de Poisson pour l'équation des ondes dans un ouvert non borné. Application à la théorie de la diffusion*, Comm.Partial Diff. Equ. **7** (1982), 905–958.
2. R.Bowen, "Equilibrium states and the ergodic theory of Anosov differomorphism," S.L.M.,470, Springer-Verlag, Berlin, 1975..
3. C.Gérard, *Asymptotique des poles de la matrice de scattering pour deux obstacles strictement convexes*, Bull.S.M.F. **116 n° 31** (1989).
4. M.Ikawa, *On the poles of the scattering matrix for two strictly convex obstacles*, J.Math.Kyoto Univ. **23** (1983), 127–194.
5. M.Ikawa, *Precise informations on the poles of the scattering matrix for two strictly convex obstacles*, J.Math.Kyoto Univ. **27** (1987), 69–102.
6. M.Ikawa, *Trapping obstacles with a sequence of poles of the scattering matrix converging to the real axis*, Osaka J.Math. **22** (1985), 657–689.
7. M.Ikawa, *Decay of solutions of the wave equation in the exterior of several convex bodies*, Ann.Inst. Fourier **38** (1988), 113–146.
8. M.Ikawa, *On the existence of poles of the scattering matrix for several convex bodies*, Proc.Japan Acad. **64** (1988), 91–93.
9. M.Ikawa, *Singular perturbation of symbolic flows and poles of the zeta functions*, to appear in Osaka J.Math.
10. P.D.Lax and R.S.Phillips, "Scattering theory," Academic Press, New York, 1967.
11. P.D.Lax and R.S.Phillips, *A logarithmic bound on the location of the poles of the scattering matrix*, Arch.Rational Mech.Anal. **40** (1971), 268–280.

12. R.Melrose, *Polynomial bound on the distribution of poles in scattering by an obstacle*, Journées Equations aux Dérivées Partielles, St.Jean de Monts (1984).
13. R.B.Melrose and J.Sjöstrand, *Singularities of boundary value problems*, Comm.Pure Appl.Math. **31** (1979), 593–617.
14. R.B.Melrose and J.Sjöstrand, *Singularities of boundary value problems.II*, Comm.Pure Appl.Math. **35** (1982), 129–168.
15. C.S.Morawetz, J.Ralston and W.A.Strauss, *Decay of solutions of the wave equation outside non-trapping obstacles*, Comm.Pure Appl.Math. **30** (1977), 447–508.
16. W.Parry, *Bowen's equidistribution theory and the Dirichlet density theorem*, Ergod.The.& Dynam. Sys. **4** (1984), 117–134.
17. W.Parry and M.Pollicott, *An analogue of the prime number theorem for closed orbits of Axiom A flows*, Ann.Math. **118** (1983), 537-591.
18. H.Soga, *The behavior of oscillatory integrals with degenerate stationary points*, Tsukuba J.Math. **11** (1987), 93-100.

Smoothing Effect for the Schrödinger Evolution Equations with Electric Fields

Tohru OZAWA

Research Institute for Mathematical Sciences, Kyoto University
Kyoto 606, Japan

1. Introduction

In recent years smoothing effects for the Schrödinger evolution equations have been established in various function spaces [1]-[15]. In this note we describe some recent results on smoothing effects for the Schrödinger evolution equations with electric fields [11]. The equation we consider is the following evolution equation in the Hilbelt space $L^2 = L^2(\mathbb{R}^n)$, $n \in \mathbb{N}$

$$(*) \qquad i \frac{d}{dt} u(t) = (H_0 + V(t))u(t), \quad t \in \mathbb{R},$$

with the initial condition $u(s) = \phi$, where $H_0 = -(1/2)\Delta + E\cdot x$, $E \in \mathbb{R}^n$, and $V(t)$ is a time-dependent real potential on $\mathbb{R}^n$. The operator H_0 is essentially self-adjoint on the Schwartz space $\mathcal{S}$. Our main purpose is to describe smoothing effects for the propagator for (*) in terms of the weighted Sobolev space $H^{m,s}$ defined by

$$H^{m,s} = \{\psi \in \mathcal{S}' ;\ \|\psi\|_{m,s} = \|(1+|x|^2)^{s/2}(1-\Delta)^{m/2}\psi\| < \infty\},\ m,\ s \in \mathbb{R},$$

where $\|\cdot\|$ denotes the L^2-norm. Concerning the existence of the propagator for (*) Yajima [14] showed that if $V \in L^{2p/(p-1)}_{loc}(\mathbb{R}; L^p) + L^1_{loc}(\mathbb{R}; L^\infty)$ for some $p \in [1,\infty)\cap(n/2,\infty)$, then the equation (*) has a unique unitary propagator $\{U(t,s);\ t,\ s \in \mathbb{R}\}$ on L^2 such that for each $\phi \in L^2$, $u(t) = U(t,s)\phi$ satisfies (*) in $H^{-2,0}\cap H^{0,-2}$ for a.e. $t \in \mathbb{R}$. To state our result we introduce

<u>Definition.</u> Let $k \in \mathbb{N}\cup\{0\}$ and $p \in [1,\infty)\cap(n/2,\infty)$. For a real function W on $\mathbb{R}\times\mathbb{R}^n$, we say that the condition $(W)_{k,p}$ holds if for all multi-indices α with $|\alpha| = j \le k$

$$\partial_x^{\alpha} W \in L_{loc}^{2p/(2p-n)}(\mathbb{R};\ L^{p(j)}) + L_{loc}^{1}(\mathbb{R};\ L^{\infty}),$$

where $p(0) = p$; $p(j) = np/(n+jp)$ when $n \geq 2j+2$, $j \geq 1$; $p(j)$ is a suitable number greater than $\max(np/(n+jp), 2p/(p+1))$ when $n = 2j+1$, $j \geq 1$; $p(j) = 2p/(p+1)$ when $n \leq 2j$, $j \geq 1$.

Assumption $(A)_k$, $k \geq 1$. There is a number $p \in [1,\infty)\cap(n/2,\infty)$ such that $(V)_{k,p}$ holds.

Assumption $(B)_k$, $k \geq 2$. (1) There is a number $p \in [1,\infty)\cap(n/2,\infty)$ such that $(V)_{k-1,p}$, $(x\cdot\nabla V)_{k-2,p}$, and $(\partial_t V)_{k-2,p}$ hold.
(2) For any $0 \leq j \leq k-2$ there is a number $q(j) \in [2,\infty)\cap(n/(j+2),\infty)$ such that for all α with $|\alpha| = j$

$$\partial_x^{\alpha} V \in C(\mathbb{R};\ L^{q(j)}) + C(\mathbb{R};\ L^{\infty}).$$

Our main result now reads as follows.

Theorem 1. Suppose that either $k \geq 1$ and V satisfies $(A)_k$ or $k \geq 2$ and V satisfies $(B)_k$. Then
(1) For any $t \neq s$, $U(t,s)$ maps $H^{0,k}$ continuously into $H^{k,-k}$. Moreover, the map $D\times H^{0,k} \ni ((t,s),\phi) \longmapsto U(t,s)\phi \in H^{k,-k}$ is continuous, where $D = \{(t,s) \in \mathbb{R}^2;\ t \neq s\}$.
(2) For any $\phi \in H^{0,k}$, $\lim_{|t-s|\to 0} \|(t-s)^k U(t,s)\phi\|_{k,-k} = 0$.

Theorem 1 shows that the propagator $U(t,s)$ improves the regularity, at least locally in space, of the initial state ϕ at s even when V and ϕ have local singularities. Moreover, we can measure the regularity gained by the propagator in terms of the decay in space of the initial state. The formulation of smoothing effects by means of the weighted Sobolev spaces was first established by Jensen [6], where $E = 0$ and the boundedness of the derivatives of time-independent potentials were assumed.

As a typical example of time-dependent singular potentials, we consider $V_1(t,x) = \sum_{j=1}^{N} Z_j|x - y_j(t)|^{-\gamma_j}$, where $Z_j \in \mathbb{R}$, $\gamma_j > 0$, $y_j \in C^1(\mathbb{R};\ \mathbb{R}^n)$. Let $\gamma = \max_{1\leq j\leq N} \gamma_j$ and $n \geq 2$. Then the conclusions of Theorem 1 hold in the following cases. (1) $0 < \gamma < \min(2, n/2)$, $k \le 2$, $n \ge 2$. (2) $0 < \gamma < \min(2, n/2-k+2)$, $k \ge 3$, $n \ge 2k-3$.

To obtain more precise estimates of $U(t,s)$ in $H^{k,-k}$ we introduce the following Assumption $(C)_k$, which is stronger than $(A)_k$.

Assumption $(C)_k$, $k \ge 1$. For any $0 \le j \le k$ there is a number $r(j)$ such that $r(0) \in [1,\infty]\cap(n/2,\infty]$, $r(j) \in [2,\infty]\cap(n/j,\infty]$ for $j \ge 1$, and $\partial_x^\alpha V \in L^\infty(\mathbb{R}; L^{r(j)}) + L^\infty(\mathbb{R}; L^\infty)$ for all α with $|\alpha| = j \le k$.

Theorem 2. Let $k \ge 1$ and let V satisfy $(C)_k$. Let $a(k) = \max_{1\le j\le k} n/(jr(j))$. Then there is a constant $C(k)$ such that

$$\|U(t,s)\phi\|_{k,-k} \le C(k)(|t-s|^{-k}+|t-s|^{k/(1-a(k))})\|\phi\|_{0,k} \tag{1.1}$$

for all $t \ne s$ and $\phi \in H^{0,k}$.

When $r(j) = \infty$ for all $1 \le j \le k$, (1.1) becomes

$$\|U(t,s)\phi\|_{k,-k} \le C(k)(|t-s|^{-k}+|t-s|^{k})\|\phi\|_{0,k}. \tag{1.2}$$

When $E = 0$ and V is time-independent, (1.2) is the same estimate given by Jensen [6]. If, in addition, $V + (1/2)x\cdot\nabla V$ satisfies some other conditions, this estimate is improved to

$$\|U(t,s)\phi\|_{k,-k} \le C(k)(|t-s|^{-k} + 1)\|\phi\|_{0,k}$$

(see [10]). When $E \ne 0$, however, (1.2) is optimal with respect to the growth rate in time.

Theorem 3. Let $k \ge 1$ and let $V = 0$. Then there is a constant $\tilde{C}(k)$ such that for any $\phi \in H^{0,k}$

$$\tilde{C}(k)^{-1}|E|^k\|\phi\|_{0,-k} \le \liminf_{|t-s|\to\infty} |t-s|^{-k}\|U(t,s)\phi\|_{k,-k}$$
$$\le \limsup_{|t-s|\to\infty} |t-s|^{-k}\|U(t,s)\phi\|_{k,-k} \le \tilde{C}(k)|E|^k\|\phi\|_{0,-k}.$$

The next theorem gives a sufficient condition in order that every wave function becomes smooth and all singularities in the initial state ϕ at time s vanish instantly for $t \ne s$, provided that ϕ decays rapidly at infinity.

Theorem 4. Let V satisfy $(A)_k$ for all $k \geq 1$. Then for any $t \neq s$ $U(t,s)$ maps $H^{0,\infty}$ continuously into C^∞, where $H^{0,\infty} = \bigcap_{k\geq 1} H^{0,k}$ and $H^{0,\infty}$ is topologized as projective limit.
Moreover, the map $D\times H^{0,\infty} \ni ((t,s),\phi) \longmapsto U(t,s)\phi \in C^\infty$ is continuous.

This result is obtained in [10] when $E = 0$ and V is time-independent.

2. Proof

We use the following notations. For an interval I,
$L^{q,\theta}(I) = L^\theta(I;\, L^q(\mathbb{R}^n))$ and $|||\cdot|||_{q,\theta;I}$ denotes the $L^{q,\theta}(I)$-norm;
$U(t) = \exp(-itH_0)$; $S(t) = \exp(i|x|^2/2t)$; $J(t) = U(t)xU(-t) = x+it\nabla$
$= (J_1(t),\cdots,J_n(t))$, $|J|^m(t) = U(t)|x|^m U(-t)$, $J_s(t) = J(t-s)$,
$|J|_s^m(t) = |J|^m(t-s)$, $J_s^\alpha(t) = \prod_{k=1}^n J_k^{\alpha_k}(t-s)$ for $\alpha \in (\mathbb{N}\cup\{0\})^n$;
$(J_s^\alpha u)(t) = J_s^\alpha(t)u(t)$; $K_s(t) = |J|_s^2(t) + 2(t-s)^2 L$,
$L = i\partial_t + (1/2)\Delta - E\cdot x$; q' denotes the index conjugate to q;
$\delta(q) = n/2-n/q$, $\theta(q) = 2/\delta(q)$, $r = 2p/(2p-n)$. We denote by G_{t_0} the integral operator

$$(G_{t_0}v)(t) = \int_{t_0}^t U(t-\tau)v(\tau)d\tau, \quad t_0,\ t \in \mathbb{R}.$$

For an interval I and q satisfying $0 \leq \delta(q) < 1$ we denote by $X_q(I)$ the Banach space

$$X_q(I) = \{v \in C(I;\, L^2)\cap L^{q,\theta(q)}(I);$$
$$|||v|||_{X_q(I)} = |||v|||_{2,\infty;I} + |||v|||_{q,\theta(q);I} < \infty\}.$$

We denote by $I(t_0;a)$ the interval $[t_0-a, t_0+a]$ with $t_0 \in \mathbb{R}$ and $a > 0$.

Lemma 1. Let q and θ satisfy $0 \leq \delta(q) < 1$ and $\theta = \theta(q)$. Then there is a constant C_0 such that for any $t_0 \in \mathbb{R}$

$$|||U(\cdot)\phi|||_{X_q(\mathbb{R})} \leq C_0\|\phi\|, \quad \phi \in L^2,$$

$$|||G_{t_0} v|||_{X_q(\mathbb{R})} \le C_0 |||v|||_{2,1;\mathbb{R}}, \quad v \in L^{2,1}(\mathbb{R}),$$

$$|||G_{t_0} v|||_{X_q(\mathbb{R})} \le C_0 |||v|||_{q',\theta';\mathbb{R}}, \quad v \in L^{q',\theta'}(\mathbb{R}).$$

Proof. See Ginibre and Velo [3] and Yajima [13].

We consider the integral equation associated with (*)

$$(**) \qquad u(t) = U(t-s)\phi - i\int_s^t U(t-\tau)V(\tau)u(\tau)d\tau$$

and its regularized equation

$$(\#\#) \qquad u_{\varepsilon,j}(t) = U(t-s)\phi_\varepsilon - i\int_s^t U(t-\tau)V_{\varepsilon,j}(\tau)u_{\varepsilon,j}(\tau)d\tau.$$

Here $\phi_\varepsilon \in \mathcal{S}$ $(\varepsilon > 0)$ tends to ϕ as $\varepsilon \longrightarrow 0$ in the sense specified later and $V_{\varepsilon,j}$ $(\varepsilon > 0,\ j \in \mathbb{N})$ is defined as

$$V_{\varepsilon,j}(t,x) = \int_{\mathbb{R}^n} \rho_j(x-y)V_\varepsilon(t,y)\,dy, \qquad V_\varepsilon(t,x) = \int_{\mathbb{R}} \zeta_\varepsilon(t-\tau)V(\tau,x)\,d\tau,$$

where $\zeta_\varepsilon(t) = \varepsilon^{-1}\zeta(\varepsilon^{-1}t)$, $\rho_j(x) = j^n\rho(jx)$ with nonnegative functions $\zeta \in C_0^\infty(\mathbb{R})$, $\rho \in C_0^\infty(\mathbb{R}^n)$ satisfying $\operatorname{supp}\zeta \subset [-1,1]$, $\operatorname{supp}\rho \subset \{x \in \mathbb{R}^n;\ |x| \le 1\}$, and $\int_{\mathbb{R}} \zeta = \int_{\mathbb{R}^n} \rho = 1$.

For $W \in L^r_{loc}(\mathbb{R};\ L^p) + L^1_{loc}(\mathbb{R};\ L^\infty)$ with $p \in [1,\infty)\cap(n/2,\infty)$ and a compact interval I we set

$$[W]_I = \inf\{|||W^{(1)}|||_{p,r;I} + |||W^{(2)}|||_{\infty,1;I}\ ;\ W = W^{(1)} + W^{(2)}\}.$$

Lemma 2. Suppose that $V \in L^r_{loc}(\mathbb{R};\ L^p) + L^1_{loc}(\mathbb{R};\ L^\infty)$ for some $p \in [1,\infty)\cap(n/2,\infty)$. Let $q = 2p/(p-1)$. Let $s \in \mathbb{R}$, let $\phi \in L^2$ and let $\phi_\varepsilon \in \mathcal{S}$ satisfy $\phi_\varepsilon \longrightarrow \phi$ in L^2 as $\varepsilon \longrightarrow 0$. Then

(1) For any $T > 0$ (**) has a unique solution $u \in X_q(I(s;T))$. Moreover, u satisfies $\|u(t)\| = \|\phi\|$, $t \in \mathbb{R}$.

(2) For any $\varepsilon > 0$ and $j \in \mathbb{N}$ (##) has a unique solution $u_{\varepsilon,j} \in C^1(\mathbb{R};\ \mathcal{S})$. Moreover, $u_{\varepsilon,j}$ satisfies $\|u_{\varepsilon,j}(t)\| = \|\phi_\varepsilon\|$, $t \in \mathbb{R}$, and

$$(\#) \qquad i\frac{d}{dt}u_{\varepsilon,j}(t) = (H_0 + V_{\varepsilon,j}(t))u_{\varepsilon,j}(t),\ t \in \mathbb{R},\ u_{\varepsilon,j}(s) = \phi_\varepsilon.$$

(3) For any $T > 0$ there are constants $C(T)$, $\varepsilon_T > 0$ such that

for any $\varepsilon \in (0,\varepsilon_T]$ and $j \in \mathbb{N}$

$$|||u_{\varepsilon,j}|||_{X_q(I(s;T))} \le C(T)\|\phi_\varepsilon\|.$$

(4) For any $T > 0$

$$\lim_{\varepsilon\to 0}\lim_{j\to\infty} |||u_{\varepsilon,j} - u|||_{X_q(I(s;T))} = 0.$$

Proof. See Ozawa [11] and Yajima [13],[14].

Proof of Theorem 1. We prove Theorem 1 by making a reduction of the probrem to the case $E = 0$. By the transformation

$$v(t,x) = (R_s u)(t,x) = \exp(i(t-s)E\cdot x - i(t-s)^3|E|^2/6)u(t,x-((t-s)^2/2)E),$$

the equation (*) becomes a new equation for v

$$i\frac{d}{dt}v(t) = (-(1/2)\Delta + \tilde{V}_s(t))v(t),\ t \in \mathbb{R},$$

with the same initial condition, where $\tilde{V}_s(t,x) = V(t,x-((t-s)^2/2)E)$. (The use of the transformation R_s was suggested by the referee.) The transformation R_s has the advantage that it gauges the electric field away and that Assumptions $(A)_k$ and $(B)_k$ are invariant under the corresponding transformation $V \longrightarrow \tilde{V}_s$. Moreover, a straightforward calculation shows that R_s is a homeomorphism on $H^{k,-k}$ for any k and that we only have to prove Theorem 1 when $E = 0$. Therefore we assume here $E = 0$, so that $H_0 = -(1/2)\Delta$. We first treat the case where V satisfies $(A)_k$ for $k \ge 1$. Let $\phi \in H^{0,k}$ and let $\phi_\varepsilon \in \mathcal{S}$ satisfy $\phi_\varepsilon \longrightarrow \phi$ in $H^{0,k}$ as $\varepsilon \longrightarrow 0$. Let $u_{\varepsilon,j}$ be the solution of (#). Let $|\alpha| = k$. Since $u_{\varepsilon,j}$ satisfies the integral equation

$$u_{\varepsilon,j}(t) = U(t-t_0)u_{\varepsilon,j}(t_0) - i(G_{t_0}V_{\varepsilon,j}u_{\varepsilon,j})(t),\ t,\ t_0 \in \mathbb{R}, \quad (2.1)$$

we have

$$(J_s^\alpha u_{\varepsilon,j})(t) = U(t-t_0)(J_s^\alpha u_{\varepsilon,j})(t_0) - i\int_{t_0}^t U(t-\tau)V_{\varepsilon,j}(\tau)(J_s^\alpha u_{\varepsilon,j})(\tau)\,d\tau$$

$$- i\sum_{\beta \lneq \alpha}\int_{t_0}^t U(t-\tau)((i(\tau-s)\partial)^{\alpha-\beta}V_{\varepsilon,j}(\tau))(J_s^\alpha u_{\varepsilon,j})(\tau)\,d\tau. \quad (2.2)$$

Let $T > 0$. There is a constant $\varepsilon_T \in (0,1]$ such that

$$C_0[V]_{I(t_0;2\varepsilon_T)} \le 1/2 \quad \text{for all} \quad t_0 \in I(s;T). \tag{2.3}$$

By the Gagliardo-Nirenberg inequality with the relation $J_s^\alpha(t) = S(t-s)(i(t-s)\partial)^\alpha S(-(t-s))$, we have for $|\beta+\gamma| = k$, $\gamma \neq 0$,

$$|||(\cdot-s)^{|\gamma|}J_s^\beta u_{\varepsilon,j}|||_{1/(1-1/q-1/p(|\alpha|)),\theta;I}$$

$$\le C\Big(\sum_{|\delta|=k} |||J_s^\delta u_{\varepsilon,j}|||_{q,\theta;I} + |||(\cdot-s)^k u_{\varepsilon,j}|||_{q,\theta;I}\Big)^\mu |||(\cdot-s)^k u_{\varepsilon,j}|||_{q,\theta;I}^{1-\mu}, \tag{2.4}$$

$$|||(\cdot-s)^{|\gamma|}J_s^\beta u_{\varepsilon,j}|||_{2,\infty;I}$$

$$\le C\Big(\sum_{|\delta|=k} |||J_s^\delta u_{\varepsilon,j}|||_{q,\theta;I} + |||(\cdot-s)^k u_{\varepsilon,j}|||_{q,\theta;I}\Big)^{\tilde\mu} |||(\cdot-s)^k u_{\varepsilon,j}|||_{q,\theta;I}^{1-\tilde\mu}, \tag{2.5}$$

for some μ, $\tilde\mu \in [0,1)$, where $I = I(t_0;\varepsilon_T)$. By Lemma 1, Assumption $(A)_k$, (2.2)-(2.5), and the Hölder inequality, we obtain

$$|||J_s^\alpha u_{\varepsilon,j}|||_{X_q(I)} \le C\|(J_s^\alpha u_{\varepsilon,j})(t_0)\| \tag{2.6}$$

$$+ C(T)\Big(\sum_{|\delta|=k} |||J_s^\delta u_{\varepsilon,j}|||_{X_q(I)} + |||(\cdot-s)^k u_{\varepsilon,j}|||_{X_q(I)}\Big)^\lambda |||(\cdot-s)^k u_{\varepsilon,j}|||_{X_q(I)}^{1-\lambda}$$

for any $\varepsilon \in (0,\varepsilon_T]$, $j \in \mathbb{N}$, $t_0 \in I(s;T)$, where $\lambda \in [0,1)$ and $q = 2p/(p-1)$. We note here that $[V_{\varepsilon,j}]_I \le [V]_{I(t_0;2\varepsilon_T)}$. By (2.6) and part (3) of Lemma 2,

$$\sum_{|\alpha|=k} |||J_s^\alpha u_{\varepsilon,j}|||_{X_q(I(t_0;\varepsilon_T))} \le C(T)\Big(\|\phi_\varepsilon\| + \sum_{|\alpha|=k} \|(J_s^\alpha u_{\varepsilon,j})(t_0)\|\Big), \tag{2.7}$$

where $C(T)$ is independent of ε and j. In particular,

$$\sum_{|\alpha|=k} |||J_s^\alpha u_{\varepsilon,j}|||_{X_q(I(s;\varepsilon_T))} \le C(T)\Big(\|\phi_\varepsilon\| + \sum_{|\alpha|=k} \|x^\alpha \phi_\varepsilon\|\Big), \tag{2.8}$$

$$\sum_{|\alpha|=k} |||J_s^\alpha u_{\varepsilon,j}|||_{X_q(I(s+\ell\varepsilon_T;\varepsilon_T))} \le C(T)\Big(\|\phi_\varepsilon\| + \sum_{|\alpha|=k} \|(J_s^\alpha u_{\varepsilon,j})(t_0)\|\Big) \tag{2.9}$$

for any $\ell \in \mathbb{Z}$ such that $|\ell\varepsilon_T| \le T$. By (2.8) and (2.9), we obtain

$$\sum_{|\alpha|=k} |||J_s^\alpha u_{\varepsilon,j}|||_{X_q(I(s;T))} \le C(T)\|\phi_\varepsilon\|_{0,k}. \tag{2.10}$$

By part (4) of Lemma 2, (2.10) and a standard compactness argument, we conclude that $J_s^\alpha u \in L^{q,\theta(q)}(I(s;T)) \cap L^{2,\infty}(I(s;T))$ for all α

with $|\alpha| \le k$ and

$$\sum_{|\alpha|\le k} (|||J_s^\alpha u|||_{q,\theta(q);I(s;T)} + |||J_s^\alpha u|||_{2,\infty;I(s;T)}) \le C(T)\|\phi\|_{0,k}. \quad (2.11)$$

By (2.11), $J_s^\alpha u\colon \mathbb{R} \longrightarrow L^2$ is weakly continuous for all α with $|\alpha| \le k$ since $u \in C(\mathbb{R};\, L^2)$. Moreover it follows from (2.11) and Lemma 1 that $J_s^\alpha u \in X_q(s;T)$ and

$$J_s^\alpha u(t) = U(t-s)x^\alpha\phi - i\int_s^t U(t-\tau)J_s^\alpha(V\cdot u)(\tau)\, d\tau.$$

Part (1) now follows from the fact that $u(t) = U(t,s)\phi$ and

$$(i(t-s)\partial)^\alpha U(t,s)\phi \quad (2.12)$$

$$= \sum_{\beta\le\alpha}\sum_{\gamma\le\beta/2}\binom{\alpha}{\beta}\frac{\beta!(-1)^{|\beta+\gamma|}}{\gamma!(\beta-2\gamma)!}\, 2^{-|\gamma|}(i(t-s))^{|\gamma|}x^{\beta-2\gamma}J_s^{\alpha-\beta}(t)U(t,s)\phi.$$

Part (2) follows from (2.12) since the R.H.S. of (2.12) is written as

$$\sum_{\beta\le\alpha}\binom{\alpha}{\beta}(-1)^{|\beta|}x^\beta(J_s^{\alpha-\beta}(t)U(t,s)\phi - x^{\alpha-\beta}\phi)$$

$$+ \sum_{\beta\le\alpha}\sum_{0\ne\gamma\le\beta/2}\binom{\alpha}{\beta}\frac{\beta!(-1)^{|\beta+\gamma|}}{\gamma!(\beta-2\gamma)!}2^{-|\gamma|}(i(t-s))^{|\gamma|}x^{\beta-2\gamma}J_s^{\alpha-\beta}(t)U(t,s)\phi.$$

We next treat the case where V satisfies $(B)_k$ for $k \ge 2$. We have for $|\alpha| = k-2$

$$(J_s^\alpha K_s u_{\varepsilon,j})(t) = U(t-t_0)(J_s^\alpha K_s u_{\varepsilon,j})(t_0)$$

$$- i\int_{t_0}^t U(t-\tau)V_{\varepsilon,j}(\tau)J_s^\alpha K_s u_{\varepsilon,j}(\tau)\, d\tau$$

$$- i\sum_{\beta\lneq\alpha}\binom{\alpha}{\beta}\int_{t_0}^t U(t-\tau)((i(\tau-s)\partial)^{\alpha-\beta}V_{\varepsilon,j}(\tau))J_s^\beta K_s u_{\varepsilon,j}(\tau)\, d\tau$$

$$+ \sum_{\beta\le\alpha}\binom{\alpha}{\beta}\int_{t_0}^t U(t-\tau)(\tau-s)((i(\tau-s)\partial)^{\alpha-\beta}MV_{\varepsilon,j}(\tau))J^\beta u_{\varepsilon,j}(\tau)\, d\tau, \quad (2.13)$$

where $(MV_{\varepsilon,j})(t) = 4V_{\varepsilon,j}(t) + 2x\cdot\nabla V_{\varepsilon,j}(t) + 2(t-s)\partial_t V_{\varepsilon,j}(t)$.

In the same way as above, we have $J_s^\alpha K_s u \in X_q(I(s;T))$. By the Gagliardo-Nirenberg inequality, this implies that $J_s^\beta u\colon \mathbb{R} \longrightarrow L^2$ is weakly continuous for all β with $|\beta| \le k$. Moreover, in the same way as in the proof of part (4) of Lemma 2, we have for all β with

$|\beta| \le k$, $J_s^\beta u_{\varepsilon,j} \longrightarrow J_s^\beta u$ in $L^{2,\infty}(I(s;T))$ as $j \longrightarrow \infty$ and $\varepsilon \longrightarrow 0$, and therefore $J^\beta u \in C(\mathbb{R}; L^2)$. Theorem 1 now follows in the same way as before. Q.E.D.

We give a brief outline of the proof of Theorems 2-4. Details will be published elsewhere [11],[10].

Proof of Theorem 2. Let $\phi \in H^{0,k}$ and let $\phi_\varepsilon \in \mathcal{S}$ satisfy $\phi_\varepsilon \longrightarrow \phi$ in $H^{0,k}$ as $\varepsilon \longrightarrow 0$. Let $u_{\varepsilon,j}$ be the solution of (#). We set

$$Z(t) = \sum_{|\alpha|=k} \|J_s^\alpha u_{\varepsilon,j}(t)\|^2 + |t-s|^{2k}\|\phi_\varepsilon\|^2.$$

Since L and J_s^α are commutative, we have by the Gagliardo-Nirenberg inequality,

$$\left|\frac{d}{dt} Z(t)\right| \le C|t-s|^{1-a(k)}\|\phi_\varepsilon\|^{1-b(k)} Z(t)^{(1+b(k))/2} + C|t-s|^{-a(k)}\|\phi_\varepsilon\|$$

where $b(k) = 1 - 1/k + a(k)/k$. By this differential inequality,

$$Z(t)^{1/2} \le C\|\phi_\varepsilon\|_{0,k} + C(|t-s|^{k(2-a(k))/(1-a(k))} + |t-s|^k)\|\phi_\varepsilon\|.$$

By a limiting argument and a formula similar to (2.12), we obtain (1.1). Q.E.D.

Proof of Theorem 3. Without loss of generality, we may assume $s = 0$ since $U(t,s) = U(t-s)$ when $V = 0$. The result follows from the relation $J^\alpha(t)U(t) = U(t)x^\alpha$ and a formula similar to (2.12). Q.E.D.

Proof of Theorem 4. The proof is essentially the same as that of [10], since it relies only on the continuity of the map $\mathbb{R}^2 \times H^{0,k} \ni ((t,s),\phi) \mapsto J_s^\alpha(t)U(t,s)\phi \in L^2$ and the relation $J_s^\alpha(t) = S(t-s)(i(t-s)\partial)^\alpha S(-(t-s))$. Q.E.D.

Acknowledgments. The author is grateful to the referee for suggesting the use of the transformation R_s and for other useful comments.

References.

[1] M. Ben-Artzi and A. Devinatz, Local smoothing and convergence properties of Schrödinger-type equations, preprint, 1989.

[2] P. Constantin and J. C. Saut, Local smoothing properties of dispersive equations, J. Amer. Math. Soc., 1, 1988, 413-439.

[3] J. Ginibre and G. Velo, Scattering theory in the energy space for a class of nonlinear Schrödinger equations, J. Math. pures et appl, 64, 1985, 363-461.

[4] N. Hayashi, K. Nakamitsu, and M. Tsutsumi, On solutions of the initial value problem for the nonlinear Schrödinger equation, J. Funct. Anal., 71, 1987, 218-245.

[5] N. Hayashi and T. Ozawa, Smoothing effect for some Schrödinger equations, J. Funct. Anal., 85, 1989, 307-348.

[6] A. Jensen, Commutator methods and a smoothing property of the Schrödinger evolution group, Math. Z., 191, 1986, 53-59.

[7] T. Kato, On nonlinear Schrödinger equations, Ann. Inst. Henri Poincaré, Physique théorique, 46, 1987, 113-129.

[8] C. E. Kenig, G. Ponce, and L. Vega, Oscillatory integrals and regularity of dispersive equations, preprint, 1989.

[9] K. Nakamitsu, Smoothing effects for Schrödinger evolution groups, Tokyo Denki Univ. Kiyo, 10, 1988, 49-52.

[10] T. Ozawa, Smoothing effects and dispersion of singularities for the Schrödinger evolution group, to appear in Arch. Rat. Mech. Anal.

[11] T. Ozawa, Space-time behavior of propagators for Schrödinger evolution equations with Stark effect, to appear in J. Funct. Anal.

[12] G. Ponce, Regularity of solutions to nonlinear dispersive equations, J. Differential Equations, 78, 1989, 122-135.

[13] K. Yajima, Existence of solutions for Schrödinger evolution equations, Commun. Math. Phys., 110, 1987, 415-426.

[14] K. Yajima, Schrödinger evolution equations with magnetic fields, preprint, 1989.

[15] M. Yamazaki, On the microlocal smoothing effect of dispersive partial differential equations, I: Second-order linear equations, in "Algebraic Analysis," ed. M. Kashiwara and T. Kawai, Academic Press, San Diego, 1988.

Blow-up of Solutions for the Nonlinear Schrödinger Equation with Quartic Potential and Periodic Boundary Condition

TAKAYOSHI OGAWA† AND YOSHIO TSUTSUMI‡

† Department of Pure and Applied Sciences
University of Tokyo
Komaba, Meguro-ku, Tokyo 153, Japan

‡ Faculty of Integrated Arts and Sciences
Hiroshima University
Higashisenda-machi, Naka-ku, Hiroshima 730, Japan

§1 Introduction and Theorems.

In the present paper we consider the blow-up of solutions for the nonlinear Schrödinger equation with quartic self-interaction potential and periodic boundary condition:

$$(1.1) \qquad i\frac{\partial u}{\partial t} + D^2 u = -|u|^4 u, \quad t \geq 0, \quad x \in I,$$

$$(1.2) \qquad u(0,x) = u_0(x), \quad x \in I,$$

$$(1.3) \qquad u(t,-2) = u(t,2), \quad t \geq 0,$$

where $D = \frac{\partial}{\partial x}$ and $I = (-2,2)$.

The equation (1.1) is of physical interest, because (1.1) describes the collapse of a plane plasma soliton (see [3]). The blow-up problem of (1.1)-(1.2) in the whole real line $\mathbb{R}$ has been studied by many authors (see, e.g., [3],[6],[10],[12]-[18] and [22]-[27]). For example, it is already proved that if u_0 is in $H^1(\mathbb{R})$ and u_0 has negative energy, then the solution $u(t)$ of (1.1)-(1.2) with $I = \mathbb{R}$ blows up in finite time (see Glassey [6], M.Tsutsumi [23] and Ogawa and Y.Tsutsumi [18]). Furthermore, it is known that in the case of $I = \mathbb{R}$, there exists a blow-up solution $u(t,x)$ of (1.1)-(1.2) whose L^2 -density approaches $c_0\delta(x)$ or $\sum_{j=1}^{N} c_j\delta(x-a_j)$ near blow-up time, where $\delta(x)$ is the Dirac δ-function (see Weinstein [26], [27], Nawa and M.Tsutsumi [16], Merle [13] and Nawa [15]).

However, there seems to be few papers concerning the blow-up problem of (1.1)-(1.2) with periodic boundary condition (1.3). In [20], C.Sulem, P.L.Sulem and Frish made a numerical experiment for (1.1)-(1.3). Their result suggests that for certain initial data the solutions of (1.1)-(1.3) might blow up in finite time like the problem in

‡Present Address: Dept. of Math., Nagoya Univ., Chikusa-ku, Nagoya 464-01, Japan

$\mathbf{R}$. In [9] Kavian treated the blow-up problem for (1.1)-(1.3). But the oddness of the solution is assumed in [9], which implies that the solution satisfies the zero Dirichlet boundary condition. For the case of $I = \mathbf{R}$, the pseudo-conformal conservation law or its variant is used to show the blow-up of solutions for (1.1)-(1.2) (see Glassey [6], M.Tsutsumi [23] and Kavian [9]). However, the influence of the boundary values is not negligible for the periodic boundary condition case, which makes the blow-up problem for (1.1)-(1.3) complicated.

In this paper, we first show a sufficient condition for the blow-up of solutions to (1.1)-(1.3), following the former paper [18] by the authors. Next we prove that for (1.1)-(1.3), there also exists a blow-up solution such that $|u(t,x)|^2$ approaches $\|u_0\|^2_{L^2(I)}\delta(x)$ near blow-up time like the problem in $\mathbf{R}$. We will construct such a blow-up solution by the perturbation argument due to Merle [13].

Before we state the main results in this paper, we define several notations. Let $\phi(x)$ be a real valued function such that $\phi(x) = -\phi(-x)$, $D^j\phi \in L^\infty(\mathbf{R})$ $(j = 0, 1, 2, 3)$,

$$\phi(x) = \begin{cases} x, & 0 \le x < 1, \\ x - (x-1)^3, & 1 \le x < 1 + \frac{1}{\sqrt{3}}, \\ \text{smooth}, & 1 + \frac{1}{\sqrt{3}} \le x < 2, \\ 0, & 2 \le x, \end{cases} \tag{1.4}$$

and $\phi'(x) \le 0$ for $1 + \frac{1}{\sqrt{3}} \le x$. We put

$$\Phi(x) = \int_0^x \phi(y)dy. \tag{1.5}$$

For $0 \le a_0 \le 2$ we write

$$\Phi_{a_0}(x) = \begin{cases} \Phi(x - a_0), & a_0 - 2 \le x < 2, \\ \Phi(x - a_0 + 4), & -2 < x < a_0 - 2, \end{cases}$$

and for $-2 \le a_0 < 0$,

$$\Phi_{a_0}(x) = \begin{cases} \Phi(x - a_0 - 4), & 2 + a_0 \le x < 2, \\ \Phi(x - a_0), & -2 < x < 2 + a_0. \end{cases}$$

$E(v)$ denotes the energy functional associated with (1.1), that is,

$$E(v) = \|Dv\|^2_{L^2(I)} - \frac{1}{3}\|v\|^6_{L^6(I)}$$

and we put $E_0 = E(u_0)$.

We have the following two theorems.

THEOREM 1.1. *Let $u_0 \in H^1(I), u_0(-2) = u_0(2)$ and $E_0 < 0$. In addition, we assume that*

$$\eta = -2E_0 - 80(1+M)^2\|u_0\|^6_{L^2(I)} - \frac{M}{2}\|u_0\|^2_{L^2(I)} > 0 \tag{1.6}$$

and for some $a_0 \in [-2, 2]$,

$$\left(\int_I \Phi_{a_0} |u_0|^2 dx\right)^{\frac{1}{2}} \left(\frac{2}{\eta}\|Du_0\|^2_{L^2(I)} + 1\right)^{\frac{1}{2}} \le \frac{1}{4}, \tag{1.7}$$

where $M = \sum_{j=1}^{3} \|D^j\phi\|_{L^\infty(I)}$. *Then, the solution* $u(t)$ *in* $H^1(\mathbf{R})$ *of (1.1)-(1.3) blows up in finite time, that is, for some finite* $T > 0$,

$$\|Du(t)\|_{L^2(I)} \to \infty \text{ as } t \to T.$$

Remark 1.1. (1) The unique local existence theorem in $H^1(I)$ for (1.1)-(1.3) is already established: If $u_0 \in H^1(I)$ and $u_0(-2) = u_0(2)$, then there exists a unique solution $u(t)$ of (1.1)-(1.3) in $C([0,T); H^1(I))$ for some $T > 0$ and $u(t)$ satisfies two conservation laws of the L^2 norm and the energy

$$\|u(t)\|_{L^2(I)} = \|u_0\|_{L^2(I)}, \tag{1.8}$$

$$E(u(t)) = E_0 \tag{1.9}$$

for $0 \le t < T$. Moreover, $T = \infty$ or $T < \infty$ and $\|Du(t)\|_{L^2(I)} \to \infty$ as $t \to T$. For the details, see, e.g., Reed [19], Ginibre and Velo [5] and Strauss [22] (see also Kato [7], and Cazenave and Weissler [2]). We note that the nonlinear term $-|u|^4u$ is locally Lipschitz continuous from $H^1(I)$ to $H^1(I)$.

(2) In the case of $I = \mathbf{R}$, the condition $E_0 < 0$ is sufficient to show that the solution of (1.1)-(1.2) blows up in finite time. But in the periodic boundary condition case, it does not assure the blow- up of solutions. For example, we consider the initial value problem of the ordinary differential equation:

$$i\frac{d}{dt}z = -|z|^4 z, \quad t > 0, \tag{1.10}$$

$$z(0) = z_0. \tag{1.11}$$

We easily see that for any $z_0 \in \mathbf{C}$, (1.10)-(1.11) has a unique global solution. The solution $z(t)$ of (1.10)-(1.11) is also a solution of (1.1)-(1.3) with $E_0 < 0$, which is homogeneous in the spatial variable x.

(3) We note that there actually exists a $u_0(x)$ satisfying all the assumptions in Theorem 1.1. In fact, we choose $v(x) \in H^1(\mathbf{R})$ satisfying $v(x) = v(-x)$ and

$$\|Dv(t)\|^2_{L^2(\mathbf{R})} - \frac{1}{3}\|v\|^6_{L^6(\mathbf{R})} < 0.$$

We put $v_\varepsilon(x) = \varepsilon^{-\frac{1}{2}} v(\frac{x}{\varepsilon})$ for $\varepsilon > 0$. Then, we can choose ε so small that the restriction of v_ε on I satisfies all the assumptions in Theorem 1.1.

THEOREM 1.2. *Let* $Q(x)$ *be a solution in* $H^1(\mathbf{R})$ *of the equation*

$$-D^2Q + Q - |Q|^4Q = 0 \quad \text{in} \quad \mathbf{R} \tag{1.12}$$

such that $Q(x) = Q(-x)$, $Q(x) > 0$, $Q(x) \in C^\infty(\mathbf{R})$ *and its derivatives up to the second decays exponentially as* $|x| \to \infty$. *Then, there exists a solution* $u(t)$ *of (1.1)-(1.3) in* $C([0,T]; H^1(I))$ *for some finite* $T > 0$ *such that*

$$\|Du(t)\|_{L^2(I)} = O((T-t)^{-1}) \quad (t \to T), \tag{1.13}$$

$$\|u(t)\|_{L^\infty(I)} = O((T-t)^{-\frac{1}{2}}) \quad (t \to T), \tag{1.14}$$

$$|u(t,x)|^2 \to \quad \|Q\|^2_{L^2(\mathbf{R})}\delta(x) \quad (t \to T) \quad in \quad \mathcal{D}'(I) \tag{1.15}$$

and for any $0 < r < 2$,

$$\|u(t)\|_{H^1(I\setminus[-r,r])} \quad \to 0 \quad (t \to T). \tag{1.16}$$

Remark 1.2. (1) It is already known that there exists a non-trivial solution of (1.12) (see, e.g., Berestycki and P.L.Lions [1] and Strauss [21]).

(2) In the case of $I = \mathbf{R}$, the existence of blow-up solutions for (1.1)-(1.2) satisfying (1.13)-(1.16) is well known (see Weinstein [26],[27], Nawa and M.Tsutsumi [16] and Merle [13]).

Our plan in this paper is as follows. In Section 2, we give several lemmas and a proposition needed for the proofs of Theorems. In Sections 3 and 4, we state the proofs of Theorems 1.1 and 1.2, respectively.

Finally we conclude this section by giving several notations. Throughout this paper we omit the integral region in x, when it is the interval I. We denote the norms of $L^q(I)$ and $L^q(\mathbf{R})$ by $\|\cdot\|_q$ and $\|\cdot\|_{L^q(\mathbf{R})}$, respectively. We abbreviate $L^q(I)$ and $H^m(I)$ to L^q and H^m, respectively. For a positive integer m, we write

$$H^m_{prd} = \left\{v \in H^m(I); \quad D^j v(-2) = D^j v(2), \quad j = 0, 1, \cdots, m-1\right\}.$$

For $z \in \mathbf{C}$, we denote the complex conjugate of z by $\bar{z}$. In the course of calculations below, various constants are simply denoted by C.

§2. Lemmas and Proposition.

In this section we state several lemmas and one proposition needed for the proofs of Theorem 1.1 and Theorem 1.2.

The following three lemmas are used in the proof of Theorem 1.1.

LEMMA 2.1. *Let* $u \in H^1_{prd}$ *and let* ρ *be a real-valued function in* L^∞ *such that* $D\rho \in L^\infty$ *and* $\rho(-2) = \rho(2)$. *Then, we have*

$$\begin{aligned} \|\rho u\|_{L^\infty(1<|x|<2)} \leq & \sqrt{2}\|u\|^{\frac{1}{2}}_{L^2(1<|x|<2)} \left[2\|\rho^2 Du\|_{L^2(1<|x|<2)}\right. \\ & \left. + \sqrt{2}\|\rho^2 u\|_{L^2(1<|x|<2)} + \|uD\rho^2\|_{L^2(1<|x|<2)}\right]^{\frac{1}{2}}. \end{aligned} \tag{2.1}$$

Proof. We can extend $\rho(x)$ from I to $\mathbf{R}$ as a periodic function on $\mathbf{R}$. We also denote it by $\rho(x)$. We first show that for $v \in H^1(\mathbf{R})$,

$$\begin{aligned} & \|\rho v\|_{L^\infty(1<|x|)} \\ & \quad \leq \|v\|^{\frac{1}{2}}_{L^2(1<|x|)} \left[2\|\rho^2 Dv\|_{L^2(1<|x|)} + \|vD\rho^2\|_{L^2(1<|x|)}\right]^{\frac{1}{2}}. \end{aligned} \tag{2.2}$$

For $x > 1$ we have

$$(2.3) \qquad |v(x)|^2\rho(x)^2 = -\int_x^\infty D(|v(y)|^2\rho(y)^2)dy \\ \leq 2\int_1^\infty |\rho^2 vDv|dy + \int_1^\infty |v|^2|D\rho^2|dy.$$

For $x < -1$ we also have

$$(2.4) \qquad |v(x)|^2\rho(x)^2 \leq 2\int_{-\infty}^{-1} |\rho^2 vDv|dy + \int_{-\infty}^{-1} |v|^2|D\rho^2|dy.$$

Combining (2.3) and (2.4), we obtain (2.2).
We next define $\Psi(x)$ by

$$\Psi(x) = \begin{cases} 1, & |x| < 2, \\ 3 - |x|, & 2 < |x| < 3, \\ 0, & 3 < |x|. \end{cases}$$

We can regard $u \in H^1_{prd}$ as a periodic function in $H^1_{loc}(\mathbf{R})$ with the period of 4. If we choose $v = \Psi u$, then we obtain (2.1) from (2.2) and the periodicity of u. ∎

LEMMA 2.2. *Let $0 < T \leq \infty$ and $u(t)$ be a solution of (1.1)-(1.3) in $C([0,T); H^1_{prd})$. Let $\phi(x)$ and $\Phi(x)$ be defined in (1.4) and (1.5), respectively. Then we have*

(2.5)
$$-Im\int \phi u(t)D\bar{u}(t)dx + Im\int \phi u_0 D\bar{u}_0 dx \\ = \int_0^t \left[2\int D\phi|Du(s)|^2dx - \frac{2}{3}\int D\phi|u(s)|^6dx - \frac{1}{2}\int D^3\phi|u(s)|^2dx\right] ds,$$

(2.6)
$$\int \Phi|u(t)|^2dx = \int \Phi|u_0|^2dx - 2\int_0^t \left[Im\int \phi u(s)D\bar{u}(s)dx\right] ds$$

for $0 \leq t < T$.

Proof. We briefly state a sketch of the proof. For the details, see Lemma 2.2 and (4.9) in [17] (see also Kavian [9, Lemma 2.9]).

We suppose that $u(t)$ is a strong solution of (1.1)-(1.3) in $C([0,T); H^2_{prd}) \cap C^1([0,T); L^2)$. We first show (2.5). We multiply (1.1) by $\phi D\bar{u}$ and take the real part to obtain

$$(2.7) \qquad -\frac{d}{dt}Im\int \phi u(t)D\bar{u}(t)dx - Im\int D\phi u(t)\bar{u}_t(t)dx \\ = \int D\phi|Du(t)|^2dx + \frac{1}{3}\int D\phi|u(t)|^6dx.$$

We next multiply the complex conjugate of (1.1) by $D\phi u$ and take the real part to obtain

$$Im\int D\phi u(t)\bar{u}_t(t)dx = \int D\phi|Du(t)|^2dx - \int D\phi|u(t)|^6dx - \frac{1}{2}\int D^3\phi|u(t)|^2dx. \tag{2.8}$$

Since ϕ is compactly supported in I, we note that in the calculations of (2.7) and (2.8), the influence of the boundary values vanishes.

Substituting (2.8) into (2.7) and integrating the both sides of (2.7) over $(0,t)$, we obtain (2.5) for the strong solution.

We next multiply the complex conjugate of (1.1) by Φu and take the imaginary part to obtain (2.6). Since $u(t) \in H^2_{prd}$ and Φ is an even function, we note that in the calculation of (2.6) the influence of the boundary values causes no problem.

For a solution $u(t)$ in $C([0,T);H^1_{prd})$, we can justify the above calculations, if we approximate $u(t)$ by a sequence of the strong solutions. ∎

LEMMA 2.3. *Let ϕ and Φ be defined as in (1.4) and (1.5), respectively. Let $0 < T \leq \infty$ and $u(t)$ be a solution of (1.1)-(1.3) in $C([0,T);H^1_{prd})$. If $u(t)$ satisfies*

$$\|u(t)\|_{L^2(1<|x|<2)} \leq \frac{1}{2}, \quad 0 \leq t < T, \tag{2.9}$$

then we have

$$\begin{aligned} -Im\int \phi u(t)D\bar{u}(t)dx + Im\int \phi u_0 D\bar{u}_0 dx &\leq \left[2E_0 + 80(1+M)^2\|u_0\|_2^6 + \frac{M}{2}\|u_0\|_2^2\right]t, \\ &\quad 0 \leq t < T, \end{aligned} \tag{2.10}$$

where M is the same one defined in Theorem 1.1.

Proof. Since we have from the energy conservation law (1.9)

$$\|Du(t)\|^2_{L^2(|x|<1)} = E_0 - \|Du(t)\|^2_{L^2(1<|x|<2)} + \frac{1}{3}\|u(t)\|_6^6,$$

we obtain from (2.5) and (1.4),

$$\begin{aligned} &-Im\int \phi u(t)D\bar{u}(t)dx + Im\int \phi u_0 D\bar{u}_0 dx \\ &= \int_0^t \Big[2\int D\phi|Du(s)|^2dx - \frac{2}{3}\int D\phi|u(s)|^6dx - \frac{1}{2}\int D^3\phi|u(s)|^2dx\Big]ds \\ &= \int_0^t \Big[2E_0 - 2\int_{1<|x|<2}(1-D\phi)|Du(s)|^2dx \\ &\quad + \frac{2}{3}\int_{1<|x|<2}(1-D\phi)|u(s)|^6dx - \frac{1}{2}\int D^3\phi|u(s)|^2dx\Big]ds. \end{aligned} \tag{2.11}$$

We choose $\rho(x) = (1 - D\phi(x))^{\frac{1}{4}}$ in Lemma 2.1 and use Lemma 2.1 with Hölder's inequality to obtain

(2.12)

$$\begin{aligned}\int_{1<|x|<2} (1-D\phi)|u|^6 dx \leq & \|u\|^2_{L^2(1<|x|<2)} \|\rho u\|^4_{L^\infty(1<|x|<2)} \\ \leq & 4\|u\|^4_{L^2(1<|x|<2)} \left[2\|\rho^2 Du\|_{L^2(1<|x|<2)} + \sqrt{2}\|\rho^2 u\|_{L^2(1<|x|<2)} \right. \\ & \left. + \|uD\rho^2\|_{L^2(1<|x|<2)}\right]^2 \\ \leq & 32\|u\|^4_{L^2(1<|x|<2)} \|\rho^2 Du\|^2_{L^2(1<|x|<2)} + 32\|u\|^4_{L^2(1<|x|<2)} \|\rho^2 u\|^2_{L^2(1<|x|<2)} \\ & + 16\|u\|^6_{L^2(1<|x|<2)} \|D\rho^2\|^2_{L^\infty(1<|x|<2)}.\end{aligned}$$

On the other hand, we have from (1.4) and the definition of ρ,

$$|D\rho(x)^2| \leq \sqrt{3}|D(|x|-1)| = \sqrt{3}$$

for $1 < |x| < 1 + \frac{1}{\sqrt{3}}$. For $1 + \frac{1}{\sqrt{3}} < |x| < 2$, we have

$$\begin{aligned}|D\rho(x)^2| &= |D(1 - D\phi(x))^{\frac{1}{2}}| \\ &= \frac{1}{2}|(1 - D\phi(x))^{-\frac{1}{2}} D^2\phi(x)| \\ &\leq \frac{1}{2}\|D^2\phi\|_\infty \leq \frac{1}{2}M.\end{aligned}$$

Accordingly, we have

$$\|D\rho(x)^2\|_{L^\infty(1<|x|<2)} \leq \sqrt{3}(1+M) \tag{2.13}$$

for $1 < |x| < 2$.

(2.12) and (2.13) give us

(2.14)

$$\begin{aligned}\int_{1<|x|<2} (1-D\phi)|u|^6 dx \leq & 32\|u\|^4_{L^2(1<|x|<2)} \|\rho^2 Du\|^2_{L^2(1<|x|<2)} \\ & + 32(1+M)\|u\|^6_{L^2(1<|x|<2)} + 48(1+M)^2\|u\|^6_{L^2(1<|x|<2)}.\end{aligned}$$

Since $\rho^4(x) = 1 - D\phi(x)$, (2.11) and (2.14) yield

(2.15)

$$\begin{aligned}-Im & \int \phi u(t) D\bar{u}(t) dx + \int \phi u_0 D\bar{u}_0 dx \\ \leq & \int_0^t \left[2E_0 - 2\{1 - \frac{32}{3}\|u(s)\|^4_{L^2(1<|x|<2)}\} \int_{1<|x|<2} (1-D\phi)|Du(s)|^2 dx \right. \\ & \left. + 80(1+M)^2\|u(s)\|^6_{L^2(1<|x|<2)} + \frac{M}{2}\|u(s)\|^2_{L^2(1<|x|<2)}\right] ds.\end{aligned}$$

By (1.8), (2.9) and (2.15) we obtain (2.10). ∎

The following proposition and lemmas are used in the proof of Theorem 1.2. We first state the result concerning the existence and the properties of nontrivial solutions for (1.12).

PROPOSITION 2.4. *There exists a solution $Q(x)$ of (1.12) in $H^1(\mathbf{R})$ such that $Q(x) \in C^\infty(\mathbf{R})$, $Q(x) > 0$ for $x \in \mathbf{R}$, $Q(x) = Q(-x)$, and for some $C_0, \delta > 0$*

$$|D^j Q(x)| \le C_0 e^{-\delta|x|}, \quad x \in \mathbf{R}, \quad j = 0,1,2. \tag{2.16}$$

For Proposition 2.4, see Berestycki and P.L. Lions [1].

By using $Q(x)$ in Proposition 2.4 and the pseudo-conformal invariance of (1.1), we can construct the explicit blow-up solution for the problem (1.1)-(1.2) in $\mathbf{R}$.

LEMMA 2.5. *Let $Q(x)$ be a solution of (1.12) given by Proposition 2.4 and let $Q_\varepsilon(x) = \varepsilon^{-\frac{1}{2}} Q(\frac{x}{\varepsilon}), \varepsilon > 0$. We put*

$$R_\varepsilon(t,x) = t^{-\frac{1}{2}} e^{\frac{4-x^2}{4it}} Q_\varepsilon(\frac{x}{t}) \tag{2.17}$$

for $t \in \mathbf{R} \setminus \{0\}$. Then, $R_\varepsilon(t)$ is a solution of (1.1)-(1.2) in $C(\mathbf{R} \setminus \{0\}; H^2(\mathbf{R})) \cap C^1(\mathbf{R} \setminus \{0\}; L^2(\mathbf{R}))$ and

$$\|DR_\varepsilon(t)\|_{L^2(-1,1)} = O(|t|^{-1}) \quad (t \to 0), \tag{2.18}$$

$$\|R_\varepsilon(t)\|_{L^\infty(-1,1)} = O(|t|^{-\frac{1}{2}}) \quad (t \to 0), \tag{2.19}$$

$$|R_\varepsilon(t,x)|^2 \to \|Q\|_2^2 \delta(x) \quad (t \to 0) \quad in \quad \mathcal{D}'(\mathbf{R}), \tag{2.20}$$

$$\|R_\varepsilon(t,x)\|_2 = \|Q\|_{L^2(\mathbf{R})}, \quad t \neq 0. \tag{2.21}$$

In addition, there exists a constant $K > 0$ independent of ε such that

$$\|DR_\varepsilon(t)\|_{L^2(\mathbf{R})} \le K(\varepsilon^{-1}|t|^{-1} + \varepsilon), \tag{2.22}$$

$$\|R_\varepsilon(t)\|_{L^\infty(\mathbf{R})} \le K\varepsilon^{-\frac{1}{2}}|t|^{-\frac{1}{2}}, \tag{2.23}$$

for $0 < t \le 1$. Moreover, for any $\varepsilon > 0$, there exists $T > 0$ such that for $|t| \le T$

$$\|D^j R_\varepsilon(t)\|_{L^2(1<|x|)} \le e^{-\frac{\delta}{2\varepsilon|t|}}, \quad j = 0,1,2, \tag{2.24}$$

where δ is defined in (2.16) of Proposition 2.4.

Proof. We only show (2.22)-(2.24). For the rest of proof of Lemma 2.5, see Weinstein [26], Nawa and M.Tsutsumi [16] and Merle [13].

By (2.16) we have

$$\begin{aligned} |t|^{-\frac{1}{2}} |D^j Q_\varepsilon(\frac{x}{t})| \\ \le & C_0(\varepsilon|t|)^{-j}\{(\varepsilon|t|)^{-\frac{1}{2}} e^{-\frac{\delta|x|}{\varepsilon|t|}}\}, \quad j = 0,1,2 \end{aligned} \tag{2.25}$$

for $x \in \mathbf{R}, t \in \mathbf{R} \setminus \{0\}$, where C_0 is independent of ε. On the other hand,

$$\begin{aligned} |DR_\varepsilon(t,x)| \le & |t|^{-\frac{1}{2}} |DQ_\varepsilon(\frac{x}{t})| + C|t|^{-\frac{1}{2}} |\frac{x}{t}| |Q_\varepsilon(\frac{x}{t})| \\ \le & C(\varepsilon|t|)^{-1}\{(\varepsilon|t|)^{-\frac{1}{2}} e^{-\frac{\delta|x|}{\varepsilon|t|}}\} + C\varepsilon\{(\varepsilon|t|)^{-\frac{1}{2}} |\frac{x}{\varepsilon t}| e^{-\frac{\delta|x|}{\varepsilon|t|}}\}. \end{aligned} \tag{2.26}$$

Since the $L^2(\mathbf{R})$ norms of $(\varepsilon|t|)^{-\frac{1}{2}}|\frac{x}{\varepsilon t}|^j e^{-\frac{\delta|x|}{\varepsilon|t|}}$, $j = 0, 1$ are independent of ε, (2.26) gives us (2.22). In the same way we can prove (2.23).

We next show (2.24). We have from (2.25)

$$
\begin{aligned}
|D^2 R_\varepsilon(t,x)| \leq & |t|^{-\frac{1}{2}}|D^2 Q_\varepsilon(\frac{x}{t})| \\
& + C|t|^{-\frac{1}{2}}|\frac{x}{t}||DQ_\varepsilon(\frac{x}{t})| + C|t|^{-\frac{1}{2}}|\frac{x}{t}|^2|Q_\varepsilon(\frac{x}{t})| \\
\leq & C(\varepsilon|t|)^{-2}\{(\varepsilon|t|)^{-\frac{1}{2}}e^{-\frac{\delta|x|}{4\varepsilon|t|}}\}e^{-\frac{3\delta}{4\varepsilon|t|}} \\
& + C\varepsilon(\varepsilon|t|)^{-1}\{(\varepsilon|t|)^{-\frac{1}{2}}|\frac{x}{\varepsilon t}|e^{-\frac{\delta|x|}{4\varepsilon|t|}}\}e^{-\frac{3\delta}{4\varepsilon|t|}} \\
& + C\varepsilon^2\{(\varepsilon|t|)^{-\frac{1}{2}}|\frac{x}{\varepsilon t}|^2 e^{-\frac{\delta|x|}{4\varepsilon|t|}}\}e^{-\frac{3\delta}{4\varepsilon|t|}}
\end{aligned}
\tag{2.27}
$$

for $1 < |x|$. Since the $L^2(\mathbf{R})$ norms of $(\varepsilon|t|)^{-\frac{1}{2}}|\frac{x}{\varepsilon t}|^j e^{-\frac{\delta|x|}{4\varepsilon|t|}}$, $j = 0, 1, 2$ are independent of ε, we obtain

$$
\|D^2 R_\varepsilon(t)\|_{L^2(1<|x|)} \leq C\{(\varepsilon|t|)^{-2} + \varepsilon^2\}e^{-\frac{3\delta}{4\varepsilon|t|}}, \tag{2.28}
$$

where C is independent of ε and t. If we choose $T > 0$ so small in (2.28) that for $|t| < T$

$$
C\{(\varepsilon|t|)^{-2} + \varepsilon^2\}e^{-\frac{\delta}{4\varepsilon|t|}} < 1,
$$

then by (2.28) we obtain (2.24) for $j = 2$. The proof of (2.24) for $j = 0, 1$ is the same. ∎

We conclude this section by giving the following elementary lemma.

LEMMA 2.6. *Let $L > 0$. Then*

$$
\int_0^t s^{-2}e^{-\frac{L}{s}}\,ds = \frac{1}{L}e^{-\frac{L}{t}}, \quad t > 0. \tag{2.29}
$$

Proof. The exchange of variables $s = -\frac{1}{r}$ immediately yields (2.29). ∎

§3. Proof of Theorem 1.1.

In this section we give the proof of Theorem 1.1. We remark that it is sufficient to show Theorem 1.1 for the case $a_0 = 0$ in (1.7), because the equation (1.1) has the translation invariance. In the case $a_0 \neq 0$, we can consider the initial data $u_0(x - a_0)$ periodically instead of $u_0(x)$.

Proof of Theorem 1.1. We suppose that the solution $u(t)$ of (1.1)-(1.3) exists for all $t \geq 0$ and derive a contradiction.

We first show that if the initial data u_0 satisfies (1.6) and (1.7), then $u(t)$ satisfies (2.9) for all $t \geq 0$. We prove this by contradiction. Since $\eta > 0$ and $1 \leq 2\Phi(x)$ for $x \in (-2,-1) \cup (1,2)$, we have from (1.7)

$$
\|u_0\|_{L^2(1<|x|<2)} \leq \frac{1}{2\sqrt{2}}. \tag{3.1}
$$

We define T_0 as follows:

$$T_0 = \sup\{t > 0; \|u(s)\|_{L^2(1<|x|<2)} \le \frac{1}{2}, \quad 0 \le s < t\}. \tag{3.2}$$

By (3.1) and the continuity in L^2 of $u(t)$ we note that $T_0 > 0$. If $T_0 = \infty$, then the desired conclusion holds. Suppose that $T_0 < \infty$. By the continuity in L^2 of $u(t)$ we have

$$\|u(T_0)\|_{L^2(1<|x|<2)} = \frac{1}{2}. \tag{3.3}$$

On the other hand, $u(t)$ satisfies all the assumptions in Lemma 2.3 on $[0, T_0)$. Therefore, by (2.6), (1.6) and Lemma 2.3 we have

$$\begin{aligned} \int \Phi|u(t)|^2 dx \le& \int \Phi|u_0|^2 dx \\ &- 2tIm \int \phi u_0 D\bar{u}_0 dx - \eta t^2, \quad 0 \le t < T_0. \end{aligned} \tag{3.4}$$

(3.4) and (1.6) yield

$$\begin{aligned} \int \Phi|u(t)|^2 dx =& -\eta(t + \frac{1}{\eta} Im \int \phi u_0 D\bar{u}_0 dx)^2 \\ &+ \frac{1}{\eta}(Im \int \phi u_0 D\bar{u}_0 dx)^2 + \int \Phi|u_0|^2 dx \\ &\le \frac{1}{\eta}\|\phi u_0\|_2^2 \|Du_0\|_2^2 + \int \Phi|u_0|^2 dx, \\ &0 \le t < T_0. \end{aligned} \tag{3.5}$$

Noting that $\phi^2 \le 2\Phi$, we obtain from (3.5)

$$\int \Phi|u(t)|^2 dx \le (\frac{2}{\eta}\|Du_0\|_2^2 + 1) \int \Phi|u_0|^2 dx, \qquad 0 \le t < T_0. \tag{3.6}$$

Since $1 \le 2\Phi(x)$ for $1 < |x| < 2$, (1.7) and (3.6) show

$$\|u(t)\|_{L^2(1<|x|<2)} \le (2 \int \Phi|u(t)|^2 dx)^{\frac{1}{2}} \le \frac{1}{2\sqrt{2}}, \qquad 0 \le t < T_0.$$

This and the continuity in L^2 of $u(t)$ give us

$$\|u(T_0)\|_{L^2(1<|x|<2)} \le \frac{1}{2\sqrt{2}},$$

which is a contradiction to (3.3). Thus, if the initial data u_0 satisfies (1.6) and (1.7), then $u(t)$ satisfies (2.9) for all $t \ge 0$.

Therefore, since all the assumptions in Lemma 2.3 hold with $T = \infty$, $u(t)$ satisfies (3.4) with $T_0 = \infty$, that is, for all $t \ge 0$. (3.4) implies that $\int \Phi|u(t)|^2 dx$ becomes negative in finite time. This is a contradiction, because $\Phi(x) > 0$ except $x = 0$. Hence, if the initial data u_0 satisfies (1.6) and (1.7), then $u(t)$ must blow up in finite time. ∎

§4. Proof of Theorem 1.2.

In this section, we prove Theorem 1.2 by the perturbation argument due to Merle [13]. From Lemma 2.5 we know the explicit blow-up solution for the Cauchy problem (1.1)-(1.2) in $\mathbf{R}$. By adding a perturbation to that solution, we can construct a blow-up solution of the periodic boundary condition case (1.1)-(1.3).

We define a cut-off function $\Psi \in C_0^\infty(\mathbf{R})$ as follows:

$$\Psi = \begin{cases} 1, & |x| < 1, \\ \text{nonnegative}, & 1 < |x| < 2, \\ 0, & 2 < |x|. \end{cases} \tag{4.1}$$

Put

$$v_\varepsilon(t,x) = \Psi(x) R_\varepsilon(t,x),$$

where $R_\varepsilon(t,x)$ is an explicit blow-up solution defined as in (2.17). Since $R_\varepsilon(t,x)$ satisfies (1.1)-(1.2), we see that v_ε satisfies

$$i\frac{\partial}{\partial t} v_\varepsilon + D^2 v_\varepsilon = -|R_\varepsilon|^4 v_\varepsilon + 2D\Psi DR_\varepsilon + R_\varepsilon D^2\Psi, \quad x \in I, \tag{4.2}$$

$$v_\varepsilon(t,-2) = v_\varepsilon(t,2) = 0, \tag{4.3}$$

for $t \in \mathbf{R} \setminus \{0\}$. Thus if we can find a time local solution $w(t,x)$ in $C([0,T]; H^1_{prd})$ for some $T > 0$ satisfying

$$\begin{aligned} i\frac{\partial}{\partial t} w + D^2 w = & -|w + v_\varepsilon|^4 (w + v_\varepsilon) \\ & + |R_\varepsilon|^4 v_\varepsilon - 2D\Psi DR_\varepsilon - R_\varepsilon D^2 \Psi, \quad t > 0, \quad x \in I \end{aligned} \tag{4.4}$$

$$w(t,-2) = w(t,2), \quad t > 0, \tag{4.5}$$

$$w(0,x) = 0, \tag{4.6}$$

we have only to put

$$u(t,x) = \overline{w(T-t,x) + v_\varepsilon(T-t,x)} \tag{4.7}$$

in order to prove Theorem 1.2. Then, $u(t,x)$ is a blow-up solution of (1.1)-(1.3) satisfying (1.13)-(1.16).

Proof of Theorem 1.2. We use the contraction mapping principle to obtain a local solution of the problem (4.4)-(4.6). We define for some positive constants T, ε, γ

$$X_T = \{\omega \in C([0,T]; H^1_{prd}); |||\omega||| < \infty\},$$

where

$$|||\omega||| = \sup_{[0,T]} \{ e^{\frac{\gamma}{\varepsilon t}} \|\omega(t)\|_2 + e^{\frac{\gamma}{2\varepsilon t}} \|D\omega(t)\|_2 \},$$

and T, ε and γ will be determined later.

Let $U(t)$ be the evolution operator of the free Schrödinger equation with periodic boundary condition (4.5). Then the solution w of (4.4)-(4.6) can be written as follows:

$$
\begin{aligned}
(4.8) \qquad w(t) =& F_0(R_\varepsilon, \Psi)(t) \\
&+ i \int_0^t U(t-s)\{|w(s)+v_\varepsilon(s)|^4(w(s)+v_\varepsilon(s))\}ds,
\end{aligned}
$$

where

$$
F_0(R_\varepsilon, \Psi)(t) = i\int_0^t U(t-s)\{2D\Psi DR_\varepsilon + R_\varepsilon D^2\Psi - |R_\varepsilon|^4 v_\varepsilon\}ds.
$$

We divide the second term of the right hand side of (4.8) as

$$
N_0(v_\varepsilon) + N_1(w, v_\varepsilon) + N_2(w, v_\varepsilon),
$$

where

$$
\begin{aligned}
N_0(v_\varepsilon) =& i\int_0^t U(t-s)|v_\varepsilon(s)|^4 v_\varepsilon(s)ds, \\
N_1(w, v_\varepsilon) =& i\int_0^t U(t-s)\{2|v_\varepsilon|^2 v_\varepsilon^2 \bar{w} + 3|v_\varepsilon|^4 w\}ds, \\
N_2(w, v_\varepsilon) =& i\int_0^t U(t-s)\{v_\varepsilon^3 \bar{w}^2 + 3v_\varepsilon^2 w(2\bar{v}_\varepsilon \bar{w} + \bar{w}^2) \\
&+ (3v_\varepsilon w^2 + w^3)(\bar{v}_\varepsilon^2 + 2\bar{v}_\varepsilon \bar{w} + \bar{w}^2)\}ds.
\end{aligned}
$$

We put $F(R_\varepsilon, v_\varepsilon, \Psi)(t) = F_0(R_\varepsilon, \Psi) + N_0(v_\varepsilon)$. Noting that

$$
-|R_\varepsilon|^4 v_\varepsilon + |v_\varepsilon|^4 v_\varepsilon = -|R_\varepsilon|^4(1-\Psi^4)v_\varepsilon
$$

and (2.24) of Lemma 2.5, we can easily obtain that for $\varepsilon > 0$ and $0 < \gamma < \frac{\delta}{2}$ there exists a $T_0 > 0$ such that

$$
(4.9) \qquad \|F(t)\|_2 + \|DF(t)\|_2 \leq \frac{1}{2} t e^{-\frac{\gamma}{\varepsilon |t|}}
$$

for $0 < t \leq T_0$, where δ is the same one defined in Proposition 2.4. Now we fix γ.

Let $A = \{v \in X_T; |||v||| \leq 1\}$. For $w \in X_T$ we define the nonlinear map

$$
N[w] = F(R_\varepsilon, \Psi, v_\varepsilon) + N_1(w, v_\varepsilon) + N_2(w, v_\varepsilon).
$$

We show that for sufficiently large $\varepsilon > 0$ and small $T > 0$, the nonlinear mapping $N[\cdot]$ is a contraction from A into A. We first choose $T > 0$ and $\varepsilon > 0$ such that

$$
(4.10) \qquad T < \min(1, T_0), \quad \varepsilon > 1,
$$

where T_0 is defined in (4.9). Then, by (4.9), (4.10) and Lemmas 2.5 and 2.6 we have

$$
(4.11) \qquad \|N[w]\|_2
$$

$$
\begin{aligned}
&\le \|F(t)\|_2 + \|N_1(w,v_\varepsilon)\|_2 + \|N_2(w,v_\varepsilon)\|_2 \\
&\le \frac{1}{2} t e^{-\frac{\gamma}{\varepsilon|t|}} + C\sum_{j=0}^{4}\int_0^t K^{4-j}\varepsilon^{-2+\frac{j}{2}} s^{-2+\frac{j}{2}} \|w(s)\|_2^{1+\frac{j}{2}} \\
&\quad \times \{\|Dw(s)\|_2 + \|w(s)\|_2\}^{\frac{j}{2}}\, ds \\
&\le \frac{1}{2} t e^{-\frac{\gamma}{\varepsilon|t|}} \\
&\quad + C\sum_{j=0}^{4} K^{4-j} t^{\frac{j}{2}} \varepsilon^{-2+\frac{j}{2}} |||w(s)|||^{1+j} \int_0^t s^{-2} e^{-(1+\frac{3j}{4})\frac{\gamma}{\varepsilon s}}\, ds \\
&\le \frac{1}{2} t e^{-\frac{\gamma}{\varepsilon|t|}} + C\sum_{j=0}^{4} K^{4-j} t^{\frac{j}{2}} \varepsilon^{-1+\frac{j}{2}} \gamma^{-1} e^{-\frac{(1+\frac{3j}{4})\gamma}{\varepsilon t}} |||w|||^{1+j} \\
&\le \{\frac{1}{2} t + C_1 \sum_{j=0}^{4} K^{4-j} (\varepsilon t)^{\frac{j}{2}} (\varepsilon\gamma)^{-1}\} e^{-\frac{\gamma}{\varepsilon t}}, \quad 0 \le t \le T
\end{aligned}
$$

for $w \in A$, where K is defined in (2.22) and (2.23) of Lemma 2.5 and C_1 is independent of ε and t. We choose $\varepsilon > 1$ so large that

$$
C_1 \sum_{j=0}^{4} K^{4-j} (\gamma\varepsilon)^{-1} < \frac{1}{4} \tag{4.12}
$$

and choose $T > 0$ such that

$$
T < \min\{\varepsilon^{-1}, \frac{1}{2}\}. \tag{4.13}
$$

Then we obtain from (4.11)-(4.13)

$$
\|N[w](t)\|_2 \le \frac{1}{2} e^{-\frac{\gamma}{\varepsilon t}}, \quad 0 \le t \le T, \tag{4.14}
$$

for $w \in A$.

Since $DU(t)v = U(t)Dv$ for $v \in H^1_{prd}$, we also have from (4.9),(4.10) and Lemma 2.5

(4.15)

$$
\begin{aligned}
\|DN[w]\|_2 \le& \frac{1}{2}\int_0^t e^{-\frac{\gamma}{\varepsilon s}}\, ds \\
&+ C\sum_{j=0}^{3}\int_0^t K^{4-j}(\varepsilon^{-1}s^{-1} + \varepsilon)\varepsilon^{-\frac{3-j}{2}} s^{-\frac{3-j}{2}} \\
&\quad \times \|w(s)\|_2^{\frac{1+j}{2}} \{\|Dw(s)\|_2 + \|w(s)\|_2\}^{\frac{1+j}{2}}\, ds \\
&+ C\sum_{j=0}^{4}\int_0^t K^{4-j}\varepsilon^{-2+\frac{j}{2}} s^{-2+\frac{j}{2}}
\end{aligned}
$$

$$\times \|Dw(s)\|_2 \|w(s)\|_2^{\frac{j}{2}} \{\|w(s)\|_2 + \|Dw(s)\|_2\}^{\frac{j}{2}} ds$$
$$\leq \frac{1}{2} t e^{-\frac{\gamma}{\varepsilon t}} + C \sum_{j=0}^{3} K^{4-j} \varepsilon^{-\frac{5}{2}+\frac{j}{2}} t^{\frac{j}{2}} \int_0^t s^{-\frac{5}{2}} e^{-\frac{3+3j}{4}\frac{\gamma}{\varepsilon s}} ds$$
$$+ C \sum_{j=0}^{3} K^{4-j} \varepsilon^{-1+\frac{j}{2}} t^{\frac{j}{2}} (\varepsilon t)^{\frac{1}{2}} \int_0^t s^{-2} e^{-\frac{3+3j}{4}\frac{\gamma}{\varepsilon s}} ds$$
$$+ C \sum_{j=0}^{4} K^{4-j} \varepsilon^{-2+\frac{j}{2}} t^{\frac{j}{2}} \int_0^t s^{-2} e^{-\frac{2+3j}{4}\frac{\gamma}{\varepsilon s}} ds$$

for $w \in A$. Here we choose $T > 0$ so small that

$$t^{-\frac{1}{2}} e^{-\frac{\gamma}{4\varepsilon t}} < 1, \quad 0 < t \leq T. \tag{4.16}$$

Then, by (4.15), (4.16) and the fact $\varepsilon > 1$ we have

(4.17)

$$\|DN[w](t)\|_2 \leq \{\frac{1}{2} t + C \sum_{j=0}^{3} K^{4-j} \gamma^{-1} (\varepsilon t)^{\frac{j}{2}} \varepsilon^{-\frac{3}{2}}$$
$$+ C \sum_{j=0}^{3} K^{4-j} \gamma^{-1} (\varepsilon t)^{1+\frac{j}{2}} e^{-\frac{\gamma}{4\varepsilon t}} + C \sum_{j=0}^{4} K^{4-j} \gamma^{-1} (\varepsilon t)^{\frac{j}{2}} \varepsilon^{-1} \} e^{-\frac{\gamma}{2\varepsilon t}}$$
$$\leq [C_2 \sum_{j=0}^{4} K^{4-j} \gamma^{-1} (\varepsilon t)^{\frac{j}{2}} \varepsilon^{-1}$$
$$+ \{\frac{1}{2} t + C_2 \sum_{j=0}^{3} K^{4-j} \gamma^{-1} (\varepsilon t)^{1+\frac{j}{2}} e^{-\frac{\gamma}{2\varepsilon t}} \}] e^{-\frac{\gamma}{2\varepsilon t}},$$
$$0 < t \leq T$$

for $w \in A$, where C_2 is independent of ε and t. We choose ε so large that

$$C_2 \sum_{j=0}^{4} K^{4-j} (\gamma\varepsilon)^{-1} < \frac{1}{4}, \tag{4.18}$$

and we choose $T > 0$ so small that T satisfies (4.13) and

$$\frac{1}{2} t + C_2 \sum_{j=0}^{3} K^{4-j} \gamma^{-1} e^{-\frac{\gamma}{4\varepsilon t}} < \frac{1}{4}. \tag{4.19}$$

Then, by (4.17)-(4.19) and (4.10) we obtain

$$\|DN[w](t)\|_2 \leq \frac{1}{2} e^{-\frac{\gamma}{2\varepsilon t}}, \quad 0 < t \leq T \tag{4.20}$$

for $w \in A$. (4.14) and (4.20) show

$$|||N[w]||| \leq 1, \quad w \in A,$$

which implies that $N[\cdot]$ is a mapping from A into A.

In the same way as above we can show that

$$|||N[w_1] - N[w_2]||| \leq \frac{1}{2}|||w_1 - w_2|||, \quad w_1, w_2 \in A \tag{4.21}$$

for sufficiently large $\varepsilon > 0$ and small $T > 0$. (4.21) shows that $N[\cdot]$ is a contraction mapping from A to A. Therefore, there exists a unique fixed point $w(t)$ of the nonlinear mapping N in A for some $\varepsilon, T > 0$. $w(t)$ is the desired local solution of (4.4)-(4.6). Accordingly, if we define $u(t,x)$ as in (4.7), we can easily verify that $u(t)$ is a blow-up solution of (1.1)-(1.3) satisfying (1.13)-(1.16). ∎

References

1. H.Berestycki, P.L.Lions, *Existence of a ground state in nonlinear equations of the Klein-Gordon type*, in "Variational Inequalities and Complementarity Problems," R.W.Cottle, F.Giannessi and J.L.Lions eds., J.Wiley, New York, 1980, pp. 37–51.
2. T.Cazenave, F.B.Weissler, *The Cauchy problem for the nonlinear Schrödinger equation in H^1*, Manuscripta Math. **61** (1988), 477–498.
3. T.Cazenave, F.B.Weissler, *The structure of solution to the pseudoconformally invariant nonlinear Schrödinger equation*, Preprint.
4. L.M.Degtyarev, V.E.Zakharov, L.I.Rudakov, *Two examples of Langmuir wave collapse*, Soviet Phys. JETP **41** (1975), 57–61.
5. J.Ginibre, G.Velo, *The global Cauchy problem for the nonlinear Schrödinger equation revisited*, Ann. Inst. Henri Poincaré, Analy. Non Linéaire **2** (1985), 309–327.
6. R.T.Glassey, *On the blow-up of solution to the Cauchy problem for the nonlinear Schrödinger equation*, J. Math. Phys. **18** (1977), 1794–1797.
7. T.Kato, *On nonlinear Schrödinger equations*, Ann. Inst. Henri Poincaré, Phys. Theor. **46** (1987), 113–129.
8. T.Kato, *Nonlinear Schrödinger equations*, Preprint.
9. O.Kavian, *A remark on the blow-up of solutions to the Cauchy problem for nonlinear Schrödinger equations*, Trans. A.M.S. **299** (1987), 193–203.
10. B.LeMesurier, G.Papanicolaou, C.Sulem, P.L.Sulem, *The focusing singularity of the nonlinear Schrödinger equation*, in "Direction in Partial Differential Equations," Academic Press, New York, 1987, pp. 159–201.
11. J.E.Lin, W.A.Strauss, *Decay and scattering of solutions of a nonlinear Schrödinger equation*, J. Funct. Anal. **30** (1978), 245–263.
12. F.Merle, *Limit of the solution of the nonlinear Schrödinger equation at the blow-up time*, J. Funct. Anal. **84** (1989), 201–214.
13. F.Merle, *Construction of solutions with exactly k blow-up points for the Schrödinger equation with critical power nonlinearity*, Preprint.

14. F.Merle, Y.Tsutsumi, L^2 *concentration of blow-up solutions for the nonlinear Schrödinger equation with the critical power nonlinearity*, Preprint.
15. H.Nawa, *"Mass concentration" phenomenon for the nonlinear Schrödinger equation with the critical power nonlinearity 2*, Preprint.
16. H.Nawa, M.Tsutsumi, *On blow-up for the pseudo-conformally invariant Schrödinger equation*, Funk. Ekvac **32** (1989), 417–428.
17. T.Ogawa, Y.Tsutsumi, *Blow-up of* H^1 *solution for the nonlinear Schrödinger equation*, J. Diff. Equ. (to appear).
18. T.Ogawa, Y.Tsutsumi, *Blow-up of* H^1 *solution for the one dimensional nonlinear Schrödinger equation with critical power nonlinearity*, Preprint.
19. M.C.Reed, "Abstract Nonlinear Wave Equations," Springer Lecture Notes in Math. 507, Springer-Verlag, Berlin, New York, Heidelberg, 1976.
20. C.Sulem, P.L.Sulem, H.Frish, *Tracing complex singularities with spectral methods*, J. Compt. Phys. **50** (1983), 138–161.
21. W.A.Strauss, *Existence of solitary waves in higher dimensions*, Comm. Math. Phys. **55** (1977), 149–162.
22. W.A.Strauss, "Nonlinear Wave Equations," to appear in CBMS Regional Conference Series in Math.
23. M.Tsutsumi, *Nonexistence of global solutions to the Cauchy problem for the damped nonlinear Schrödinger equation*, SIAM J.Math. Anal. **15** (1984), 357–366.
24. Y.Tsutsumi, *Rate of* L^2 *concentration of blow-up solutions for the nonlinear Schrödinger equation with critical power nonlinearity*, Nonlinear Anal., T.M.A. (to appear).
25. M.I.Weinstein, *Nonlinear Schrödinger equations and sharp interpolation estimates*, Comm. Math. Phys. **87** (1983), 567–576.
26. M.I.Weinstein, *On the structure and formation of singularities in solutions to nonlinear dispersive evolution equations*, Comm. P.D.E. **11** (1986), 545–565.
27. M.I.Weinstein, *The nonlinear Schrödinger equation– Singularity formation stability and dispersion*, Contemporary Mathematics **99** (1989) (to appear).

List of Recent Publications of Tosio Kato

An almost complete list of publications of Professor Kato up to 1986 appeared in the Journal of Mathematical Analysis and Application, **127** (1987), 303-311, together with his brief chronological record. The following are some recent publications not included in that list.

Abstract Differential Equations and Nonlinear Mixed Problems, Lezioni Fermiane[Fermi Lectures], Accademia Nazionale dei Lincei, Scuola Normale Superiore, 1985, 87pp.

On nonlinear Schrödinger equations, Ann. Inst. H. Poincaré Phys. Théor. **46**(1987), 113–129.

Variation of discrete spectra, Comm. Math. Phys. **111**(1987), 501–504.

With G. Ponce, On nonstationary flows of viscous and ideal fluids in $L^p_s(\mathbb{R}^2)$, Duke Math. J. **55**(1987), 487–499.

With G. Ponce, Commutator estimates and the Euler and Navier-Stokes equations, Comm. Pure Appl. Math. **41**(1988), 891–907.

Nonlinear Schrödinger equations, Schrödinger Operators, H. Holden and A. Jensen (Eds.), Lecture Notes in Physics 345, Springer 1989, 218–263.

With K. Yajima, Some examples of smooth operators and the associated smoothing effect, Rev. Math. Phys. **1**(1989), 481–496.

Lecture Notes in Mathematics

Edited by A. Dold, B. Eckmann and F. Takens

Editorial Policy

for the publication of proceedings of conferences and other multi-author volumes

---

Lecture Notes aim to report new developments – quickly, informally and at a high level. The following describes criteria and procedures for multi-author volumes. For convenience we refer throughout to "proceedings" irrespective of whether the papers were presented at a meeting.
The editors of a volume are strongly advised to inform contributors about these points at an early stage.

§ 1. One (or more) expert participant(s) should act as the scientific editor(s) of the volume. They select the papers which are suitable (cf. §§ 2 – 5) for inclusion in the proceedings, and have them individually refereed (as for a journal). It should not be assumed that the published proceedings must reflect conference events in their entirety. The series editors will normally not interfere with the editing of a particular proceedings volume – except in fairly obvious cases, or on technical matters, such as described in §§ 2 – 5. The names of the scientific editors appear on the cover and title page of the volume.

§ 2. The proceedings should be reasonably homogeneous i.e. concerned with a limited and well-defined area. Papers that are essentially unrelated to this central topic should be excluded. One or two longer survey articles on recent developments in the field are often very useful additions. A detailed introduction on the subject of the congress is desirable.

§ 3. The final set of manuscripts should have at least 100 pages and preferably not exceed a total of 400 pages. Keeping the size below this bound should be achieved by stricter selection of articles and NOT by imposing an upper limit on the length of the individual papers.

§ 4. The contributions should be of a high mathematical standard and of current interest. Research articles should present new material and not duplicate other papers already published or due to be published. They should contain sufficient background and motivation and they should present proofs, or at least outlines of such, in sufficient detail to enable an expert to complete them. Thus summaries and mere announcements of papers appearing elsewhere cannot be included, although more detailed versions of, for instance, a highly technical contribution may well be published elsewhere later.
Contributions in numerical mathematics may be acceptable without formal theorems resp. proofs provided they present new algorithms solving problems (previously unsolved or less well solved) or develop innovative qualitative methods, not yet amenable to a more formal treatment.
Surveys, if included, should cover a sufficiently broad topic, and should normally not just review the author's own recent research. In the case of surveys, exceptionally, proofs of results may not be necessary.

§ 5. "Mathematical Reviews" and "Zentralblatt für Mathematik" recommend that papers in proceedings volumes carry an explicit statement that they are in final form and that no similar paper has been or is being submitted elsewhere, if these papers are to be considered for a review. Normally, papers that satisfy the criteria of the Lecture Notes in Mathematics series also satisfy this requirement, but we strongly recommend that each such paper carries the statement explicitly.

§ 6. Proceedings should appear soon after the related meeting. The publisher should therefore receive the complete manuscript (preferably in duplicate) including the Introduction and Table of Contents within nine months of the date of the meeting at the latest.

§ 7. Proposals for proceedings volumes should be sent to one of the editors of the series or to Springer-Verlag Heidelberg. They should give sufficient information on the conference, and on the proposed proceedings. In particular, they should include a list of the expected contributions with their prospective length. Abstracts or early versions (drafts) of the contributions are helpful.

General Remarks

Lecture Notes are printed by photo-offset from the master-copy delivered in camera-ready form by the authors of monographs, resp. editors of proceedings volumes. For this purpose Springer-Verlag provides technical instructions for the preparation of manuscripts. Volume editors are requested to distribute these to all contributing authors of proceedings volumes. Some homogeneity in the presentation of the contributions in a multi-author volume is desirable.
Careful preparation of manuscripts will help keep production time short and ensure a satisfactory appearance of the finished book. The actual production of a Lecture Notes volume normally takes approximately 8 weeks.
For monograph manuscripts typed or typeset according to our instructions, Springer-Verlag can, if necessary, contribute towards the preparation costs at a fixed rate.

Authors of monographs receive 50 free copies of their book. Editors of proceedings volumes similarly receive 50 copies of the book and are responsible for redistributing these to authors etc. at their discretion. No reprints of individual contributions can be supplied. No royalty is paid on Lecture Notes volumes.
Volume authors and editors are entitled to purchase further copies of their book for their personal use at a discount of 33.3 %, other Springer mathematics books at a discount of 20 % directly from Springer-Verlag. Authors contributing to proceedings volumes may purchase the volume in which their article appears at a discount of 20 %.

Commitment to publish is made by letter of intent rather than by signing a formal contract. Springer-Verlag secures the copyright for each volume.

Vol. 1290: G. Wüstholz (Ed.), Diophantine Approximation and Transcendence Theory. Seminar, 1985. V, 243 pages. 1987.

Vol. 1291: C. Mœglin, M.-F. Vignéras, J.-L. Waldspurger, Correspondances de Howe sur un Corps p-adique. VII, 163 pages. 1987

Vol. 1292: J.T. Baldwin (Ed.), Classification Theory. Proceedings, 1985. VI, 500 pages. 1987.

Vol. 1293: W. Ebeling, The Monodromy Groups of Isolated Singularities of Complete Intersections. XIV, 153 pages. 1987.

Vol. 1294: M. Queffélec, Substitution Dynamical Systems – Spectral Analysis. XIII, 240 pages. 1987.

Vol. 1295: P. Lelong, P. Dolbeault, H. Skoda (Réd.), Séminaire d'Analyse P. Lelong – P. Dolbeault – H. Skoda. Seminar, 1985/1986. VII, 283 pages. 1987.

Vol. 1296: M.-P. Malliavin (Ed.), Séminaire d'Algèbre Paul Dubreil et Marie-Paule Malliavin. Proceedings, 1986. IV, 324 pages. 1987.

Vol. 1297: Zhu Y.-l., Guo B.-y. (Eds.), Numerical Methods for Partial Differential Equations. Proceedings. XI, 244 pages. 1987.

Vol. 1298: J. Aguadé, R. Kane (Eds.), Algebraic Topology, Barcelona 1986. Proceedings. X, 255 pages. 1987.

Vol. 1299: S. Watanabe, Yu.V. Prokhorov (Eds.), Probability Theory and Mathematical Statistics. Proceedings, 1986. VIII, 589 pages. 1988.

Vol. 1300: G.B. Seligman, Constructions of Lie Algebras and their Modules. VI, 190 pages. 1988.

Vol. 1301: N. Schappacher, Periods of Hecke Characters. XV, 160 pages. 1988.

Vol. 1302: M. Cwikel, J. Peetre, Y. Sagher, H. Wallin (Eds.), Function Spaces and Applications. Proceedings, 1986. VI, 445 pages. 1988.

Vol. 1303: L. Accardi, W. von Waldenfels (Eds.), Quantum Probability and Applications III. Proceedings, 1987. VI, 373 pages. 1988.

Vol. 1304: F.Q. Gouvêa, Arithmetic of p-adic Modular Forms. VIII, 121 pages. 1988.

Vol. 1305: D.S. Lubinsky, E.B. Saff, Strong Asymptotics for Extremal Polynomials Associated with Weights on $\mathbb{R}$. VII, 153 pages. 1988.

Vol. 1306: S.S. Chern (Ed.), Partial Differential Equations. Proceedings, 1986. VI, 294 pages. 1988.

Vol. 1307: T. Murai, A Real Variable Method for the Cauchy Transform, and Analytic Capacity. VIII, 133 pages. 1988.

Vol. 1308: P. Imkeller, Two-Parameter Martingales and Their Quadratic Variation. IV, 177 pages. 1988.

Vol. 1309: B. Fiedler, Global Bifurcation of Periodic Solutions with Symmetry. VIII, 144 pages. 1988.

Vol. 1310: O.A. Laudal, G. Pfister, Local Moduli and Singularities. V, 117 pages. 1988.

Vol. 1311: A. Holme, R. Speiser (Eds.), Algebraic Geometry, Sundance 1986. Proceedings. VI, 320 pages. 1988.

Vol. 1312: N.A. Shirokov, Analytic Functions Smooth up to the Boundary. III, 213 pages. 1988.

Vol. 1313: F. Colonius, Optimal Periodic Control. VI, 177 pages. 1988.

Vol. 1314: A. Futaki, Kähler-Einstein Metrics and Integral Invariants. IV, 140 pages. 1988.

Vol. 1315: R.A. McCoy, I. Ntantu, Topological Properties of Spaces of Continuous Functions. IV, 124 pages. 1988.

Vol. 1316: H. Korezlioglu, A.S. Ustunel (Eds.), Stochastic Analysis and Related Topics. Proceedings, 1986. V, 371 pages. 1988.

Vol. 1317: J. Lindenstrauss, V.D. Milman (Eds.), Geometric Aspects of Functional Analysis. Seminar, 1986–87. VII, 289 pages. 1988.

Vol. 1318: Y. Felix (Ed.), Algebraic Topology – Rational Homotopy. Proceedings, 1986. VIII, 245 pages. 1988

Vol. 1319: M. Vuorinen, Conformal Geometry and Quasiregular Mappings. XIX, 209 pages. 1988.

Vol. 1320: H. Jürgensen, G. Lallement, H.J. Weinert (Eds.), Semigroups, Theory and Applications. Proceedings, 1986. X, 416 pages. 1988.

Vol. 1321: J. Azéma, P.A. Meyer, M. Yor (Eds.), Séminaire de Probabilités XXII. Proceedings. IV, 600 pages. 1988.

Vol. 1322: M. Métivier, S. Watanabe (Eds.), Stochastic Analysis. Proceedings, 1987. VII, 197 pages. 1988.

Vol. 1323: D.R. Anderson, H.J. Munkholm, Boundedly Controlled Topology. XII, 309 pages. 1988.

Vol. 1324: F. Cardoso, D.G. de Figueiredo, R. Iório, O. Lopes (Eds.), Partial Differential Equations. Proceedings, 1986. VIII, 433 pages. 1988.

Vol. 1325: A. Truman, I.M. Davies (Eds.), Stochastic Mechanics and Stochastic Processes. Proceedings, 1986. V, 220 pages. 1988.

Vol. 1326: P.S. Landweber (Ed.), Elliptic Curves and Modular Forms in Algebraic Topology. Proceedings, 1986. V, 224 pages. 1988.

Vol. 1327: W. Bruns, U. Vetter, Determinantal Rings. VII,236 pages. 1988.

Vol. 1328: J.L. Bueso, P. Jara, B. Torrecillas (Eds.), Ring Theory. Proceedings, 1986. IX, 331 pages. 1988.

Vol. 1329: M. Alfaro, J.S. Dehesa, F.J. Marcellan, J.L. Rubio de Francia, J. Vinuesa (Eds.): Orthogonal Polynomials and their Applications. Proceedings, 1986. XV, 334 pages. 1988.

Vol. 1330: A. Ambrosetti, F. Gori, R. Lucchetti (Eds.), Mathematical Economics. Montecatini Terme 1986. Seminar. VII, 137 pages. 1988.

Vol. 1331: R. Bamón, R. Labarca, J. Palis Jr. (Eds.), Dynamical Systems, Valparaiso 1986. Proceedings. VI, 250 pages. 1988.

Vol. 1332: E. Odell, H. Rosenthal (Eds.), Functional Analysis. Proceedings, 1986–87. V, 202 pages. 1988.

Vol. 1333: A.S. Kechris, D.A. Martin, J.R. Steel (Eds.), Cabal Seminar 81–85. Proceedings, 1981–85. V, 224 pages. 1988.

Vol. 1334: Yu.G. Borisovich, Yu. E. Gliklikh (Eds.), Global Analysis – Studies and Applications III. V, 331 pages. 1988.

Vol. 1335: F. Guillén, V. Navarro Aznar, P. Pascual-Gainza, F. Puerta, Hyperrésolutions cubiques et descente cohomologique. XII, 192 pages. 1988.

Vol. 1336: B. Helffer, Semi-Classical Analysis for the Schrödinger Operator and Applications. V, 107 pages. 1988.

Vol. 1337: E. Sernesi (Ed.), Theory of Moduli. Seminar, 1985. VIII, 232 pages. 1988.

Vol. 1338: A.B. Mingarelli, S.G. Halvorsen, Non-Oscillation Domains of Differential Equations with Two Parameters. XI, 109 pages. 1988.

Vol. 1339: T. Sunada (Ed.), Geometry and Analysis of Manifolds. Procedings, 1987. IX, 277 pages. 1988.

Vol. 1340: S. Hildebrandt, D.S. Kinderlehrer, M. Miranda (Eds.), Calculus of Variations and Partial Differential Equations. Proceedings, 1986. IX, 301 pages. 1988.

Vol. 1341: M. Dauge, Elliptic Boundary Value Problems on Corner Domains. VIII, 259 pages. 1988.

Vol. 1342: J.C. Alexander (Ed.), Dynamical Systems. Proceedings, 1986–87. VIII, 726 pages. 1988.

Vol. 1343: H. Ulrich, Fixed Point Theory of Parametrized Equivariant Maps. VII, 147 pages. 1988.

Vol. 1344: J. Král, J. Lukeš, J. Netuka, J. Veselý (Eds.), Potential Theory – Surveys and Problems. Proceedings, 1987. VIII, 271 pages. 1988.

Vol. 1345: X. Gomez-Mont, J. Seade, A. Verjovski (Eds.), Holomorphic Dynamics. Proceedings, 1986. VII, 321 pages. 1988.

Vol. 1346: O. Ya. Viro (Ed.), Topology and Geometry – Rohlin Seminar. XI, 581 pages. 1988.

Vol. 1347: C. Preston, Iterates of Piecewise Monotone Mappings on an Interval. V, 166 pages. 1988.

Vol. 1348: F. Borceux (Ed.), Categorical Algebra and its Applications. Proceedings, 1987. VIII, 375 pages. 1988.

Vol. 1349: E. Novak, Deterministic and Stochastic Error Bounds in Numerical Analysis. V, 113 pages. 1988.

Vol. 1350: U. Koschorke (Ed.), Differential Topology. Proceedings, 1987. VI, 269 pages. 1988.

Vol. 1351: I. Laine, S. Rickman, T. Sorvali, (Eds.), Complex Analysis, Joensuu 1987. Proceedings. XV, 378 pages. 1988.

Vol. 1352: L.L. Avramov, K.B. Tchakerian (Eds.), Algebra – Some Current Trends. Proceedings, 1986. IX, 240 Seiten. 1988.

Vol. 1353: R.S. Palais, Ch.-l. Terng, Critical Point Theory and Submanifold Geometry. X, 272 pages. 1988.

Vol. 1354: A. Gómez, F. Guerra, M.A. Jiménez, G. López (Eds.), Approximation and Optimization. Proceedings, 1987. VI, 280 pages. 1988.

Vol. 1355: J. Bokowski, B. Sturmfels, Computational Synthetic Geometry. V, 168 pages. 1989.

Vol. 1356: H. Volkmer, Multiparameter Eigenvalue Problems and Expansion Theorems. VI, 157 pages. 1988.

Vol. 1357: S. Hildebrandt, R. Leis (Eds.), Partial Differential Equations and Calculus of Variations. VI, 423 pages. 1988.

Vol. 1358: D. Mumford, The Red Book of Varieties and Schemes. V, 309 pages. 1988.

Vol. 1359: P. Eymard, J.-P. Pier (Eds.), Harmonic Analysis. Proceedings, 1987. VIII, 287 pages. 1988.

Vol. 1360: G. Anderson, C. Greengard (Eds.), Vortex Methods. Proceedings, 1987. V, 141 pages. 1988.

Vol. 1361: T. tom Dieck (Ed.), Algebraic Topology and Transformation Groups. Proceedings, 1987. VI, 298 pages. 1988.

Vol. 1362: P. Diaconis, D. Elworthy, H. Föllmer, E. Nelson, G.C. Papanicolaou, S.R.S. Varadhan. École d'Été de Probabilités de Saint-Flour XV–XVII, 1985–87. Editor: P.L. Hennequin. V, 459 pages. 1988.

Vol. 1363: P.G. Casazza, T.J. Shura. Tsirelson's Space. VIII, 204 pages. 1988.

Vol. 1364: R.R. Phelps, Convex Functions, Monotone Operators and Differentiability. IX, 115 pages. 1989.

Vol. 1365: M. Giaquinta (Ed.), Topics in Calculus of Variations. Seminar, 1987. X, 196 pages. 1989.

Vol. 1366: N. Levitt, Grassmannians and Gauss Maps in PL-Topology. V, 203 pages. 1989.

Vol. 1367: M. Knebusch, Weakly Semialgebraic Spaces. XX, 376 pages. 1989.

Vol. 1368: R. Hübl, Traces of Differential Forms and Hochschild Homology. III, 111 pages. 1989.

Vol. 1369: B. Jiang, Ch.-K. Peng, Z. Hou (Eds.), Differential Geometry and Topology. Proceedings, 1986–87. VI, 366 pages. 1989.

Vol. 1370: G. Carlsson, R.L. Cohen, H.R. Miller, D.C. Ravenel (Eds.), Algebraic Topology. Proceedings, 1986. IX, 456 pages. 1989.

Vol. 1371: S. Glaz, Commutative Coherent Rings. XI, 347 pages. 1989.

Vol. 1372: J. Azéma, P.A. Meyer, M. Yor (Eds.), Séminaire de Probabilités XXIII. Proceedings. IV, 583 pages. 1989.

Vol. 1373: G. Benkart, J.M. Osborn (Eds.), Lie Algebras, Madison 1987. Proceedings. V, 145 pages. 1989.

Vol. 1374: R.C. Kirby, The Topology of 4-Manifolds. VI, 108 pages. 1989.

Vol. 1375: K. Kawakubo (Ed.), Transformation Groups. Proceedings, 1987. VIII, 394 pages, 1989.

Vol. 1376: J. Lindenstrauss, V.D. Milman (Eds.), Geometric Aspects of Functional Analysis. Seminar (GAFA) 1987–88. VII, 288 pages. 1989.

Vol. 1377: J.F. Pierce, Singularity Theory, Rod Theory, and Symmetry-Breaking Loads. IV, 177 pages. 1989.

Vol. 1378: R.S. Rumely, Capacity Theory on Algebraic Curves. III, 437 pages. 1989.

Vol. 1379: H. Heyer (Ed.), Probability Measures on Groups IX. Proceedings, 1988. VIII, 437 pages. 1989

Vol. 1380: H.P. Schlickewei, E. Wirsing (Eds.), Number Theory, 1987. Proceedings. V, 266 pages. 1989.

Vol. 1381: J.-O. Strömberg, A. Torchinsky. Weighted Hardy Sp V, 193 pages. 1989.

Vol. 1382: H. Reiter, Metaplectic Groups and Segal Algebras. XI pages. 1989.

Vol. 1383: D.V. Chudnovsky, G.V. Chudnovsky, H. Cohn, M.B. Natha (Eds.), Number Theory, New York 1985–88. Seminar. V, 256 pages. 19

Vol. 1384: J. Garcia-Cuerva (Ed.), Harmonic Analysis and F Differential Equations. Proceedings, 1987. VII, 213 pages. 1989.

Vol. 1385: A.M. Anile, Y. Choquet-Bruhat (Eds.), Relativistic Fluid I mics. Seminar, 1987. V, 308 pages. 1989.

Vol. 1386: A. Bellen, C.W. Gear, E. Russo (Eds.), Numerical Metho Ordinary Differential Equations. Proceedings, 1987. VII, 136 pages. 1

Vol. 1387: M. Petković, Iterative Methods for Simultaneous Inclusi Polynomial Zeros. X, 263 pages. 1989.

Vol. 1388: J. Shinoda, T.A. Slaman, T. Tugué (Eds.), Mathematical I and Applications. Proceedings, 1987. V, 223 pages. 1989.

Vol. 1000: Second Edition. H. Hopf, Differential Geometry in the Larg 184 pages. 1989.

Vol. 1389: E. Ballico, C. Ciliberto (Eds.), Algebraic Curves and Proj Geometry. Proceedings, 1988. V, 288 pages. 1989.

Vol. 1390: G. Da Prato, L. Tubaro (Eds.), Stochastic Partial Differ Equations and Applications II. Proceedings, 1988. VI, 258 pages. 19

Vol. 1391: S. Cambanis, A. Weron (Eds.), Probability Theory on V Spaces IV. Proceedings, 1987. VIII, 424 pages. 1989.

Vol. 1392: R. Silhol, Real Algebraic Surfaces. X, 215 pages. 1989.

Vol. 1393: N. Bouleau, D. Feyel, F. Hirsch, G. Mokobodzki (E Séminaire de Théorie du Potentiel Paris, No. 9. Proceedings. VI, pages. 1989.

Vol. 1394: T.L. Gill, W.W. Zachary (Eds.), Nonlinear Semigroups, F Differential Equations and Attractors. Proceedings, 1987. IX, 233 p 1989.

Vol. 1395: K. Alladi (Ed.), Number Theory, Madras 1987. Proceeding 234 pages. 1989.

Vol. 1396: L. Accardi, W. von Waldenfels (Eds.), Quantum Probabilit Applications IV. Proceedings, 1987. VI, 355 pages. 1989.

Vol. 1397: P.R. Turner (Ed.), Numerical Analysis and Parallel Proces Seminar, 1987. VI, 264 pages. 1989.

Vol. 1398: A.C. Kim, B.H. Neumann (Eds.), Groups – Korea 1988. ceedings. V, 189 pages. 1989.

Vol. 1399: W.-P. Barth, H. Lange (Eds.), Arithmetic of Complex Mani Proceedings, 1988. V, 171 pages. 1989.

Vol. 1400: U. Jannsen. Mixed Motives and Algebraic K-Theory. XIII, pages. 1990.

Vol. 1401: J. Steprāns, S. Watson (Eds.), Set Theory and its Applica Proceedings, 1987. V, 227 pages. 1989.

Vol. 1402: C. Carasso, P. Charrier, B. Hanouzet, J.-L. Joly (E Nonlinear Hyperbolic Problems. Proceedings, 1988. V, 249 pages. 1

Vol. 1403: B. Simeone (Ed.), Combinatorial Optimization. Seminar, V, 314 pages. 1989.

Vol. 1404: M.-P. Malliavin (Ed.), Séminaire d'Algèbre Paul Dubreil et M Paul Malliavin. Proceedings, 1987 – 1988. IV, 410 pages. 1989.

Vol. 1405: S. Dolecki (Ed.), Optimization. Proceedings, 1988. V, pages. 1989.

Vol. 1406: L. Jacobsen (Ed.), Analytic Theory of Continued Fractio Proceedings, 1988. VI, 142 pages. 1989.

Vol. 1407: W. Pohlers, Proof Theory. VI, 213 pages. 1989.

Vol. 1408: W. Lück, Transformation Groups and Algebraic K-Theory 443 pages. 1989.

Vol. 1409: E. Hairer, Ch. Lubich, M. Roche. The Numerical Soluti Differential-Algebraic Systems by Runge-Kutta Methods. VII, 139 p 1989.

Vol. 1410: F.J. Carreras, O. Gil-Medrano, A.M. Naveira (Eds.), Differe Geometry. Proceedings, 1988. V, 308 pages. 1989.